POWER SEMICONDUCTOR DEVICES AND CIRCUITS

Asea Brown Boveri Symposia

1991 Power Semiconductor Devices and Circuits
Edited by André A. Jaecklin

Brown Boveri Symposia previously published:

POWER SEMICONDUCTOR DEVICES AND CIRCUITS

Edited by

André A. Jaecklin

Asea Brown Boveri Corporate Research
Baden, Switzerland

SPRINGER SCIENCE+BUSINESS MEDIA, LLC

Library of Congress Cataloging-in-Publication Data

Power semiconductor devices and circuits / edited by André A.
Jaecklin.
    p.   cm. -- (Asea Brown Boveri symposia series)
  "Proceedings of an International Symposium on Power Semiconductor
Devices and Circuits, held September 26-27, 1991, in Baden-Dättwil,
Switzerland"--T.p. verso.
  Includes bibliographical references and index.
  ISBN 978-0-306-44402-9   ISBN 978-1-4615-3322-1 (eBook)
  DOI 10.1007/978-1-4615-3322-1
  1. Power semiconductors--Congresses. 2. Electronic apparatus and
appliances--Power supply--Congresses.   I. Jaecklin, André A.
II. International Symposium on Power Semiconductor Devices and
Circuits (1991 : Baden, Switzerland, and Dättwil, Switzerland)
III. Series.
TK7871.85.P667   1992
621.317--dc20   92-42064
                                                                    CIP

Proceedings of an International Symposium on Power Semiconductor Devices
and Circuits, held September 26-27, 1991, in Baden-Dättwil, Switzerland

ISBN 978-0-306-44402-9

© 1992 Springer Science+Business Media New York
Originally published by Plenum Press, New York in 1992

FOREWORD

This symposium was the scientific-technical event of the centennial celebration of the Asea Brown Boveri Switzerland. The purpose was to assess the present state of the art as well as shaping the basis for future progress in the area of power devices and related power circuits.

The merger of Brown Boveri (BBC) with Asea to Asea Brown Boveri (ABB) three years ago gave new stimulus and enriched the technical substance of the symposium. By 1991, 100 years after the formation of BBC in Switzerland as a single company, this organization has been decentralized, forming 35 independent ABB companies. One of them - ABB Semiconductors Ltd. - directly deals with the power semiconductor business. These significant changes reflect the changes in the market place: increased competition and higher customer expectations have to be fulfilled.

In line with the core business activities of ABB and with the concept of sustainable development, it is natural for ABB to be active in the area of power devices and circuits. Increased awareness towards energy conservation is one of the main drives for these activities. User friendliness is another drive: integration of intelligent functions, e.g. protection and/or increased direct computer interfacing of the power circuits. Therefore, also the R&D activities related to the subject of this symposium will in the future be characterized by an even stronger coupling with the market needs. For the members of the R&D Laboratories this means improved customer partnership beyond operational excellence.

The symposium has been attended by 107 participants from 11 countries. It was a great pleasure to welcome leading scientist from many prominent institutions from all over the world as well as competent specialists in the field of ABB, thus providing a unique opportunity for high level discussions.

We express our sincere thanks to:
- the authors, who made major contributions to the success of this symposium and also for their effort reflected in the excellent papers contained in this volume.
- to all the other participants for their contributions to lively discussions.
- to the collaborators of ETH Zurich, Professor H. Baltes and especially to the Vice President for Research, Professor R. Hütter, for his collaboration and contributions to the success of this symposium.

- to Dr. A. Jaecklin and Dr. J. Gobrecht from ABB Corporate Research Switzerland, for their careful layout of the scientific program.

We also would like to recognize the special efforts of Mr. H. P. Kaufmann and his staff of ABB Corporate Research Switzerland for the smooth running of the administrative side of this meeting.

Maurice Campagna
Director of ABB Corp.
Research Switzerland

PREFACE

The subjects of power semiconductor devices and the circuits associated with them have had an impact going far beyond the significance of the hardware itself. Their influence is ubiquitous and involves many facets like a steady supply of undistorted electrical power and efficient, pollution-free public transportation. Our way of life has been penetrated so deeply that their impact goes unnoticed in most cases.

Twenty years ago, the activity "Power Semiconductor Devices" has been started within this Research Center. Exactly ten years ago, a similar Symposium has taken place, entitled "Semiconductor Devices for Power Conditioning". Looking back at this event reveils a tremendous amount of progress made in ten years :

- Conventional thyristors with very high switching power have been an issue at that time. By now, these devices have attained a rather mature state, and in various areas they are even gradually being replaced.

- The very first high power Gate Turn-Off thyristors (GTO) were just emerging from the laboratory, and they were not yet available to the general public. In the meantime, almost all high power applications are dominated by turn-off devices which has meant a significant step toward the utopia of an ideal switch. The consequence was that such a switch, combining turn-on and turn-off in a single unit, has triggered almost a revolution in applications of power electronics. For the first time, concepts like the inverter driven locomotive became attractive solutions, both technically and economically.

- One of the most important facts is probably that, in 1981, people were at best dreaming to combine bipolar devices with field effect control, but there was no viable proposal how to achieve that. In the meantime, such elements have grown to be a class by itself. Already, devices like the Insulated Gate Bipolar Transistor (IGBT) have found their way to the marketplace. On the other hand, there is the MOS-Controlled Thyristor (MCT) which - on a small scale still - exhibits a very promising behaviour in the laboratory. This subject is one of the important topics in the present context. The belief is that, again, we may be at the verge of new revolution of power electronics with a significant impact in the foreseeable future.

The evolution of semiconductor devices has acted as a promoter of new solutions for power electronics and for the associated applications in the past. We expect them to continue being a most valuable source of innovations. Hence, the goal of this Symposium was to assess the future potential of this field, based on the present state of the art of both power semiconductor devices and the circuits associated with them.

We had the pleasure to welcome some of the most prominent leading scientists and engineers in these fields who have presented their view and who were ready to discuss freely about the future trends expected.

Taking into account the increasing interaction between power semiconductor device concepts, their realization, and their circuit applications, the Symposium has been split into four subject areas :

The first part, entitled Modern Power Devices, starts with an illustration of the increasing need to apply Very Large Scale Integration (VLSI) technology to power devices. Subsequently, device concepts including turn-off are presented and discussed. The most radical one for high power applications, the MOS-controlled thyristor, is expected to have a very high potential.

The subject Simulation has been treated separately because advanced software tools are presently available which can give a very essential support for the realization of the final devices, thus helping to reduce the costly technological effort or to predict critical device phenomena.

In the context of Circuits and Control Concepts, the interactions between turn-off devices and converters in general and more specifically between optically fired thyristors and their applications are considered. On the low end of the power rating, this includes the integration of the control circuit on the power device.

The chapter Future Trends attempts to give an outlook on recent developments in the field of power electronics on the one hand and on hybrid integration of circuits and devices as well as on the potential of future semiconductor materials on the other hand.

An edited, slightly shortened version of the discussions following each individual paper has been added to the text.

It is a pleasure to thank all the authors for their enthusiastic participation at the Symposium and especially for the careful preparation of their contributions. The high quality of their work is reflected in this volume. Additional thanks go to the four session chairman for their competent guiding, mainly through the discussions.

The efforts of Dr. R. W. Meier who has paved the way and secured the support needed and of Dr. J. Gobrecht who has assisted in setting up the program are gratefully acknowledged. Mrs. B. Säring and Mrs. M. Gerber were responsible for the extensive secretarial work. Sincere thanks are expressed them - as well as to many other individuals - who have helped to make this Symposium a successful event.

A. A. Jaecklin
Editor

CONTENTS

PARTICIPANTS

Dr. C. Abbas	Institut für Kommunikationstechnik, CH-8092 Zürich
Prof. Dr. L. Abraham	Universität der Bundeswehr München, D-8014 Neubiberg
Mr. G. Anzalone	Ministero Difeso Marina, I-00100 Roma
Dr. H. Asal	Elektrizitäts-Gesellschaft Laufenburg AG, CH-4335 Laufenburg
Dr. B. Åstrand	ABB HAFO AB, S-175 26 Järfälla
Dr. M. Bakowski	Swedish Institute of Microelectronics, S-16421 Kista
Prof. Dr. B.J. Baliga	North Carolina State University, Raleigh NC 27 695, USA
Prof. Dr. H. Baltes	ETH Zürich, CH-8093 Zürich
Dr. F. Bauer	ABB Semiconductor AG, CH-5405 Dättwil
Dr. R. Bayerer	ABB IXYS, D-6840 Lampertheim
Mr. J.O. Boerlis	ABB Semiconductor AB, CH-5600 Lenzburg
Mr. W.U. Bohli	ABB Transportation AG, CH-8050 Zürich
Dr. F. Bonzanigo	ETH Zürich, CH-8092 Zürich
Dr. P. Bordignon	Ansaldo Industria SpA, Power Electronics and Motor Division, I-20 126 Milano
Mr. H.J. Bossi	ABB Drives AG, CH-5300 Turgi
Mr. H. Braunsdorfer	Wiener Stadtwerke/Elektrizitätswerke, A-1095 Wien
Dr. B. Breitholz	ABB Corp. Research, S-721 78 Västeras
Mr. K. Brisby	ABB Semiconductor AG, CH-5600 Lenzburg
Prof. Dr. M. Campagna	ABB Corporate Research, CH-5405 Dättwil
Mr. M. Ciappa	ETH Zürich, CH-8092 Zürich
Dr. G. Corbett	ABB Industri SpA, I-20099 Sesto S.Giovanni
Mr. S. Corsi	ENEL Centro di ricerca di automatic, I-20093 Cologno Monzese/Milan
Mr. G. Crawshaw	Brush Traction Ld., UK-Loughborough LE11 ILJ
Dr. B. Danielsson	ABB Power Systms AB, S-771 01 Ludvika
Prof. Dr. N. de Rooij	IMT Institute de Microtechnique de l'Université, CH-2007 Neuchâtel
Prof. Dr. R. Dutton	Stanford Univerity, USA-Stanford, CA 94315
Prof. Dr. A. Ekström	Royal Institute Techn. Stockholm, S-10 044 Stockholm
Mr. P. Etter	ABB Transportation, CH-5300 Turgi
Prof. Dr. W. Fichtner	ETH Zürich, CH-8092 Zürich
Dr. K. Funk	TAG Semiconductors Ltd., CH-8048 Zürich
Prof. Dr. W. Gerlach	TU Berlin, D-1000 Berlin 12
Dr. J. Gobrecht	ABB Corporate Research, CH-5405 Dättwil
Dr. H. Grüning	ABB Corp. Research, CH-5405 Dättwil
Dr. G. Güth	ABB Power Systems AG, CH-5405 Dättwil

Mr. L. Halbo	ABB Corporate Research, N-1360 Nesbru
Mr. D. Halvorsen	ABB Sigma Elektroteknisk A/S, N-1540 Vestby
Dr. M. Häusler	ABB Power Systems AG, CH-5405 Dättwil
Mr. Ch. Hauswirth	ABB Drives AG, CH-5300 Turgi
Dr. M. Held	ETH Zürich, CH-8092 Zürich
Mr. J. Holm	ABB Transportation Management & Systems Development GmbH, D-68 Mannheim 1
Prof. Dr. R. Hütter	ETH Zürich, CH-8092 Zürich
Dr. A.A. Jaecklin	ABB Corporate Research, CH-5405 Dättwil
Mr. C.M. Johnson	University of Cambridge, UK-Cambridge CB2 1 PZ
Dr. H.K. Krokoszinski	Asea Brown Boveri AG/ Corp. Research Center, D-6900 Heidelberg
Mr. J. Langer	Rhein.Westf. Technische Hochschule, D-5100 Aachen
Mr. K. Lilja	ABB Semiconductor AG, CH-5405 Dättwil
Prof. Dr. T.P. Lipo	Wisconsin University, USA-Madison WI
Dr. R.W. Meier	ABB Corporate Research, CH-5405 Dättwil
Prof. Dr. A. Menth	Oerlikon Bührle AG, CH-8021 Zürich
Mr. A. Mertens	Rhein.Westf. Technische Hochschule, D-5100 Aachen
Mr. D. Metzner	Institute of Electrical Drives, Technical University Munich, D-8000 München 2
Dr. G. Moraw	Österr. Elektrizitätswirtschafts AG, A-1010 Wien
Dr. J. Naumann	ABB IXYS, D-6840 Lampertheim
Mr. A. Nilarp	ABB Semiconductors AG, CH-5600 Lenzburg
Prof. Dr. T. Ohmi	Tohoku University, Sendai, Japan
Dr. C. Ovrén	ABB HAFO AB, S-175 26 Järfälla
Dr. P.R. Palmer	University of Cambridge, Cambridge CB2 1 PZ, UK
Mr. J.M. Peter	SGS Thomson Microelectronics, F-13 106 Rousset
Dr. P. Pfluger	CSEM Centre Suisse d'Electronique et de Microtechnique SA, CH-2007 Neuchâtel
Dr. E. Ramezani	ABB Semiconductor AG, CH-5600 Lenzburg
Mr. V. Robibero	Schindler Aufzüge AG, CH-6030 Ebikon
Dr. P. Roggwiller	ABB Semiconductor AG, CH-5405 Dättwil
Mr. W. Roos	ABB Asea Brown Boveri Ltd, CH-5401 Baden
Dr. M. Rossinelli	ABB Semiconductor AG, CH-5600 Lenzburg
Dr. H. Rüegg	Faselec AG, CH-8045 Zürich
Dr. A. Rüegg	ABB Semiconductor AG, CH-5600 Lenzburg
Mr. Ch. Ruetsch	ABB Drives AG, CH-5300 Turgi
Mr. I. Ruohonen	ABB Drives-Antriebe GmbH, D-6800 Mannheim 1
Dr. T. Salo	ABB Strömberg Research Center, SF-65101 Vaasa
Mr. R. Schäfer	ABB Kraftwerke AG, CH-5401 Baden
Dr. S. Schafir	Schindler Aufzüge AG, CH-6030 Ebikon
Mr. H. Scheibengraf	ABB Transportation AG, CH-5300 Turgi
Dr. H.J. Schötzau	Aarg. Elektrizitätswerke, CH-5001 Aarau
Dr. F. Schwab	Aare - Tessin AG, CH-4601 Olten
Mr. T. Seger	ABB Transportation AG, CH-5300 Turgi
Mr. M. Serizawa	Tohoku University/Mitsubishi International GmbH, D-4000 Düsseldorf 30

Mr. J. Setchell	ABB Control Limited, Exhall/Coventry CV7 9ND, UK
Dr. D. Sigurd	Institute of Microwave Technology, S-164 21 Kista
Prof. Dr. A. Silard	TU Bucharest, R-71 273 Bucharest
Prof. Dr. D. Silber	TU Bremen, D-2800 Bremen 33
Prof. Dr. R. Sittig	Technische Universität Braunschweig, D-3300 Braunschweig
Prof. Dr. H. Skudelny	Rhein. Westf. Technische Hochschule Aachen, D-5100 Aachen
Mr. C. Spikings	AEA Technology Colour, Culham Laboratory, Abingdon/Oxfordshire, UK
Prof. Dr. A. Steimel	Ruhr-Universität-Bochum, D-4630 Bochum
Dr. P. Steimer	ABB Drives AG, CH-5300 Turgi
Prof. Dr. H. Stemmler	ETH Zürich, CH-8092 Zürich
Dr. T. Stockmeier	ABB Semiconductor AG, CH-5405 Dättwil
Dr. P. Streit	ABB Semiconductor AG, CH-5600 Lenzburg
Prof. Dr. P. Svedberg	ABB HAFO AB, S-175 26 Järfälla
Dr. C. Tedmon	ABB Asea Brown Boveri Ltd, CH-8050 Zürich
Dr. P. Tenti	University of Padova, I-35 131 Padova
Dr. L. Ulrich	ABB Mittelspannungstechnik AG, CH-8050 Zürich
Mr. K. Vik	ABB Corporate Research, N-1360 Nesbru
Dr. J. Vitins	ABB Transportation AG, CH-8050 Zürich
Mr. J. Vorwerk	Bernische Kraftwerke AG, CH-3000 Bern
Mr. B. Voss	Fraunhofer Institut für Solare Energiesysteme, D-7800 Freiburg
Dr. G. Wachutka	ETH Zürich Institut für Quantenelektronik, CH-8093 Zürich
Dr. J. Waldmeyer	ABB Semiconductor AG, CH-5600 Lenzburg
Mr. F. Walenberg	N.V. Nederlandse Spoorwegen, NL-3500 HA Utrecht
Mr. W. Wehrle	ABB Drives AG, CH-5300 Turgi
Prof. Dr. J. Weiler	ETH Zürich, CH-8092 Zürich
Mr. P. Willi	ABB Normelec AG, CH-8953 Dietikon
Mr. M. Wimshurst	Hill Graham Control Ltd., Bucks HP12 3RB, UK
Prof. Dr. W. Zaengl	ETH Zürich, CH-8092 Zürich
Dr. H.R. Zeller	ABB Semiconductor AG, CH-5405 Dättwil
Dr. W. Zimmermann	Ascom Favag S.A., CH-2022 Bevaix
Dr. N. Zommer	ABB IXYS Corp., San Jose CA 95 131, USA

TECHNOLOGY FOR HIGH-POWER DEVICES

Tadahiro Ohmi

Tohoku University
Sendai 980, Japan

ABSTRACT

The simultaneous fulfillment of three principles, viz., **Ultra Clean Wafer Surface**, **Ultra Clean Processing Environment** and **Perfect Process-Parameter Control** is the key to high performance processes for realizing advanced high power devices.

The importance of the concept has been demonstrated by the experimental results of low-temperature silicon epitaxy by low kinetic-energy particle processes. As a result of optimizing pertinent process parameters under ultra clean conditions, high-crystallinity silicon epitaxial layers have been successfully grown at temperatures as low as 250 °C. Advanced copper metallization for large-current driving interconnect formation have also been established. Giant-grain-copper thin films also formed by low-kinetic-energy particle process exhibit very low resistivity as well as excellent reliability against electromigration failures. Employment of native-oxide-free processing allows us to form ideal metal silicon contacts without any alloying heat cycles. All these advanced process technologies realized for the first time by Ultra Clean Technology have made it possible to establish total low temperature processing which is most essential for high-performance power semiconductor devices.

1. INTRODUCTION

The guiding principles for the development of high-performance high power semiconductor devices is to increase the maximum power that can be handled in a unit area of the device and to enhance the speed performance of the device as well.

In order to enhance the switching speed, reduction of the terminal resistance and inductance is the most essential requirement. If resistance is existing in the cathode terminal, it severely degrades the current driving capability of the device. The resistance in the control-gate electrode of an IGBT or a GTO, for instance, must be eliminated as completely as possible because the multi-channels in these devices can not be turned on or off simultaneously due to the signal propagation delays occurring in the gate electrode. Such non uniform switching causes the degradation in its switching performance. Since the gate capacitance is primarily determined by the geometrical structure of the device and can not be made zero, there is no other way but to reduce the gate electrode resistance in order to minimize the RC delays. In addition, typically a current as large as 100 A must be switched during a very short period of time of e.g. 0.1 μsec which corresponds to the current switching of 1 mA for 1 psec in the regime of high speed LSI's. The effect of interconnect inductance (L) becomes enormous, because the voltage drop across the inductance increases in proportion to the time derivative of current. Therefore the inductive components in interconnects must be minimized.

All these requirements to realize high-performance power devices are only fulfilled by utilizing low-resistance metals much more abundantly in the hart of device structures. However, the use of metals in semiconductor devices has been severely limited so far, mainly due to the high temperature processes that must be carried out after such metal pattern formation. **Ultra Clean Technology** that has been developed very extensively under the leadership of Tohoku University has made it possible to establish a total low-temperature processing for semiconductor device fabrication. **Ultra Clean Technology** has been developed in order to establish high performance processes for realizing ultra-high-density integration of ultimate-small-dimension ultra-high-speed devices. However, the achievements of **Ultra Clean Technology** are all directly transferable to power devices to enhance their performances.

The purpose of this article is to discuss the principles of **Ultra Clean Technology** and present several remarkable achievements in the process and device technologies. It is shown that the creation of **ultra clean wafer surface** in combination with **low kinetic energy particle processes** is a key to establish low temperature processing technologies to form high quality thin films having ideal interface characteristics. In the first place, the growth of high-crystallinity **in situ**-doped epitaxial silicon films at temperatures as low as 250 °C is presented. Then large-electromigration-resistance copper interconnect technology and hillock-free aluminum metallization technology are described. The formation of ideal metal/silicon contacts without any alloying heat cycles is also presented. All these technologies are quite essential to realize high-performance high power semiconductor devices.

2. PRINCIPLES OF ULTRA CLEAN TECHNOLOGY FOR ULTRA LARGE SCALE INTEGRATION

We have claimed that the simultaneous fulfillment of the following three principles is most essential to establish high performance processes for fabricating ULSI's. The three principles are: **(1) ULTRA CLEAN PROCESSING ENVIRONMENT; (2) ULTRA CLEAN WAFER SURFACE; (3) PERFECT-PROCESS PARAMETER CONTROL.** In order to create **ultra clean processing environment,** tremendous efforts have been devoted to the development of Ultra Clean Technologies [1-7] such as super cleanroom technology [2,8-14], ultra clean gas technology [3,8-25], ultra pure water (UPW) technology [2,26-28], ultra clean chemicals technology [29,30] and so forth. As a result, we have achieved extremely high levels of cleanliness and purity of a cleanroom system, of gases, of ultra pure water and of chemicals. This is quite important because they directly affect the quality of wafer processing. In spite of the remarkable advancement in these clean technologies, there still exist difficulties in realizing **Ultra Clean Wafer Surface** which, we believe, is most essential in establishing high performance processes.

The **Ultra Clean Wafer Surface** is characterized as a surface which is **PARTICLE FREE, ORGANIC-CONTAMINATION FREE, METALLIC-CONTAMINATION FREE, NATIVE-OXIDE FREE, COMPLETELY HYDROGEN-TERMINATED, and ATOMIC-LEVEL PERFECTLY FLAT**. The particles, organic materials, and metallic elements are the well-recognized classical contaminations. We have achieved a good control of them, owing to the advancement of super cleanroom technology and wet chemical cleaning technology. It is important to note that we have included **native oxide** in the category of contamination in a sense that it severely deteriorates the processing integrity. It was found that the native oxide growth occurs at room temperatures only under the coexistence of O_2 and H_2O [31-33]. This is why the dissolved oxygen (DO) concentration in UPW must be reduced to a level as low as possible, at least less than 2 ppb. It is shown that the growth of native oxide is suppressed when wafers are rinsed in UPW with reduced DO concentrations. The silicon surface removed of native oxide by diluted HF, for instance, is very active and easily contaminated by adsorbed air molecules. The molecular adsorption on a cleaned Si surface also degrades the processing quality. In order to prevent such degradation, the cleaned silicon surface must be perfectly hydrogen-terminated. Finally, silicon surfaces must be perfectly flat at an atomic level. We have shown the improvement in the surface microroughness has great impacts on the integrity of thin gate oxide as well as the channel electron mobility[34].

In the following, the importance of ultra clean technology on the advancement of process and device technologies are presented and discussed.

3. IMPACT OF THREE ULTRA-CLEAN-TECHNOLOGY PRINCIPLES ON LOW TEMPERATURE SILICON EPITAXY [35-41]

The importance of three principles of ultra clean technology are demonstrated by experimental data in the following where low temperature silicon epitaxy by a low-kinetic energy particle process is taken as an example. In this process, concurrent Ar ion bombardment of a growing silicon film surface in a very low energy regime is utilized to activate the very surface layer. As a result, single crystal silicon layers, having high crystalline perfection and in situ impurity doping with 100% electrical activation, have been successfully grown at temperatures as low as 300 °C [36,39] and 250 °C [40,41]. The process has been realized by using a RF-DC coupled mode bias sputtering system[4], shown in Fig. 1, equipped with ultra high vacuum (10^{-10} Torr) and ultra clean gas delivery (1~2 ppb moisture level at a point of use) systems .

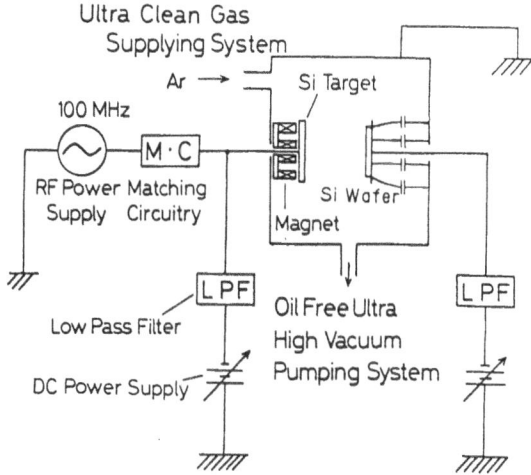

Fig. 1. Schematic of RF-DC coupled mode bias sputtering system.

Figure 2 demonstrates the crystallinity changes in epitaxial silicon films depending on the ion bombardment energy. When the energy of individual ion bombardment is precisely controlled to an optimum value of 25 eV, perfect epitaxy occurs. However, a change in the energy severely degrades the crystallinity. Bombarding ion energy dependence of the film resistivity is shown in Fig. 3. The resistivity is a very sensitive measure of the film crystallinity[37], and the best quality film was obtained at the minimum of resistivity. Important to note is that appreciable change in the resistivity, occurs by only 2 eV change in the bombarding ion energy, thus showing the importance of PERFECT PROCESS-PARAMETER CONTROL.

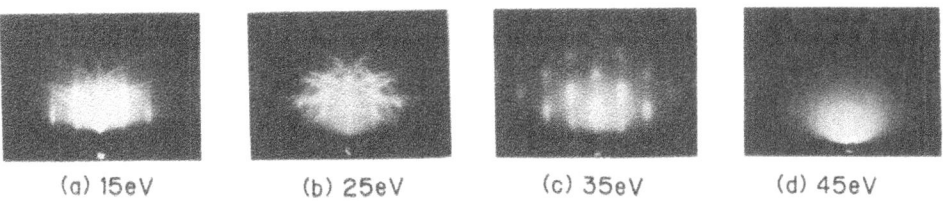

(a) 15eV   (b) 25eV   (c) 35eV   (d) 45eV

Fig. 2. Electron diffraction patterns for epitaxial silicon films formed at 300 °C under Ar ion bombardment, having four different bombarding energies such as 15 eV (a), 25 eV (b), 35 eV (c), and 45eV (d) (target bias of -120 V).

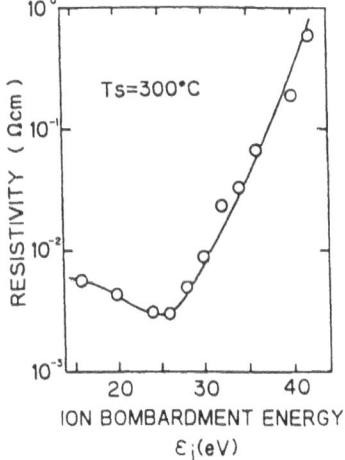

Fig. 3. Deposited Si film resistivity as a function of the ion bombardment energy.

ε_i=25eV

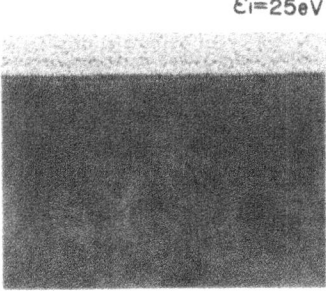

Fig. 4. The High Resolution TEM (HRTEM) image of the film cross section at the surface which was formed under the optimum epitaxial growth condition (ε_i= 25 eV). The electron beam was parallel to the <100> silicon axis.

Figure 4 shows the High Resolution TEM (HRTEM) image[38] of the film cross section at the surface for a film grown with the optimum deposition condition. A very clear lattice image in this figure demonstrates that a high quality epitaxial silicon film was successfully grown at 300 °C. Formation of a very flat surface at the atomic level is also visible.

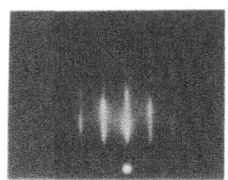   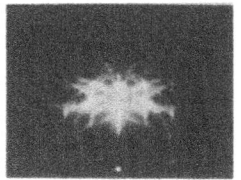

WITHOUT IN SITU   WITH IN SITU
SURFACE CLEANING   SURFACE CLEANING

Fig. 5. Electron diffraction patterns obtained from Si films, grown at optimum ion bombardment condition with (right) and without (left) in situ substrate surface cleaning.

Figure 5 shows the electron diffraction patterns obtained from low temperature grown Si films with(right) and without(left) **in situ** substrate surface cleaning process. **In situ** surface cleaning was performed by extremely low energy Ar ion bombardment having an energy of approximately 2 ~ 3 eV just before the Si deposition[35]. In this case, substrate surfaces were exposed to the clean room air for a few minutes before loading to the chamber. Although the optimum silicon deposition condition was employed in both samples, a perfect crystal was not obtained without the **in situ** surface cleaning, verifying the importance of ULTRA CLEAN WAFER SURFACE. In Fig. 6, the resistivity of the silicon film deposited with or without **in situ** substrate surface cleaning is shown as a function of air-exposure time. The wafers were wet chemically cleaned with a diluted HF etch at the final step and exposed to clean room air for a certain period of time before setting into the sputtering chamber. Then the silicon growth was carried out, using the optimum condition for epitaxy at 300 °C. Degradation in crystallinity for air-exposure time longer than 1 h is evident even with the **in-situ** cleaning. Five orders of magnitude increase in the resistivity is observed for sample without **in-situ** surface cleaning. This has resulted from moisture molecule adsorption and succeeding native oxide formation[31-33]. This fact indicates that the low-energy ion bombardment surface cleaning can remove adsorbed molecules but cannot remove native oxide. **In-situ** removal of native oxide should be conducted, using the HF gas selective etching process[42]. It has been thus confirmed that wafer surfaces change continuously with an elapse of air-exposure time. These results clearly indicate that the present open manufacturing system must be replaced by a closed manufacturing system where wafers are not exposed to the air and are transported in a clean environment such as ultra clean N_2 environment [43,44].

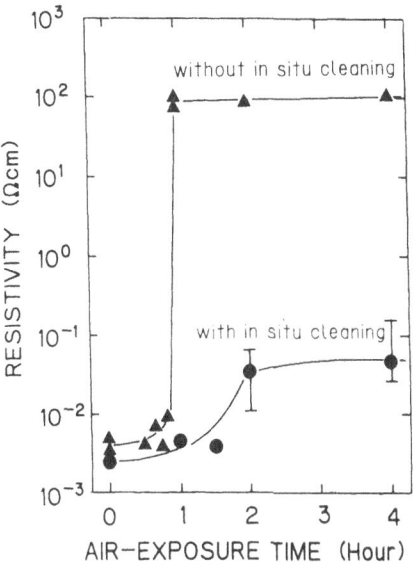

Fig. 6. Deposited Si film resistivity with and without in situ substrate surface cleaning as a function of air-exposure time.

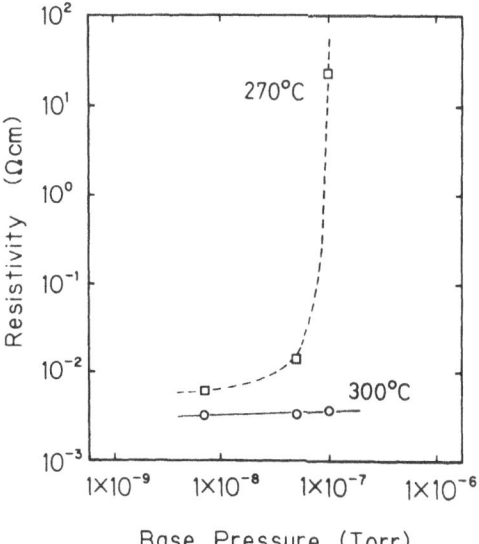

Fig. 7. Resistivity of an epitaxially grown film as a function of the base pressure before the film growth. The films were grown at 270 °C and 300 °C under ion bombardment conditions optimized at 300 °C.

Figure 7 demonstrates the importance of ULTRA CLEAN PRO-
CESSING ENVIRONMENT where the film resistivity is shown as a function of the background pressure before film growth for two different substrate temperatures. Increase in the base pressure (i.e. increase in the
contamination level) results in a severe degradation in the film crystallinity in the case of 270 °C. Thus environmental cleanliness is particularly important for lower processing temperatures.

As discussed above, the significance of the three principles on the
establishment of high performance processes has been experimentally
verified.

Figure 8 shows typical I-V characteristics of reverse biased p-n junctions[37]. The diode size was 1 x 1mm^2 and n-type epitaxial layers were
grown on p-type substrates using a phosphorus-doped target under three
different ion bombardment energies (ε_i's). The current level at a reverse
bias voltage of 5 V for the diode, formed at the optimum condition for
single-crystal growth ($\varepsilon_i = 25$ eV), is 1.88×10^{-9} A/cm^2 which is a very
low reverse current level for diodes produced at the extremely low tem-
perature of 300 °C. Other samples ($\varepsilon_i = 15$ eV and 35 eV) show much
higher reverse current density compared to that formed at $\varepsilon_i = 25$ eV,
showing a good correlation with the cristallinity of the epilayer as shown
in Fig. 2.

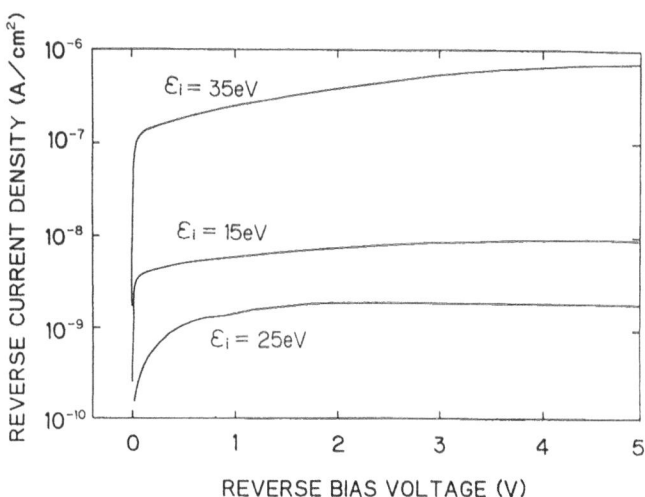

Fig. 8. Typical current-voltage characteristics of a 1 x 1 mm^2 p-n junction diode, fabricated by depositing an n-type epi-layer directly on a p-type substrate under three different ion bombardment energies.

The calculated depletion layer widths at a reverse bias voltage of 5 V
for the n region and for the p substrate are 25.6 Å and 1.46 µm, respectively, indicating that the epitaxial layer-substrate interface is right in
the midst of the depletion layer. It is worthwhile to note that such low

reverse current diodes have been realized, utilizing the original wafer surface as an interface of the p-n junction.

The experimental data, presented so far, are all obtained from silicon epitaxial layers grown at a substrate temperature of 300 °C. In the following, we would like to discuss how to reduce the epitaxial temperature below 300 °C by optimizing the ion bombardment conditions. The discussion is made in reference to a series of electron diffraction patterns shown in Fig. 9.

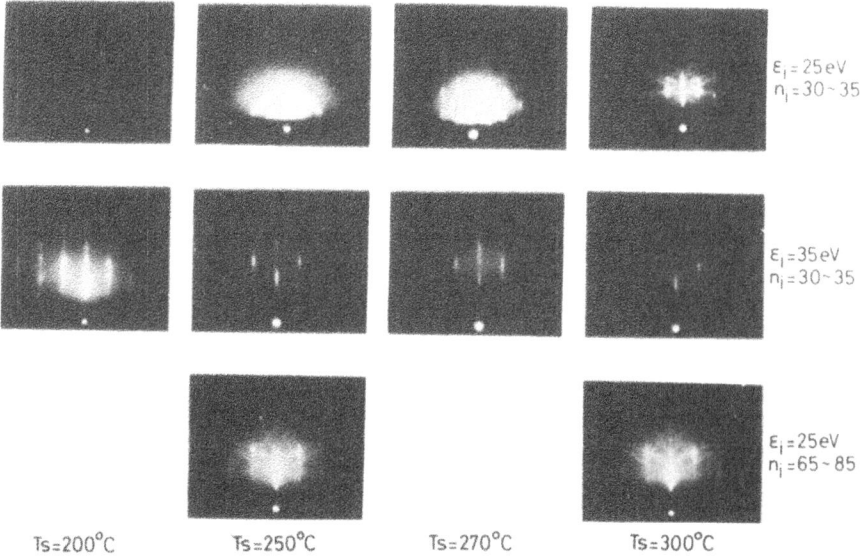

Fig. 9. Reflection electron diffraction patterns obtained from silicon films, deposited under two different ion bombardment energies ε_i and normalized ion flux densities n_i. The substrate temperature was varied from 200 to 300 °C.

The diffraction patterns shown in the top row were obtained from silicon films deposited at varying substrate temperatures, using the ion bombardment condition optimized at 300 °C. Here the energy of individual ion bombardment (ε_i) is 25 eV and the normalized ion flux n_i is 30 ~ 35. n_i is defined as a ratio of the Ar ion flux to the Si atom flux, which can be calculated from the measured data of ion current density in the substrate and the film growth rate. That is, $n_i = 30$ means 30 Ar ions are bombarding the growing film surface for a single deposited Si atom. The degradation in the film crystallinity with decrease in the substrate temperature is evident. Only hazy rings are seen at 250 °C. At 200 °C, no diffraction pattern is visible, indicating the formation of an amorphous film.

Let E_{tot} denote the total energy that a single deposited Si atom gains at the surface of a growing film. E_{tot} consists of two components, i.e. E_t, the thermal energy provided through substrate heating and E_{ion}, the energy originating from surface activation by Ar ion bombardment. It is assumed that, in order for silicon epitaxy to occur, the total energy E_{tot} must be larger than a certain threshold value, E_{epi}, viz

$$E_{tot} = E_t + E_{ion} > E_{epi} \qquad (1)$$

Here E_t will be on the order of kT. The energy transfer process among the bombarding Ar ions and the Si adatoms involves a number of complex physical processes occurring at the surface, such as the collision of an incoming Ar ion with a Si atom in the surface layer of the film and the resultant localized optical phonon excitation followed by the interaction between the phonon and the adatom, and so forth. The detailed description of the process is beyond the scope of this paper. However, E_{ion} can be reasonably approximated as

$$E_{ion} \propto n_i \bullet \varepsilon_i \qquad (2)$$

where ε_i is the bombardment energy of an individual ion and n_i the normalized ion flux. Complete disappearance of Bragg spots in the diffraction pattern at 250 °C (top row) can be interpreted by the inequality, eq. (1), not holding due to the deficit in the thermal energy E_t as a result of 50 °C reduction in the substrate temperature. In order to compensate for the deficit in the thermal energy, ion energy E_{ion} must be increased. Since E_{ion} is proportional to $n_i \bullet \varepsilon_i$, this can be achieved either by increasing n_i or ε_i. The results obtained by increasing ε_i are given in the middle row of Fig. 9.

Here the ε_i was increased by 10 eV as compared to the sample in the top low. It is interesting to see that streak patterns are obtained at 250 °C and that Bragg spots are also seen even at 200 °C. The deficit in the thermal energy has been compensated for by an increase in E_{ion} and epitaxial growth has occurred. However, it is important to note that no Kikuchi lines are observed even at 300 °C where the inequality, eq (1), definitely holds. The crystallinity degradation is due to the excess ion bombardment energy that has produced damages in the film. Therefore the right way to increase the E_{tot} for silicon epitaxy at reduced temperature is to increase E_{ion} by increasing the normalized ion flux n_i while keeping the individual ion bombardment energy ε_i at the optimum value of 25 eV. The results of silicon film growth under $\varepsilon_i = 25$ eV and $n_i = 65 \sim 85$ are given in the bottom row of Fig. 9. Perfect epitaxy occurred at 250 °C and the resistivity of the film is identical to that grown with optimum ion bombardment conditions at 300 °C.

In conclusion, the surface activation of a growing film surface under optimized ion energy and flux conditions and, of course, under **Ultra Clean** conditions has made it possible to grow device-grade epitaxial silicon films at a temperature as low as 250 °C. Such a very low-temperature epitaxial silicon growth process has allowed us to have a great deal of flexibility in the design of advanced device structures and their fabrication processes.

4. APPLICATION TO ADVANCED COPPER METALLIZATION

The interconnect formation is one of the most essential process in the fabrication of ULSI's. However, the device scaling and the resultant miniaturization of interconnect structures have presented a number of severe limitations in the performance and reliability of the conventional metallization scheme. Since the aluminum based alloys, the predominantly used interconnect material so far, are showing poor electromigration reliability and their resistivities are not low enough to operate ULSI systems at high speeds, pure copper is now drawing a considerable attention as an alternative to aluminum based alloys[45-49]. The advanced copper metallization established by ultra clean technology[50,51] is described in the following.

Approximately 1 μm-thick Cu films were grown on thermal SiO_2 with varying substrate bias voltages using a RF-DC coupled mode bias sputtering system[4] similar to the one shown in Fig. 1. The X-ray diffraction analysis revealed that majority of the film orientation was (111) and only small (200) peaks were observed for samples grown under smaller ion bombardment energies. The X-ray diffraction peak heights are plotted in   Fig. 10(a) as a function of the substrate bias voltage applied during the film growth. More negative biases indicate the larger ion bombardment energies.

Figure 10(b) demonstrates the results of X-ray diffraction analysis performed after the Cu films were thermally annealed at 450 °C for 30 min in N_2 ambience. Very interesting to note is that the films grown with biases more negative than -50 V all exhibit almost perfect transformation from (111) to (100) orientation, i.e., the appearance of (200) peaks. Such a transformation occurs at temperatures as low as 180 °C.

Energetic ion bombardment of a growing film surface tends to make (111) planes of thin films align normal to the incoming ion direction. This phenomenon can be interpreted to mean that only close-packed surfaces survive under the influence of energetic particle bombardment[52]. This is the reason why as-deposited films with substrate bias larger than -50 V exhibit an almost complete (111) orientation. However, the most stable film orientation of Cu seems to be (100) on the SiO_2 films. In other words, ion bombardment created (111)-oriented films are metastable, and the transformation from a (111)-oriented film into a more stable (100) oriented film occurs very easily.

One of the most interesting phenomena that we have observed is the growth of giant grains in the Cu film that accompanies such crystal orientation transformation. The Cu films, grown with a sufficient amount of energy deposition during the film growth, exhibit growth of grains as large as 100 μm after 450 °C thermal annealing[51,53,54].

One of the most interesting applications of such (111) to (100) transformation, accompanied by giant grain formation, is demonstrated in Fig. 11.

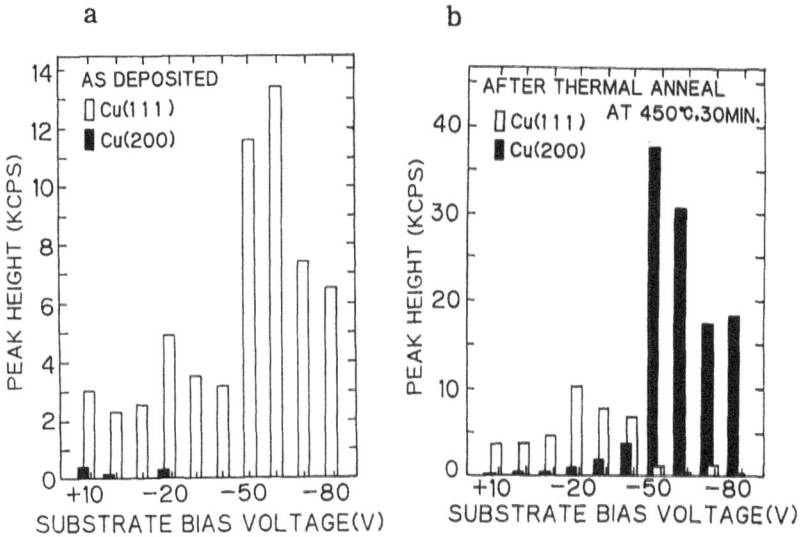

Fig. 10. Substrate bias dependence of Cu(111) and Cu(200) x-ray diffraction peak heights for films formed on SiO_2:(a) as-deposited films, (b) films annealed for 30 min at 450 °C.

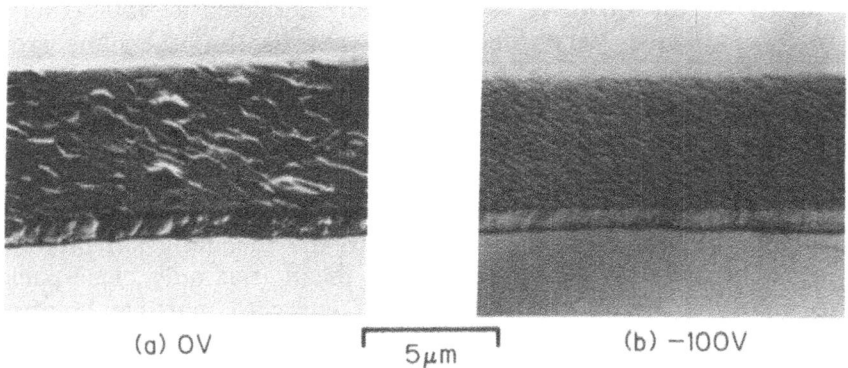

Fig. 11. Part of a 300 x 10 μm Cu island formed on SiO_2 after annealing at 600 °C for 1 h. The films were grown with substrate biases of 0 V (a) and -100 V (b).

Cu films were deposited on SiO_2 with varying substrate bias voltages, then patterned into 300 x 10 μm^2 islands. These wafers were annealed at 600 °C for 1 h. The results of SEM observation of islands after thermal annealing are presented in Fig. 11. When the Cu film was formed with $V_s = 0$ V, the island is a polycrystalline Cu film, as is evident from the high density of the grain boundary lines seen in the picture (a). In Cu samples deposited with $V_s = -100$ V, however, only a few grain boundaries were found in an island(b). In the sample shown in Fig. 11(b), the island is almost single crystal. By careful observation using the low-energy SEM technique (5 keV), we found only a few small grains (~10 μm) localized in a restricted area in the island. By optimizing the ion bombardment condition as well as the post metallization anneal process, we expect that single-crystal Cu interconnects can be formed on SiO$_2$; this study is now in progress.

The resistivity of the giant-grain Cu film is 1.76 μΩcm, which is almost the same as the bulk resistivity of 1.72 μΩcm. At 12 K, the resistivity further reduces to 18.3 nΩcm which is almost one order of magnitude smaller than that of an as-deposited film of 152 nΩcm. The temperature dependence of Cu film resistivity is shown in Fig. 12 along with the data for an Al-Si-Cu alloy film[57].

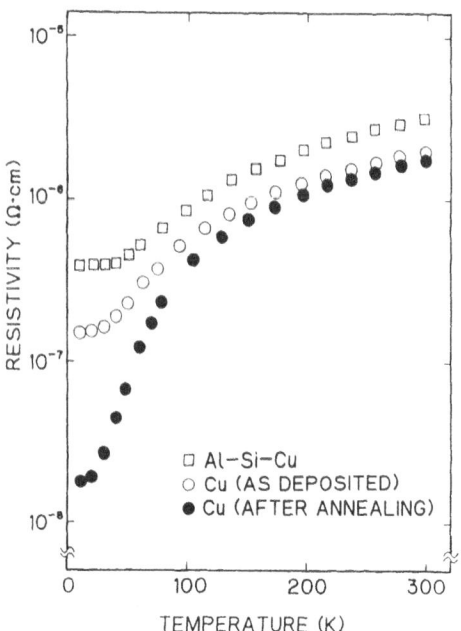

Fig. 12. Temperature dependence of film resistivities for Al-Si-Cu, as deposited Cu and giant grain Cu films.

Figure 13 represents the electromigration resistance evaluated by a new acceleration test method[58]. Four different metal interconnects were tested, namely the as-deposited Cu film, the Cu film having giant grains, Al-Si-Cu and Al-Si. The product of the electromigration lifetime and the squared current density of the current flowing through the interconnect is plotted as a function of the reciprocal temperature (1/T). Table 1 summarizes the activation energy (E_a) obtained by performing best-fit of these experimental data to the Black's formula [$\tau = A/J^2 \cdot \exp(qE_a/kT)$]. Compared with the conventional interconnect material of Al-Si-Cu alloy, it is clear that the giant-grain Cu exhibits very high electromigration resistance. Specifically the activation energy (E_a) of the giant-grain Cu is 1.25 eV, almost twice as large as that of Al-Si-Cu (0.62 eV). As the data are extrapolated to a room temperature (300 K), it is found that the giant-grain-Cu interconnect can withstand about two orders of magnitude larger current than the Al-Si-Cu alloy interconnect when the same interconnect lifetime is assumed. These data verify that the giant-grain-Cu film is a very promising material to realize ultra high-speed performance of the device. In addition, when the same current density is assumed, it is expected that the giant-grain-Cu interconnect will exhibit three to four orders of magnitude longer lifetime than Al-Si-Cu interconnects. Thus it has been demonstrated that the giant-grain Cu is the promising interconnect material featuring a low resistance as well as a sufficient reliability which enables the subhalf micron ULSI devices to operate at ultra high speed.

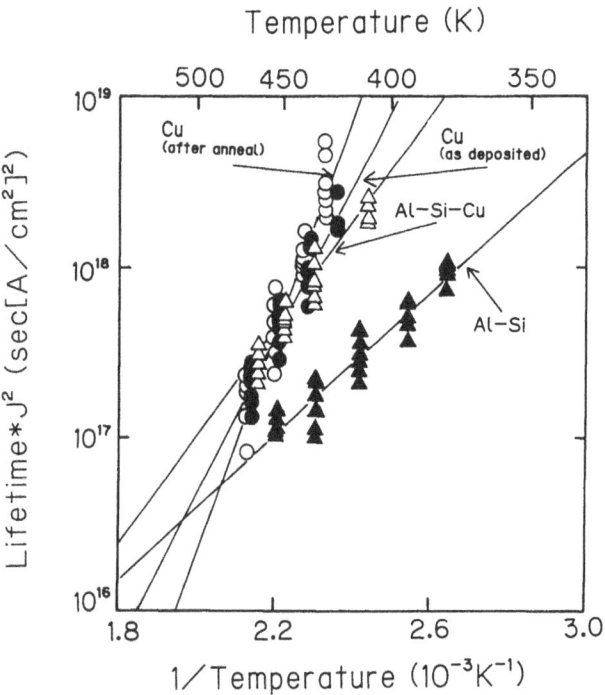

Fig. 13. The results of accelerated electromigration life tests carried out on Cu, Al-Si, Al-Si-Cu alloy interconnects. Here, the Arrhenius plot of $\tau \cdot J^2$ (τ: electromigration lifetime as defined to yield 5% resistivity increase, J: stress current density) is presented.

Table 1. Activation energies of electromigration failure for various interconnect materials, determined by a newly-developed accelerated test method[58].

INTERCONNECT MATERIALS	ACTIVATION ENERGY
Al–Si[*]	0.427eV
Al–Si–Cu[*]	0.617eV
pure Al[**]	0.591eV
Cu[**] (as deposited)	0.864eV
Cu[**] (after anneal)	1.255eV

5. PURE ALUMINUM METALLIZATION

The low kinetic energy particle process has also been successfully applied to high quality pure aluminum metallization. Some of the principal results are presented in the following[59].

Pure aluminum films, formed by this process, exhibited a number of interesting features such as the as-deposited thin film resistivity nearly equal to the bulk value, completely (111) oriented films on (100)Si, (111)Si or SiO_2. One of the most interesting, and perhaps the most important features is the formation of hillock-free films. Surface morphology changes by thermal annealing occurring in Al films are shown in Fig. 14. The Al-Si film formed by a conventional sputtering technique exhibits high-density hillock formation. On the other hand, the pure Al film formed by this process is hillock-free up to 500 °C. Although hillocks are formed at 550 °C, their density is quite low as compared to the conventional film.

The effects of low energy Ar ion bombardment on the step coverage of Al films are shown in Fig. 15. The width and depth of the hole are 2 and 0.9 μm, respectively. In the Figure, a significant shadowing effect can be observed which has resulted in poor step coverage (Fig. 15a, Vs = 0 V). In contrast, in the cross section of an Al film formed with a substrate bias of -20 V (Fig. 15b), the contact holes are filled up with a very gently sloped profile. It is important to note that the good step coverage has been achieved using the same condition that form hillock-free Al films in Fig. 14.

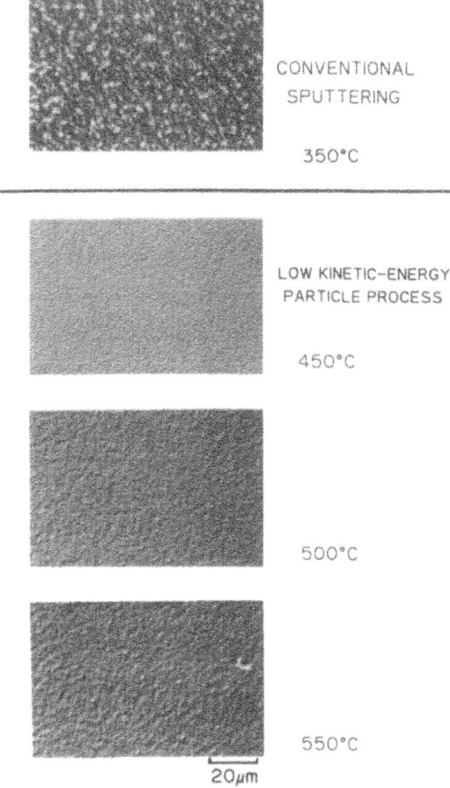

Fig. 14. Nomarsky interference contrast reflection images of Al films after thermal annealing at varying temperatures.

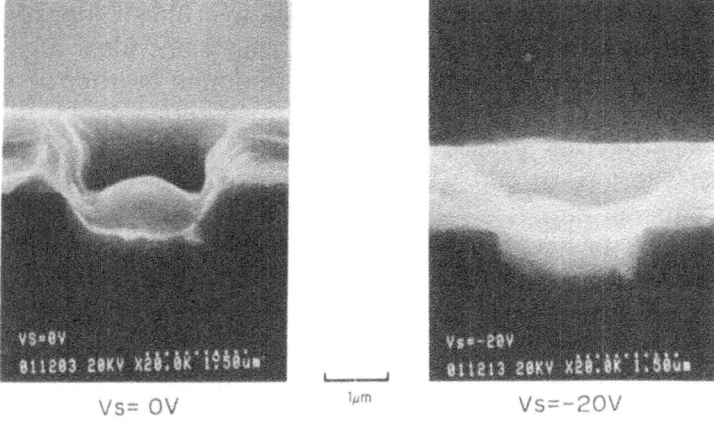

Fig. 15. Influence of substrate biasing on Al film profiles at contact holes: (a) substrate bias of 0 V; (b) substrate bias of -20 V.

6. IDEAL METAL/SILICON CONTACT FORMATION BY NATIVE-OXIDE FREE PROCESSING

One of the easiest way, to realize native-oxide free processing, is to perform all wafer processes in a clean nitrogen ambient. Since the native oxide grows only under the simultaneous presence of oxygen and moisture[31-33], native oxide-free processing is possible by replacing clean air by clean nitrogen gas. As we have already established advanced technologies for delivering ultra clean nitrogen gas, we do not worry about the wafer contamination by impurities in the nitrogen gas. When the low-kinetic-energy particle process is combined with the native-oxide free processing, ideal metal/silicon contacts are obtained without any post metallization annealing heat cycles.

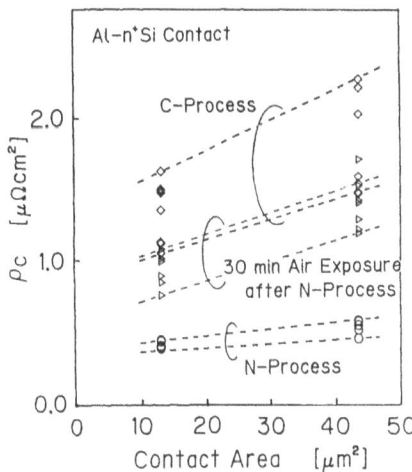

Fig. 16. Al-n+ Si contact resistance as a function of contact area for three different processes.

Contact resistance (Al-n+ Si) is plotted as a function of the contact hole area in Fig. 16 [55]. About 1 μm thick pure aluminum films were grown on an arsenic implanted n+ silicon surface by the low-kinetic-energy particle process. The n+ regions are formed by implanting 1×10^{16} cm^{-2} of arsenic at 70 keV and annealed at 1000 °C for 10 minutes in nitrogen gas ambient. The arsenic concentration at the surface was found to be approximately 5×10^{19} cm^{-3}. No alloying heat treatments were conducted after metallization. The results are plotted for three different processes. In the nitrogen-seal processing (indicated as "N-process" in the figure), all processes from the final stage of the wet chemical cleaning (i.e., the diluted HF etching, ultra pure water rinsing and drying, and the transport and loading to the vacuum chamber) were carried out in an ultra pure nitrogen ambient. While in the conventional process (C-process), the processes are all carried out in a super clean room air. The data are also shown for samples that were prepared by nitrogen-seal processing, but exposed to clean room air for 30 minutes before loading to the sputtering chamber.

Very low contact resistances of about 0.4 $\mu\Omega$ cm^2 are obtained for the N-process. Furthermore the fluctuations in the data are drastically reduced as compared to those of the C-process. Fig. 16 clearly shows that contact resistance and its fluctuations are degraded by exposing the substrate surfaces to the air after nitrogen-gas-sealed wet cleaning and drying processes.

The same nitrogen-seal process has been applied to ideal Cu/Si contact formation. Schottky diodes were fabricated by depositing Cu films on n-type Si ((100), 30 Ω cm) by the low-kinetic-energy particle process[56]. The Schottky barrier heights as determined from the forward J-V characteristics are plotted in Fig. 17 as a function of the area of contact. The date are again shown for the C-process and the N-process. Slightly lower values are obtained in the N-process as compared to those in the C-process. This is due to the native oxide existing at the Cu-Si interface in the C-process sample. The value of the barrier height is almost independent of the contact area over four orders of magnitude. Moreover, the fluctuation of the data is less than 0.005 eV.

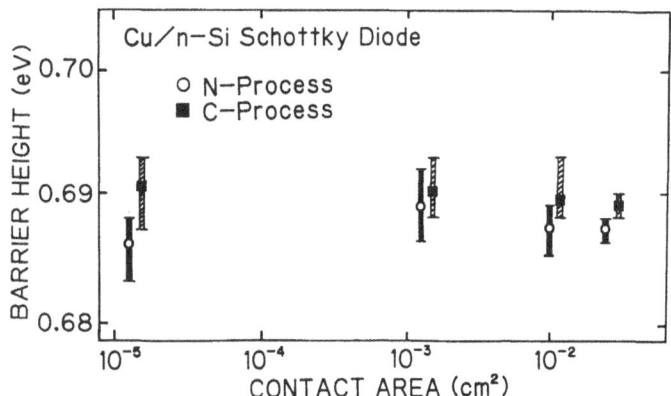

Fig. 17. Schottky barrier height of Cu/n-Si contact as a function of contact area.

In order to verify the formation of ideal metal/silicon contacts, Cu films were deposited on both n-type and p-type silicon substrates. Fig. 18 shows the current-voltage characteristics of Cu/Si Schottky diodes.

The measured data exhibit excellent agreement with the calculation by the thermionic emission theory. The deviation of measured data at large current density is due to the series resistance effect of the Si substrate. The Schottky barrier heights were obtained from the temperature variation of the saturation current density, yielding the values of Φ_{bn} and Φ_{bp} of 0.668 eV and 0.470 eV, respectively. The sum of these values, 1.138 eV, is very close to the band gap of Si at 0 °C (1.14 eV). These observations indicate the formation of ideal metal/silicon interfaces by our process[56]

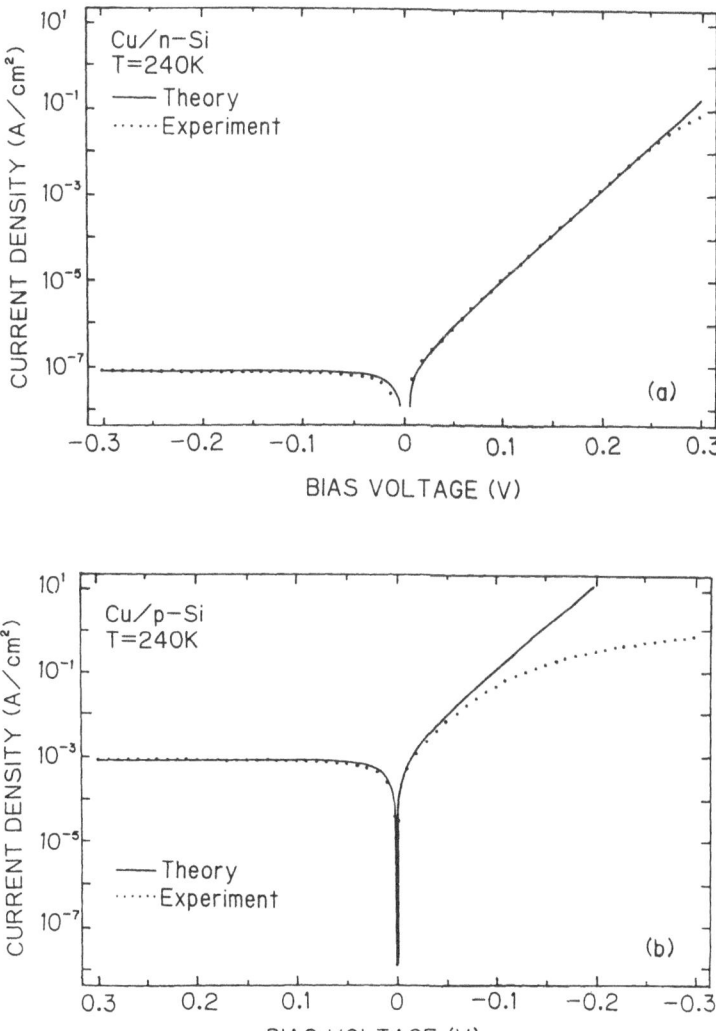

Fig. 18. J-V characteristics of Schottky diodes for Cu/n-type Si (a) and Cu/p-type Si (b) measured at 240 K. Dotted lines demonstrate measurement points and solid lines are calculated from the thermionic emission theory.

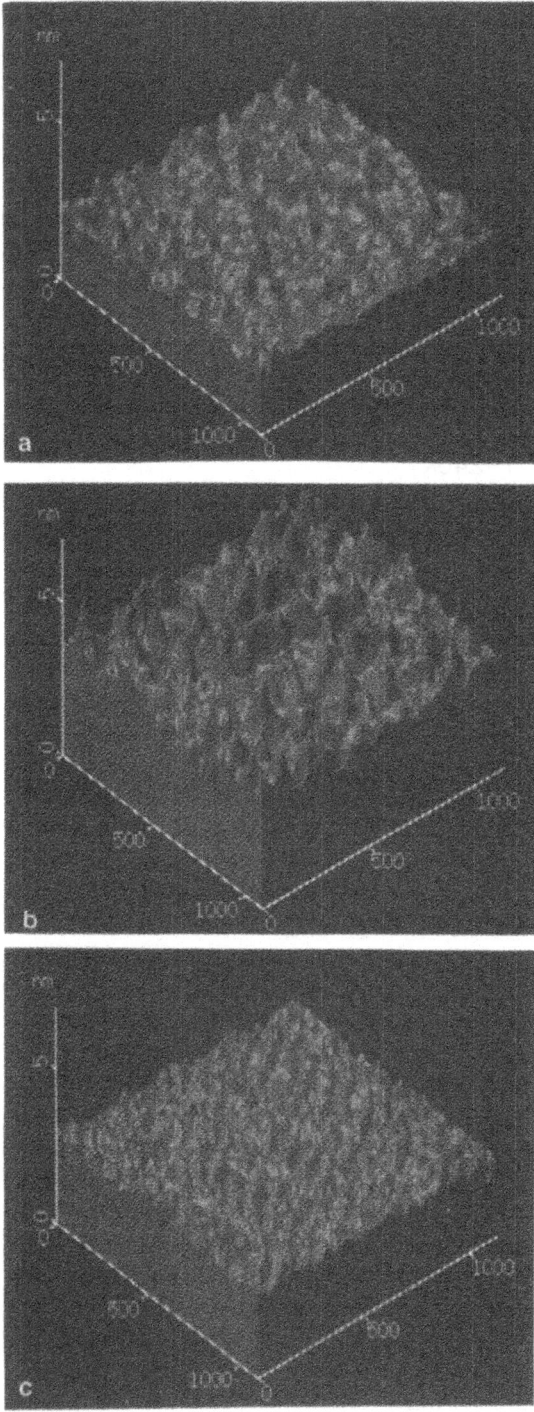

Fig. 19. Scanning tunneling microscope (STM) images of silicon wafer surfaces: (a) before cleaning; (b) after cleaning with $NH_4OH : H_2O_2 : H_2O = 1 : 1 : 5$; (c) after cleaning with $NH_4OH : H_2O_2 : H_2O = 0.05 : 1 : 5$.

7. IMPACT OF MICROROUGHNESS-FREE SURFACES ON DEVICE PERFORMANCE[34]

The results of silicon wafer surface observation by a scanning tunneling microscope (STM) are demonstrated in Fig. 19. A roughening of the original surface (a) occurs after a wafer was cleaned (b), using the conventional RCA cleaning technique[60] ($NH_4OH:H_2O_2:H_2O=1:1:5$). This is due to the non uniform etching of silicon by the $NH_4OH/H_2O_2/H_2O$ solution. The dependence of the surface microroughness, R_a, on the content of NH_4OH in the solution is given in Fig. 20 which indicates that the roughening becomes negligible when the mixing ratio is reduced to 0.05:1:5. The STM image of the silicon surface cleaned with this NH_4OH solution is shown in Fig. 19 (c).

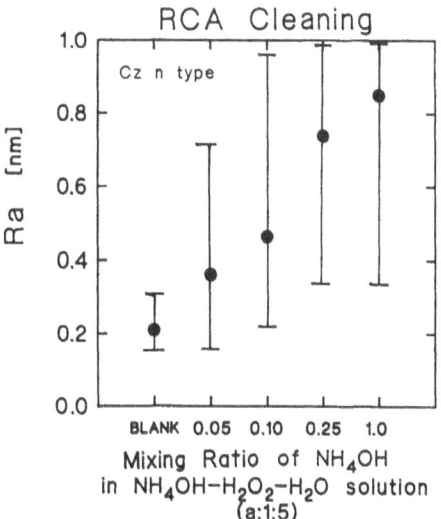

Fig. 20. Surface microroughness of Si wafer, Ra, as a function of the NH_4OH content in the cleaning solution.

The breakdown field intensity of a 100 Å SiO_2 formed on a variety of silicon wafer surfaces are summarized in Fig. 21. Here, various surface roughness samples were prepared by the NH_4OH/H_2O_2 solution cleaning having varying NH_4OH contents.

When the data are compared within the same group of n-type or p-type, the breakdown field depends only on the roughness and not on the kinds of wafers such as Epi, Cz,or Fz. Epi or Fz wafers are more resistant against roughening by NH_4OH/H_2O_2 cleaning, resulting in the larger breakdown fields. The difference between the p and n groups are due to the different barrier heights for electron injection to the gate oxide. Since aluminum electrodes were employed for MOS capacitors, electrons were injected from Al to SiO_2 in the case of p-substrates, while electron injection occurs from n-Si to SiO_2 in the case of n-substrates. The barrier height is larger for injection from Al to SiO_2.

Fig. 22 demonstrates the electron surface channel mobility as a function of the surface microroughness. It is very interesting to note that the surface mobility approaches the bulk mobility when the microroughness is reduced.

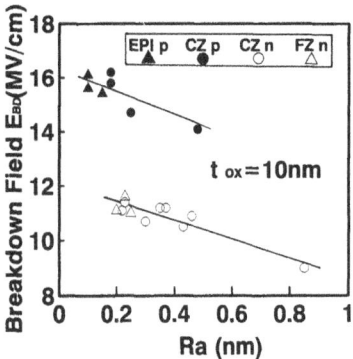

Fig. 21. Breakdown field intensity of 100 Å SiO_2 vs. microroughness, Ra, on various Si wafer surfaces.

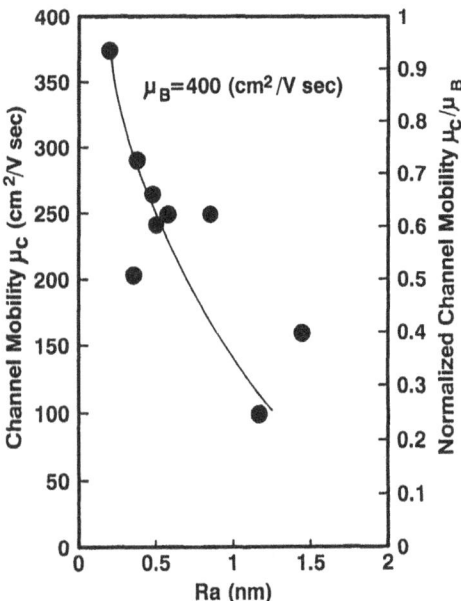

Fig. 22. Electron surface channel mobility as a function of surface microroughness.

8. CONCLUSIONS

The importance of Ultra Clean Technology in establishing advanced process technologies for deep submicron ULSI fabrication has been discussed. The most essential requirement is the simultaneous fulfillment of the three principles of **Ultra Clean Technology, i.e., Ultra Clean Wafer Surface, Ultra Clean Processing Environment, and Perfect Process-Parameter Control**. As a result of optimizing pertinent process parameters in low kinetic-energy particle processes, high-crystallinity silicon epitaxial layers have been successfully grown at temperatures as low as 250 °C. Giant-grain copper thin films have been grown also by low-kinetic energy particle processes which exhibit very low resistivity as well as excellent reliability against electromigration failures. As a result, advanced copper metallization technology has been established as a process for future ultra-high-density ultra-high-speed ULSI's. In addition, the employment of native-oxide-free processing allows us to form ideal metal silicon contacts without any alloying heat cycles. All these advanced process technologies, realized for the first time by Ultra Clean Technology, have made it possible to establish a complete low temperature processing which is most essential for realizing advanced high-power semiconductor devices.

REFERENCES

1   T. Ohmi, "Soft and clean technologies for submicron LSI," *Proc. 1986 SEMI Technology Symposium*, pp. A1-1~A1-21, Dec., 1986.

2   T. Ohmi, N. Mikoshiba, and K. Tsubouchi, "Super clean room-ultra clean technologies for submicron LSI fabrication," S. Broydo and C.M. Osburn, eds., *ULSI Science and Technology / 1987*, PV87-11 (The Electrochemical Society, Pennington, 1987), pp. 761-785.

3   T. Ohmi, J. Murota, Y. Mitsui, K. Sugiyama, T. Kawasaki, and H. Kawano, "Ultra clean gas supplying system for ULSI fabrication and its evaluation," ibid., pp. 805-821 (1987).

4   T. Ohmi, H. Kuwabara, T. Shibata and T. Kiyota, "RF-DC coupled mode bias sputtering for ULSI metallization," ibid., pp. 574-592 (1987).

5   T. Ohmi, "Ultraclean technology : ULSI processing's crucial factor," *Microcontamination*, Vol.6, No.10, pp. 49-58, Oct.1988.

6   T. Ohmi, "Future trends and applications of ultra clean technology," *Tech. Dig., 1989 International Electron Devices Meeting*, Washington D.C., pp. 49-52.

7   T. Ohmi, "Proposal for advanced semiconductor manufacturing
        equipment - an approach to automated IC manufacturing,"
        V. Akins ed., *Automated Integrated Circuits Manufacturing*, PV90-3
        (The Electrochemical Society, Pennington, 1990), pp. 3-18.

8   T. Takenami, H. Inaba, and T. Ohmi, "Total system cost
        effectiveness must keep pace with submicron manufacturing,"
        Microcontamination, Vol.7, No.8, pp. 28-32 (1989).

9   T. Ohmi and M. Yasuda, "Evaluating passive and active
        microvibration control technologies," *Microcontamination*, Vol.7,
        No.9, pp. 23-30 (1989).

10   T. Ohmi, H. Inaba, and T. Takenami, "Research on adhesion of
        particles to charged wafers critical in contamination control,"
        Microcontamination, Vol.7, No.10, pp. 29-32 (1989).

11   T. Ohmi, H. Inaba, and T. Takenami, "Preventing electromagnetic
        interference essential for ULSI E-beam performance,"
        Microcontamination, Vol.7, No.11, pp. 29-32 (1989).

12   T. Ohmi, H. Inaba, and T. Takenami, "Using water-based cooling
        systems in cleanroom environments," *Microcontamination*, Vol.7,
        No.12, pp. 27-34 (1989).

13   T. Ohmi, Y. Kasama, H. Inaba, T. Takenami, and S. Fukuda,
        "Developing a central monitoring system for the ULSI
        environment," *Microcontamination*, Vol.8, No.1, pp. 39-41 (1990).

14   T. Ohmi, Y. Kasama, K. Sugiyama, Y. Mizuguchi, Y. Yagi, H. Inaba
        and M. Kawakami, "Controlling wafer surface contamination in air
        conditioning particle removal subsystems," *Microcontamination*,
        Vol.8, No.2, pp. 45-47 (1990).

15   K. Sugiyama, F. Nakahara, T. Okumura, T. Ohmi and J. Murota,
        "Detection of sub ppb impurities in gas using atmospheric
        pressure ionization mass spectrometry," *Proceedings 9th
        International Symposium on Contamination Control*, Los Angeles,
        pp. 332-340 (1988).

16   Y. Kanno and T. Ohmi, "Development of contamination-free gas
        components and ultra clean gas supply system for ULSI
        manufacturing," ibid., pp. 345-351 (1988).

17   S. Mizogami, Y. Kunimoto and T. Ohmi, "Ultra clean gas transport
        from manufacture to users by newly developed tank lorries and
        gas storage tanks," ibid., pp. 352-359 (1988).

18   K. Sugiyama and T. Ohmi, "ULSI fabrication must being with ultraclean nitrogen system," *Microcontamination*, Vol.6, No. 11, pp. 49-54 (1988).

19   Y. Kanno and T. Ohmi, "Components key to developing contamination free gas supply," *Microcontamination*, Vol.6, No.12, pp. 23-30 (1988).

20   K. Sugiyama, T. Ohmi, T. Okumura and F. Nakahara, "Electropolished moisture-free piping surface essential for ultrapure gas system," *Microcontamination*, Vol.7, No.1, pp. 37- 65 (1989).

21   K. Sugiyama, F. Nakahara and T. Ohmi, "Designing a gas delivery system for lower submicron ULSI processes," *Microcontamination*, Vol.7, No.7, pp. 29-32 (1989).

22   Y. Mizuguchi, K. Sugiyama and T. Ohmi, "Welding technology for passivated tubing system," *Proc. Microcontamination 89 Conf.*, Anaheim, pp. 49-56 (1989).

23   H. Berger, F. Nakahara, T. Ohmi, K. Sugiyama, Y. Mizuguchi, M. Nakamura, H. Mihira and K. Sato, "High purity gas dilution system and its evaluation by APIMS," ibid., pp. 65-79 (1989).

24   T. Ohmi, Y. Kanno and S. Mizogami, "Plastic material-free and oxygen-passivated gas tubing system for ultra-clean process environment," ibid., pp. 80-90 (1989).

25   M. Nakamura, T. Ohmi, K. Sugiyama, Y. Mizuguchi, A. Ohkura and K. Kawata, "All-metal and O_2 passivation ultraclean gas delivery system for submicron VLSI manufacturing," Ext. Abstracts, *178th ECS Meeting*, Seattle, pp. 633-634 (Abst.No.433) (1990).

26   K. Yabe, Y. Motomura, H. Ichikawa, T. Mizuniwa and T. Ohmi, "Responding to the future quality demands of ultrapure water," *Microcontamination*, Vol.7, No.2, pp. 37-46 (1989).

27   K. Yabe, T. Kumagai, H. Ichikawa, S. Akiyama, T. Mizuniwa and T. Ohmi, "Evaluating equipment technologies for future monitoring demands of ultrapure water," *Microcontamination*, Vol.7, No.3, pp. 25-30 (1989).

28   Y. Kasama, Y. Yagi, T. Imaoka and T. Ohmi, "Low dissolved oxygen ultrapure water systems for native oxide free wafer processing," Ext. Abstract, *1990 Int. Conf. Solid State Devices and Materials*, Sendai, pp. 1139-1142.

29   H. Kikuyama, N. Miki, J. Takano and T. Ohmi, "Developing
      property-controlled, high-purity buffered hydrogen fluorides for
      ULSI processing," *Microcontamination*, Vol.7, No.4, pp. 25-28
      (1989).

30   S. Hashimoto, M. Kaya and T. Ohmi, "Improving and maintaining
      electronics-grade chemical quality requires technological
      advances," *Microcontamination*, Vol.7, No.6, pp. 25-28 (1989).

31   M. Morita, T. Ohmi, E. Hasegawa, M. Kawakami and K. Suma,
      "Control factor of native oxide growth on silicon in air or in
      ultrapure water," *Appl. Phys. Letters*, Vol.55, No.6, pp. 562-564
      (1989).

32   M. Morita, T. Ohmi, E. Hasegawa, M. Kawakami and M. Ohwada,
      "Growth of native oxide on a silicon surface," *J. Appl. Phys.*,
      Vol.68, No.3, pp. 1272-1281 (1990).

33   M. Morita, T. Ohmi, E. Hasegawa and A. Teramoto, "Native oxide
      growth on silicon surface," Ext. Abstracts, *1990 Int. Conf. Solid
      State Devices and Materials*, Sendai, pp. 1063-1066.

34   M. Miyashita, M. Itano, T. Imaoka, I. Kawanabe and T. Ohmi,
      "Dependence of thin oxide films quality on surface micro-
      roughness," Dig. Technical Papers, *1991 Symp. VLSI Technology*,
      Oiso, pp. 45-46, May. 1991.

35   T. Ohmi, T. Ichikawa, T. Shibata, K. Matsudo and H. Iwabuchi, "In
      situ substrate-surface cleaning for very low temperature silicon
      epitaxy by low-kinetic-energy particle bombardment," *Appl. Phys.
      Letters*, Vol.53, No.1, pp. 45-47 (1988).

36   T. Ohmi, K. Matsudo, T. Shibata, T. Ichikawa and H. Iwabuchi,
      "Low-temperature silicon epitaxy by low-energy bias sputtering,"
      Appl. Phys. Letters, Vol.53, No.5, pp. 364-366 (1988).

37   T. Ohmi, H. Iwabuchi, T. Shibata and T. Ichikawa, "Electrical
      Characterization of epitaxial silicon films formed by a low kinetic
      energy particle process," *Appl. Phys. Letters*, Vol.54, No.3, pp.
      253-255 (1989).

38   T. Ohmi, T. Ichikawa, T, Shibata and H. Iwabuchi, "Crystal
      structure analysis of epitaxial silicon films formed by a low kinetic
      energy particle process," *Appl. Phys. Letters*, Vol.54, No.6, pp.
      523-525 (1989).

39   T. Ohmi, T. Ichikawa, H. Iwabuchi and T. Shibata, "Formation of device-grade epitaxial silicon films at extremely low temperatures by low energy bias sputtering," *J. Appl. Phys.* Vol.66, No.10, pp. 4756-4766 (1989).

40   T. Ohmi, K. Hashimoto, M. Morita and T. Shibata, "In situ doped epitaxial silicon film growth at 250 °C by an ultra-clean low-energy bias sputtering," Tech. Dig., *1989 Int.Electron Devices Meeting*, Washington D.C., pp. 53-56.

41   T. Ohmi, K. Hashimoto, M. Morita and T. Shibata, "Study on further reducing the epitaxial silicon temperature down to 250 °C in low-energy bias sputtering," *J. Appl. Phys.* Vol.69, No.4, pp. 2062-2071(1991).

42   N. Miki, H. Kikuyama, I. Kawanabe, M. Miyashita and T. Ohmi, "Gas-phase selective etching of native oxide," *IEEE Trans. Electron Devices*, Vol.37, No.1, pp. 107-115 (1990).

43   T. Ohmi, "Closed system essential for high-quality processing in advanced semiconductor lines," *Microcontamination*, Vol.8, No.6, pp. 27-32 (1990).

44   T. Ohmi and T. Shibata, "Developing a fully automated, closed wafer manufacturing system," *Microcontamination*, Vol.8, pp. 25-32 (1990).

45   N. Awaya and Y. Arita, "Selective chemical vapor deposition of copper," Digest of Technical Papers, *Symposium on VLSI Technology*, Kyoto, Japan, May 1989, p. 103.

46   N. Awaya and Y. Arita, "High rate deposition copper CVD," Digest of Technical Papers, *1991 Symposium on VLSI Technology*, Oiso, May, 1991, p. 37.

47   J.S.H. Cho, H-K. King, M.A. Beiley and S.S. Wong, "Copper interconnection with tungsten cladding for ULSI," Digest of Technical Papers, *1991 Symposium on VLSI Technology*, Oiso, May 1991, p. 39 .

48   M. Sosnowski and I. Yamada, "Deposition of thin films of copper on silicon substrates at low temperature by the ICB Method," *Nucl. Instrum. Master. Phys. Res.*, B37/38, 874 (1989).

49   G.H. Takaoka, J. Ishikawa and T. Takagi, "Surface and interface characteristics of Cu films deposited by ionized cluster beam," *J. Vac. Sci. Technol.*, A8, 840 (1990).

50   T. Ohmi, T. Saito, T. Shibata and T. Nitta, "Room-temperature copper metallization for ultra large scale integrated circuits by a low-kinetic-energy particle process," *Appl. Phys. Letters*, Vol.52, No.26, pp. 2236-2238 (1988).

51   T. Ohmi, T. Sato, M. Otsuki, T. Shibata and N. Nitta, "Formation of copper thin films by a low kinetic energy particle process," *J. Electrochem. Soc.*, Vol.138, No.4, pp. 1089-1097 (1991).

52   M. Kiuchi, K. Fujii, T. Tanaka, M. Sato and F. Fujimoto, "Microstructure of titanium nitride films produced by the dynamic mixing method," *Proc. 12th Int. Conf. Atomic Collisions in Solids*, Okayama, pp. 649-652 (1987).

53   K. Tsubouchi, K. Masu, M. Tanaka, Y. Hiura, T. Ohmi, N. Mikoshiba, S. Hayashi, T. Marui, A. Teramoto, T. Kajiyama and H. Soejima, "Development of scanning μ-RHEED microscopy for imaging poly-crystal grain structure in LSI," Extended Abstracts, Tokyo 1989, p. 217.

54   K. Tsubouchi, S. Masu, M. Tanaka, Y. Hiura, T. Ohmi, N. Mikoshiba, S. Hayashi, T. Marui, A. Teramoto, T. Kajikawa and H. Soejima, "Development of scanning, μ-RHEED microscopy for imaging poly-crystal grain structure in LSI," *Jpn. J. Appl. Phys.* 28 (1989) L2075.

55   M. Miyawaki, S. Yoshitake, T. Sato, S. Saito and T. Ohmi, "Improvement of aluminum-Si contact performance in native-oxide-free processing," *IEEE Electron Device Letters*, Vol.11, No.10, pp. 448-450 (1990).

56   H. Kuwabara, M. Otsuki and T.Ohmi, "Ideal metal silicon contact formation by clean-nitrogen-seal processing," Extended Abstracts, the *179th Electrochemical Society Meeting*, Washington D.C., May 1991, Abstract No.311, pp. 463-464 (1991).

57   M. Otsuki, T. Takewaki, H. Kuwabara, T. Shibata, T. Ohmi and T. Nitta, "High performance copper metallization for ULSI interconnects," Extended Abstracts, *1991 Int. Conf. Solid State Device and Materials*, Yokohama, August 1991, pp. 186-188

58   T. Nitta, T. Ohmi, T. Shibata, T. Hoshi, S. Sakai, K. Sakaibara and S. Imai, Submitted to *JECS*.

59   T. Ohmi, H. Kuwabara, S. Saitho and T. Shibata, "Formation of high quality pure aluminum films by low kinetic energy particle bombardment," *J. Electrochem. Soc.*, Vol.137, No.3, pp. 1008-1016 (1991).

60   W. Kern and D.A. Poutien, "Cleaning solutions based on hydrogen peroxide for use in silicon semiconductor technology," *RCA Review*, Vol.31, pp. 187-205, June (1970).

DISCUSSION

M. Sommer (ABB-IXYS, San Jose, CA, USA)

Did you find any indication that changes occur if localized areas in a power device might exceed the processing temperature ?

T. Ohmi

There is no data as yet, but we are planning to implement a metal-electrode in order to test high power devices. After metallization, I like to insist that all processing must be carried out at less than about 500 °C. In order to improve high speed performance, resistance and inductance must be reduced. Therefore, metallization technology becomes very important.

M. Campagna (ABB Corp. Research, Baden)

Have you looked into other refractory metals except aluminium and copper ?

T. Ohmi

For the metallization, the most important characteristic of performance is low resistance. So far, everybody is using refractory metals. Because VLSI fabrication requires processing temperatures of over 1000 °C, only high melting temperature refractory metals are acceptable for present production. According to our research results, however, all processing temperatures are decreasing down to less than about 600 °C. So other metals like copper or pure aluminium will become attractive. The most important points are very low resistivity and high electron migration resistance.

T. Stockmeier (ABB Corp. Research, Baden)

In the metallization, both the electrical and the mechanical properties are very important. Can you comment on the adhesion of a copper metallization on large areas of silicon ? Is it feasible to produce very thick metal layers with the technique you have presented here?

T. Ohmi

The adhesion capabilities of metals to silicon oxide films is a first requirement. In general, the copper films do not adhere on the silicon dioxide films. In order to improve this situation, we have eliminated the moisture molecules from the silicon dioxide film surface. The resulting

The adhesion capabilities of metals to silicon oxide films is a first requirement. In general, the copper films do not adhere on the silicon dioxide films. In order to improve this situation, we have eliminated the moisture molecules from the silicon dioxide film surface. The resulting adhesion capabilities are very similar to those of aluminium alloy metals. All processing has been carried out at very low temperature. Thus, the mechanical strength is established and the copper metallization becomes acceptable. Furthermore, a large area copper metallization exhibits less than 1 or 2 % non-uniformities.

Low energy ion bombardment processing during the film growth aids in realizing a very tight adhesion of films not only to metals. The stress problems, related for example to very hard metals, can be overcome not only for copper but also for refractory metals; 1 to 2 μm of metallization will be acceptable.

W. Fichtner (ETHZ, Zürich)

Did you do any investigations of the quality of thin oxides, grown on materials that have been processed at high temperatures like 1100 to 1200 °C ?

T. Ohmi

I have described the importance of the surface smoothness. Its degradation results from the wet chemical cleaning steps as well as from high temperature processing, particularly if performed in an inert gas atmosphere. If you increase the temperature for the CZ wafers in an inert gas ambient to greater than 800 or 900 °C, the surface smoothness is degraded very drastically. E.g. if the temp. is increased to 1000 °C or 1100 °C in an argon or nitrogen gas ambiance, the surface microroughness is increased up to 100 Å or greater.

J. Gobrecht (ABB Corp. Research, Baden)

The surface roughness you have demonstrated of 1 or 2 Å corresponds basically to the natural atomic roughness. How is this achieved and how can you maintain such a value over large areas ?

T. Ohmi

In Japan, the improvement of substrate surface polishing technologies has led to average surface microroughnesses of just about 1 Å for wafers up to six inch diameter. Average surface microroughness of 1 Å means that the peak to peak value is corresponding to 10 Å . This is a very advanced wafer surface polishing technology which we are improving even more in order to guarantee eventually atomic order flatness.

THE MOS CONTROLLED THYRISTOR AND ITS LIMITS

Friedhelm Bauer

Asea Brown Boveri Corporate Research
Baden, Switzerland

ABSTRACT

The MOS Controlled Thyristor (MCT) is turned on and off by the action of a high impedance MOS gate electrode. Due to a cellular, fine-structured cathode, this new generation of thyristors opens the way to using simple gate drive circuits in high power applications and offers considerable improvements of the dynamic behaviour as compared to conventional Gate Turn-Off thyristors (GTO's).

This paper gives an introduction to the basic MCT operation and device concept. Special emphasis is put on the physical effects which limit the maximum turn-off capability of small size MCT's, i.e. single cell devices as well as devices with up to 20,000 cells. MCT's with only a few cells are able to control extremely large current densities of some thousand A/cm^2 at small anode voltages. This level of performance decreases strongly in larger devices with several thousands of cells due to inhomogeneous current density distributions which eventually end up in current filamentation during turn-off. Despite this fact, current densities of up to 70 A cm^{-2} have been turned off under snubberless, clamped inductive load conditions into anode voltages of 2000 V. This places the MCT among the premier candidates for an efficient and simple switch in future high power applications.

Power Semiconductor Devices and Circuits, Edited by A.A. Jaecklin
Plenum Press, New York, 1992

31

1. INTRODUCTION

In recent years, progress in the field of power electronics has been closely related to the emergence of intelligent or "smart" silicon power devices. This trend has been particularly strong in low and medium power applications[1] and will become even stronger in the future. It is also predicted that intelligent switches will start penetrating the field of high power applications such as traction and industrial drives. Electronic intelligence has always been intimately related to the way of controlling conducting and non-conducting states of a semiconductor device. For about two decades, the rapid progress of Very Large Scale Integrated (VLSI) circuits and in particular that of Dynamic Random Access Memory (DRAM) chips has demonstrated year after year the extraordinary capabilities of the Metal Oxide Semiconductor Field Effect Transistor (MOSFET)[2]. It is therefore not surprizing that power device designers have adopted similar techniques to integrate MOS control into bipolar power devices in order to turn them on and off with simple and cheap gate drive units. The first member of this new familly of power devices being produced in very large quantities is the Insulated Gate Bipolar Transistor (IGBT)[3,4]. Its success is not only based on the simple gate drive requirements. Due to its extremely high degree of ruggedness, the IGBT can operate without passive protection components, commonly needed for the safe operation of conventional power switches. Both factors contribute to the emergence of simple, inexpensive, small and light-weight power electronic systems.

As the next logical step in this development, device designers try to integrate MOS control into thyristors; to date, only thyristors provide sufficiently low losses as required for switches to be used in high power applications. The low on-state loss of thyristors is a result of double sided injection. Their action relies on coupling two bipolar transistors which leads to regenerative feedback, called latching. However, a latched device is considerably more difficult to control as a non-latching device such as the IGBT. It may be for this reason, that the development of the MOS controlled thyristor (MCT) is still in an infant state although research and development activities have been going on for some time[5,6].

This paper attempts to provide a quantitative description of important characteristics and limits of the MCT. The data reported here were measured at an early stage of our ongoing device development. To gain an understanding of the physical mechanisms governing the device operation, including turn-off failure modes, extensive use was made of two-dimensional process and device simulation tools, running in a supercomputer environment[7]. Concerning MCT performance limits, it should be stressed, that the data presented in this paper have a preliminary nature: throughout the paper it will become clear that practical performance limits are strongly dependent on the maturity of available semiconductor manufacturing technologies. With the rapid progress in this field, important improvements of MCT performance limits can be expected in the future.

Among the large number of electrical, thermal and mechanical limits of a power semiconductor device, we will focus on the most fundamental issue for a gate turn-off power switch: the maximum anode current which can be safely turned off as well as the physical mechanisms leading to a turn-off failure. Although MCT's are expected to have inherent advantages in the forward voltage drop for any application involving voltages exceeding 300 V, we will concentrate on high voltage implementations of the MCT. With respect to the blocking capability, the MCT is a typical thyristor: the breakdown voltage is determined by the thickness and the doping of the silicon wafer as well as the particular high voltage junction termination structure chosen. The devices investigated in this paper were able to block anode voltages as high as 3.5 kV; however, this value is not the result of a physical limit. Today, silicon power devices are able to block voltages approaching 10 kV. Some of our MCT's were electron-irradiated to set their on-state voltage to the level of comparable GTO's. Since the discussion of MCT limits concerning minimum device losses is beyond the scope of this paper, we will not return to blocking and conduction properties furtheron in this paper. In addition, concepts for lower voltage lateral MCT's[8] which may open the way to new power electronic applications requiring bilateral MOS controlled switches, are not considered in this paper.

The following chapter reviews the basic concept of the MCT and its different implementations are presented. Technologically based arguments are given, which have led us to the decision to utilize a DMOS approach for the realization of a high voltage MCT[9]. From the study of turn-off failures occuring in small MCT test samples we shall develop a picture of the physical effects leading to the failure in the next chapter. This understanding is of prime importance for the identification of the nature of turn-off failures in larger MCT devices with up to 20,000 elementary cells. It is also useful for finding an interpretation for the dependence of the maximum anode turn-off current on DMOS cell size as well as the active device area. In the final discussion, we attempt to estimate the absolute turn-off performance limit for power devices; it will be shown, that this limit has already been reached with smaller MCT devices; however, this task becomes increasingly more difficult, when larger devices are to be manufactured. The chance of realizing very large MCT's with spacially homogeneous turn-off characteristics may be actually rather limited; however, MOS power device technologies, which are basically the same for power MOSFET's, IGBT's, MCT's as well as smart power integrated circuits, are strongly affected by the development of VLSI technologies. Thus, DMOS technology will advance and become ever more sophisticated in the future. Together with the physical understanding gained to date, the MOS controlled thyristor will soon reach ratings of current, voltage, and power permitting its use in many traditional as well as new high power applications.

2. THE BASIC MCT DEVICE CONCEPT

At the beginning of this chapter, we have first to face the fact that it is more or less impossible to define one unique or one "basic MCT device concept". "MCT" must be understood as a term describing a large family of semiconductor devices, whose conduction states are controlled by the action of one or more integrated MOSFET's. At the moment, we are far from having identified all members of this device family. In the following, some members of the MCT family are mentioned. As a first criterion, MCT's may be classified as emitter switched or emitter shunted thyristors. Emitter switched thyristors (EST) incorporate a MOSFET in series with the four-layer thyristor structure[10]. As outlined below, the emitter may also be shunted such that the current takes an alternative path in the device. Throughout this paper, we will refer to the latter device structure, being the oldest known MCT concept, as the "MCT". Combinations of emitter shunted and emitter switched MCT's, which might have extremely low switching losses, have not yet been discussed in the literature.

The operation of thyristors in the conducting on-state is commonly explained by the regenerative feedback of two coupled complementary bipolar transistor structures. Both n^+ and p^+ emitters inject minority carriers into their respective p- and n-bases; electrons and holes form a characteristic, well known distribution of the carrier plasma, where the plasma density attains maximum values at both emitters and reaches a minimum in between. Coupled to the minority carrier injection is an emitter-to-base voltage which increases with increasing densities of injected minority carriers according to the Boltzmann relation. In the following, we will refer to this potential drop as the junction voltage. Characteristic values are in the range of 700 to 800 mV for anode current densities at and above 100 A/cm^2. In order to inhibit the injection of minority carriers from the emitter effectively, the junction voltage must be reduced to approximately 200 to 300 mV in an emitter shunted MCT; this is a consequence of the exponential nature of the Boltzmann relation. Once the injection of minority carriers from the emitters ceases, the latched thyristor on-state is interrupted and the device is able to support high anode voltages in the blocking state.

The ability of emitter shunted MCT's to turn-off large anode currents relies on the behaviour outlined above; integrating a MOSFET between any emitter and its associated base region permits the shorting or shunting of the injecting emitter to its base by just switching the integrated MOSFET into the conducting state. Several important aspects should be noted at this point: although the integrated emitter short will immediately cut off the emission of minority carriers once it is activated by a MOS gate signal, anode current continues to flow through the device for some time since it is enforced by the external circuit. If a clamped inductive load is switched, using an unprotected or unsnubbered power switch, the device has to support the full anode current until the anode voltage has reached the supply voltage. Obviously, at high currents and voltages large amounts of heat can be generated during these transients. The device must be able to handle these dynamic power losses as well as the static losses safely. For the MCT this means that during a turn-off transient the activated MOS shorts have to conduct the full anode current until the clamping voltage level is reached. Taking into account the low value of the effective drain voltage, being identical to the junction voltage, quite severe restrictions are imposed on the resistive components of the internal current commutation path.

At this point, we can already formulate some general design guidelines for an MCT. All resistive components included in the current shorting path must be reduced to a minimum. The carriers injected from the opposite non-shorted emitter constitute the anode current which continues to flow for some time during the turn-off transient. The charge carriers will develop a potential drop across the resistive components of the FET shorting path. The value of this potential difference determines the efficiency of shorting the emitters. In other words, if the potential drop across the FET path is larger than 200 to 300 mV, the junction voltage will remain at these high levels and the minority carrier injection of the emitter is not effectively eliminated. A turn-off failure may result from such a condition. It is simple to calculate the maximum current allowed to flow in the FET shorting path once the total shorting resistance is known.

Among the resistive components in the FET path we can easily identify the resistance of the inversion layer of the integrated MOS short. With respect to the device design, MOS channel lengths as short as possible should be used. There are additional resistive drops associated with the current path in the p-base during turn-off. In particular, the lateral and vertical specific resistance of the base layer associated with the shorted emitter must be minimized. In principle, this can be accomplished by increasing the doping density of the base layer. There are, however, practical limits. First, the firing sensitivity of the thyristor degrades severely at high base doping levels. In addition, integrating the MOS short into a base layer requires base doping levels, which permit sufficiently high inversion layer mobilities and resonably small threshold voltages. In a realistic case, the surface doping density of the base layer should not exceed 10^{17} cm^{-3}. The corresponding specific base resistance just underneath the emitter is very high; as a consequence, low resistance current shunting paths in the base layer must be realized, creating very small emitter islands and placing the MOS controlled shorts as close as possible to the emitters. It is obvious, that VLSI technologies are best suited to provide the resolution of such small feature sizes.

3. EMITTER SHUNTED MCT IMPLEMENTATIONS

So far, the discussion has been general with respect to the emitter chosen for the integration of the MOS shorts. It is clear that the concept can be applied to either the cathode or the anode emitter. Integrating MOS controlled emitter shorts on the cathode as well as on the anode side of the device is conceptually possible. However, equipment cost and complexity of the processing hamper a realization in practice. If shorts to the anode or p^+ emitter are chosen, the gate control signal has to be supplied independent of the instantaneous high positive potential of the anode. This complicates the gate drive circuitry considerably. A simple gate drive is possible if shorts to the n^+ emitter of the cathode side are chosen. In this case, the potential of the cathode is always fixed at the ground level.

The discussion in the last chapter has shown, that a small resistance of the MOS channel is extremely important for achieving large turn-off current densities. The preferred MOSFET concept would thus be an n-channel device, providing a roughly threefold improvement in inversion layer mobility compared to a p-channel device of equal channel length. In principle, the MOS short may be integrated as a planar or a vertically Double Diffused MOS (DMOS) structure. With this additional degree of freedom, four fundamental emitter shunted MCT structures can be described: n^+ emitter shorting is obtained by integrating either a p-channel DMOS structure[11] or a planar n-channel MOSFET[5,12] into the p-base. Both concepts have been experimentally investigated. For p^+ emitter shorts, either an n-channel DMOSFET[6] or a p-channel planar short-structure may be used. The combination of p^+ emitter shorts, accomplished by a p-channel lateral MOS, has not been reported in the literature so far.

MCT implementations with planar shorts are a last important item. All results available to date clearly point out that lateral MCT's are considerably more susceptible to parasitic effects[12] as compared to DMOS type MCT's. Suppressing these parasitic structures would require an addition of complicated processing steps to a technology, which is already more difficult to control as state-of-the-art DMOS technologies; for the integration of short-channel planar MOS shorts into a thyristor structure, very small feature sizes in the micron range must be resolved. Although VLSI technologies are able to provide this range of resolution, using micron size features in an MCT chip will limit the chip size to approximately 1 cm^2. This is due to the fact that small features are more susceptible to process related defects as compared to larger structures. DMOS type structures with their uncritical larger features experience this yield limit only at considerably larger chip sizes Moreover, MCT's with planar shorts are bound to a double layer metallization scheme. This is not the case for DMOS type devices. For these technological reasons, it is considerably more simple to fabricate DMOS type MCT's; only such devices are considered in the remainder of this paper.

4. OPTIMIZED HIGH VOLTAGE DMOS MCT STRUCTURES

For the manufacture of high voltage power devices, usually neutron transmutation doped n-type float zone silicon wafers are used. Today, this is the only silicon material offering the tight tolerances in dopant density fluctuations, required to achieve high manufacturing yields for large area power devices. In the finished device, the lowly doped substrate constitutes the thick n⁻ base layer which blocks the high anode potential as long as the device is turned off.

Shorting of anode versus cathode emitter shall now be considered in terms of the physical effects at turn-off of an MCT. Starting with a DMOS type MCT by shorting the p⁺ emitter, the injection of holes from the p⁺ emitter will be stopped immediately after applying the MOS gate signal. As already discussed, the current continues to flow for some time in the external circuit. Since the supply of holes has vanished, this current is supported solely by electrons injected from the cathode emitter. Once the anode voltage reaches a high level, these electrons may create additional carriers by impact ionization. Eventually, a carrier avalanche is generated, resulting in a turn-off failure. In the case of an equivalent MCT with MOS shorts on the cathode side, holes are now the carriers triggering impact ionization. Compared to electrons, holes have substantially smaller ionization rates. As a consequence, the MCT with n⁺ emitter shorts is much less susceptible to the occurence of catastrophic avalanche breakdown as compared to its counterpart with p⁺ emitter shorts.

Having thus defined the optimum MCT structure for high voltage applications, the device is shown schematically in Fig. 1. The cathode consists of a large number of elementary cells; each cell accomodates an individual emitter completely surronded by the p⁺ short, which corresponds to the drain of the integrated MOSFET. The polysilicon gate covers the p-base between the n⁺ emitters. P-type channels are created by applying a negative potential to the gate electrode, resulting in the formation of an inversion layer at the surface of the vertically diffused n-well regions. Neglecting the n⁻ base layer, the device structure corresponds perfectly to that of a p-channel DMOS power MOSFET. The reasons leading to the ever increasing number of cells per silicon area in the race for minimizing the on-resistance of low-voltage power MOSFETs can be directly traced to the performance optimization of DMOS type MCT's.

For the fabrication of emitter shunted MCT's, we have utilized a simple planar junction termination with a lateral doping gradient[13] and a high voltage blocking capability of 3.5 kV. The n-type silicon wafers had a thickness of 500 μm and a resistivity of 200 Ωcm. MCT's with even higher blocking voltages of 4.2 kV are reported by Stoisiek et al.[14]. As for gate turn-off thyristors (GTO's), the level of blocking voltage of an MCT is determined by the thickness and resistivity of the silicon wafers used. The process steps with the highest thermal budgets - junction termination drive-in and formation of the p⁺ emitter - were carried out simultaneously at the beginning of the process sequence. After implantation and drive-in of the p-base, the MCT process is similar to a p-channel power DMOSFET process.

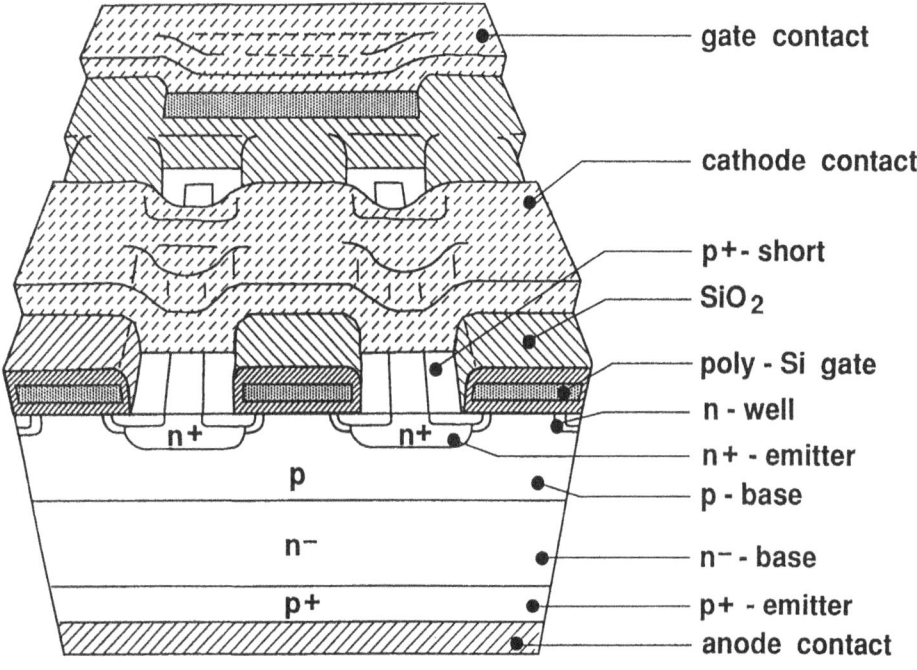

Fig. 1. Three-dimensional cellular type MCT structure.

As mentioned earlier, it is imperative for a high performance MCT to keep channel lengths of the integrated shorts as short as possible. The combined formation of the submicrometer MOS device (L = 0.5 μm) integrated with the thyristor structure required a trade-off analysis of several complex process parameters. The use of results obtained from numerical process and device simulation studies in one and two dimensions[7] has been a helpful guide toward an efficient and successful device design. By properly adjusting implantation dose and annealing treatments, it was possible to obtain a high-performance, DMOS device with a threshold voltage of - 3 V. The influence of the parasitic bipolar p-n-p transistor inherent to the DMOS MCT structure resulted in an almost negligible increase of the thyristor on-state voltage by 200 mV.

5. MCT TURN-OFF FAILURE MODES

The development of modern GTO's has shown, that the performance limit, i.e. , the maximum anode turn-off current, does not scale linearly with the number of cathode fingers[15]. This is due to a current redistribution taking place during the turn-off transient. As a result, the turn-off capability degrades with increasing size of the device. The phenomenon may occur in all four-layer devices characterized by a latched or regenerative mode in the on-state. The MCT also falls into this category of power devices; it has been known for some time that large area MCT's suffer from current redistributions during turn-off[14]. However, no clear picture exists concerning the influence of MCT design and layout on the extent of the nonlinearity as well as the nature of the physical effects which are its cause.

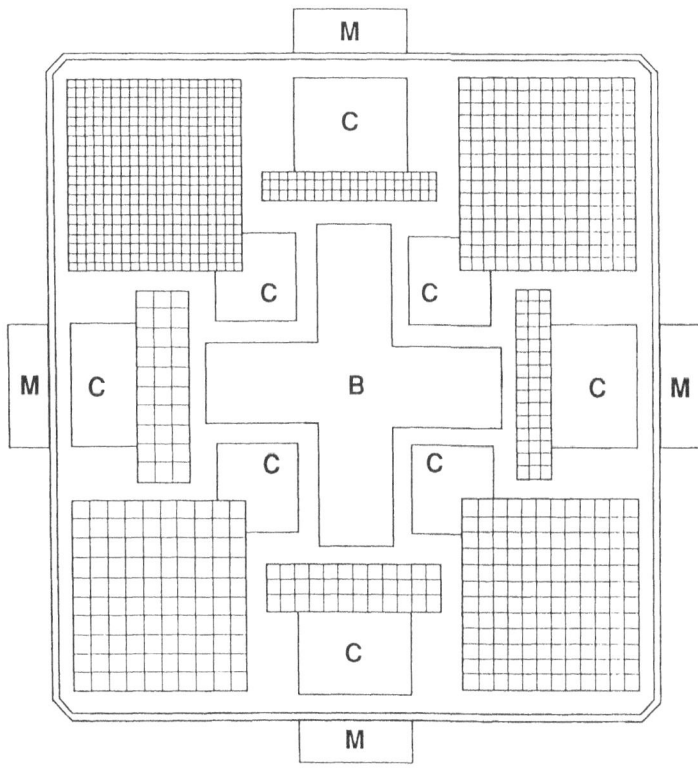

Fig. 2. MCT test device layout: triple cell lines and square cell arrays with common MOS turn-off pads (M), bipolar turn-on gate pads (B), and individual cathode pads (C); cell pitch between 15 and 30 μm.

In order to fully investigate the MCT concept, we have designed a large variety of different test structures. Device structures were fabricated as single cells, one-dimensional ensembles (single and multiple cell lines), and two-dimensional configurations. Figure 2 shows an MCT test device layout with four triple cell lines and four square cell arrays with an active cathode area of 9 • 10^{-4} cm^2 each. In this particular case, cell pitches vary from 15 to 30 μm. A typical MCT cell layout is shown in Fig. 3. For larger MCT's with active cathode areas up to 0.56 cm^2, square cells with a pitch of 20 μm were utilized. The total number of cells for such devices exceeds 100,000. Because of the large variations in device size, turn-off currents span a range from several mA up to several A. We used a test set-up with a clamped inductive load to determine the maximum controllable anode current as a function of the applied anode voltage. Since we were interested in the safe operating area of the snubberless MCT device, passive protection circuits were not employed during these experiments. A schematic picture of the test set-up is shown in Fig. 4. The same test arrangement was used for the study of turn-off failures. A sophisticated pulse sequencer, integrated in the gate drive unit, made sure that a device undergoing a turn-off failure was turned on again a few microseconds after an unsuccessful attempt to turn it off. Between successive turn-off events, ample time was made provided to dissipate the heat generated during the turn-off failure.

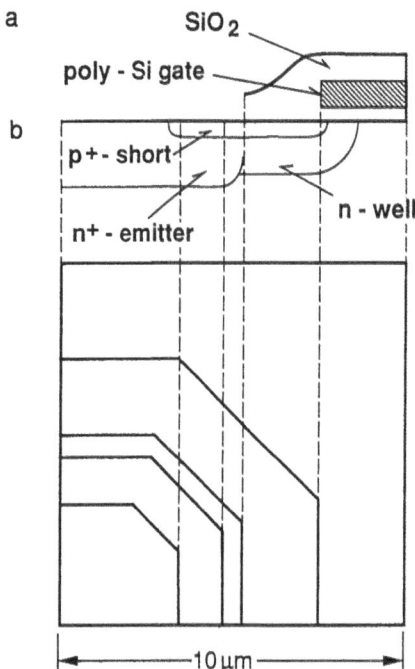

Fig. 3. MCT square cell layout schematic: a) cross-section; b) top view.

We will first focus our discussion on small MCT test devices, which consist of only a few cells and do not suffer from a spatially inhomogeneous current distribution during turn-off. Figure 5 shows an overlay of five turn-off events (I_a in steps of 0.1 A from 0.1 to 0.5 A). The MCT under consideration has an active area of $9 \cdot 10^{-4}$ cm^2; thus the current density increases in steps of 110 A/cm^2 up to a level of 550 A/cm^2. The anode clamping voltage is fixed to 500 V. Looking at the corresponding voltage traces, it can be seen that the anode voltage increases faster for larger anode current levels. This is a common behaviour of all power devices: the current flowing in the on-state provides the driving force for the anode voltage built-up and therefore determines the speed at which sweep-out of carriers from the growing depletion layer occurs. Recombination in the emitter layers is also enhanced at higher forward current levels. In Fig. 6, we show a series of turn-off events, where the current density is increased to even larger values (370 to 1100 A/cm^2). The MCT consists of 60 cells with a pitch of 15 μm and the anode voltage is limited to 50 V. At a critical level of anode current, dV_a/dt saturates. When the anode current is further increased, dV_a/dt starts to decrease again and more time is required for the anode voltage to reach the clamping level. A physical interpretation of this behavior was found using two-dimensional device simulation. Beyond current densities of approximately 400 A/cm^2 (for a cell pitch of 20 μm), the voltage drop across the current shorting path is too high to completely eliminate the emission of electrons from the n+ emitter (see the discussion in Chapter 2). This means that a reduced supply of electrons to the p-base and the depletion layer exists during the turn-off event. As a consequence, more conductivity modulated charge is present in the device and more holes must flow via the shorting path: the build-up of the anode voltage takes considerably more time, since more charge has to be extracted.

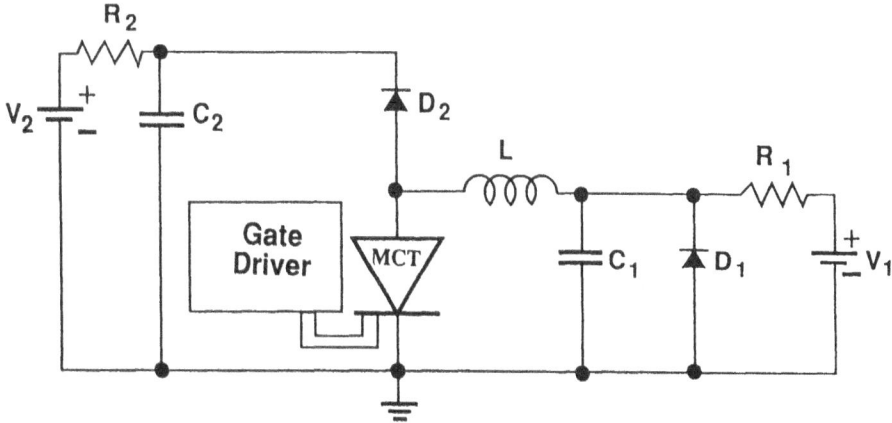

Fig. 4. Test arrangement for snubberless switching of a clamped inductive load.

So far, the only result of the increase of dV_a/dt during turn-off of high values of the anode current results in increased switching losses. However, a device failure must not necessarily occur as long as the anode voltage remains at a low level and the losses can be extracted. In experiments using single cell MCT's we have found, that a cell with a pitch of 15 μm is able to handle current densities of 9,000 A/cm^2 before the amount of charge injected from the n+ emitter is equal to the charge extracted via the activated MOS shorts. It was observed, that devices with an actice area of 1.6 mm^2 were able to turn-off currents in excess of 18 A. The corresponding current density exceeds 1,000 A/cm^2.

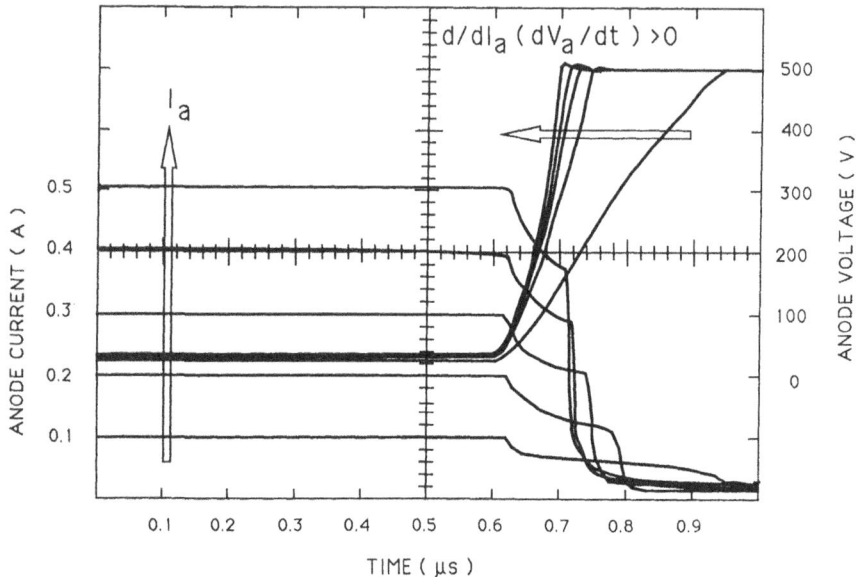

Fig. 5. Snubberless clamped inductive turn-off transients using a 400 cell MCT array (active cathode area: $9 \bullet 10^{-4}$ cm^2); the anode current is increased in steps of 0.1 A (110 A/cm^2) from 0.1 to 0.5 A (550 A/cm^2); the anode voltage is 500 V.

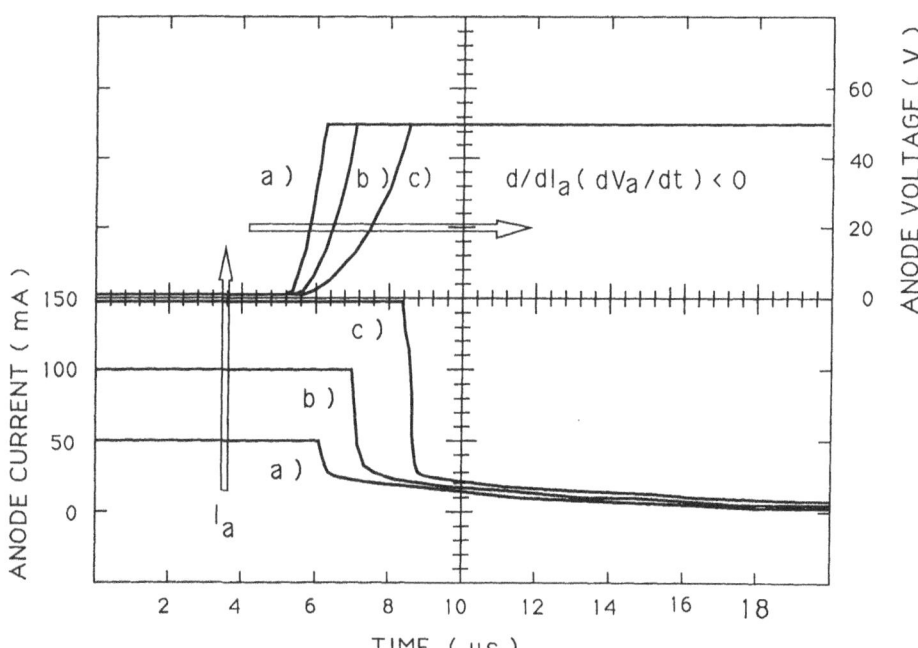

Fig. 6. Snubberless clamped inductive turn-off transients using an MCT with 60 cells (15 μm pitch; active device area: $1.35 \bullet 10^{-4}$ cm^2); a) I_a = 50 mA, b) I_a = 100 mA, c) I_a = 150 mA. For c), the current density reaches 1100 A/cm^2. Stepper lithography was used for the fabrication of this device.

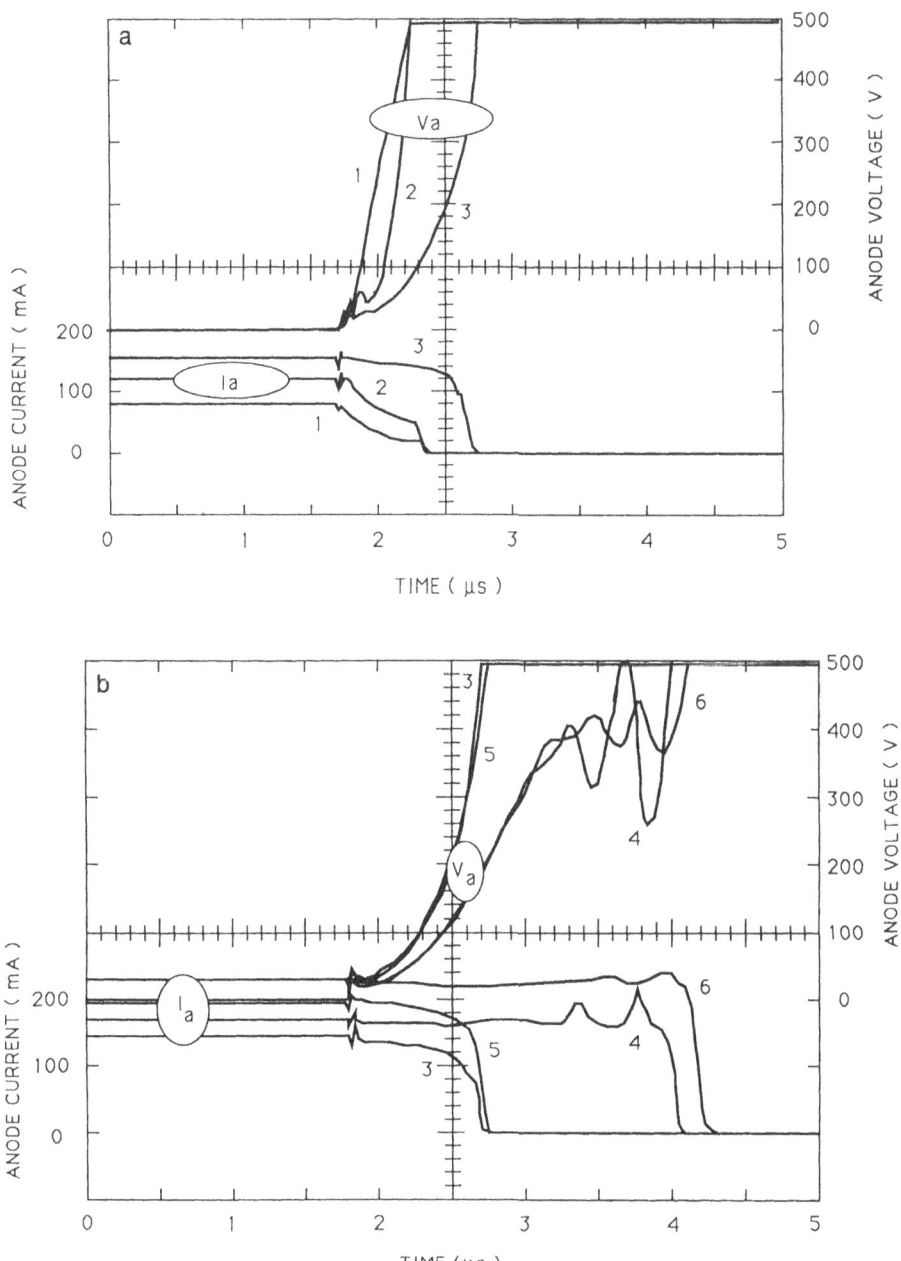

Fig. 7. Snubberless clamped inductive turn-off transients using an MCT with 60 cells (15 μm pitch; active device area: $1.35 \cdot 10^{-4}$ cm^2)
a) 1: I_a = 90 mA;   2: I_a = 110 mA;   3: I_a = 150 mA
b) 3: I_a = 150 mA;   4: I_a = 170 mA;   5: I_a = 190 mA;   6: I_a = 230 mA.

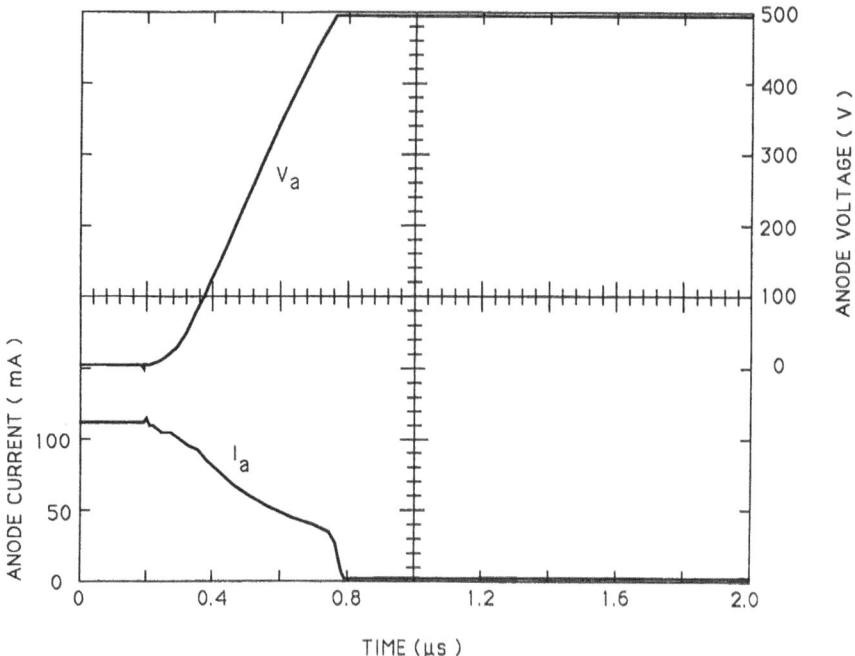

Fig. 8. Snubberless clamped inductive turn-off transient using a 400 cell MCT array (active device area: $9 \cdot 10^{-4}$ cm^2). The device was fabricated using conventional lithography.

We will consider turn-off events at higher anode voltage (mostly 500 V) in the following. For this purpose, the same device already referred to in Fig. 6, is used once again. In Fig. 7a, voltage and current traces labeled 1 correspond to the anode current (I_a = 90 mA or J_a = 700 A/cm^2) resulting in the highest value of dV_a/dt. For anode currents of 110 mA and 150 mA, the delayed increase of anode voltage can be seen clearly. Nevertheless, after the initial delay phase, the anode voltage rises quickly up to the clamping voltage of 500 V. When the anode current is further increased (see Fig. 7b, labels 4 through 6), this is no longer the case: uncontrollable oscillations of the anode voltage may occur, once a level of 300 to 400 V has been reached. Once again, two-dimensional device simulation helped to find an interpretation for this behavior. As the anode current continues to flow at its full on-state value, carriers (in this particular case, electrons from the still injecting n$^+$ emitter and holes from the p$^+$ emitter) travel with saturation drift velocity across the emerging depletion layer. Some of them attain energies sufficient to ionize neutral silicon atoms. Thus, there is a second additional source of charge generated in the depletion layer. If the integrated MOS shorts are still sufficiently conductive to carry the current corresponding to this additional charge, the anode potential may eventually reach the level of the clamping voltage. This is the case for all turn-off transients shown in Fig. 7b. If, however, the rate of ionization is too high, an avalanche is generated in the space charge region. The anode current continues to flow unaltered and the anode voltage cannot rise to the clamping voltage. Safe operation of an MCT under such conditions is impossible; this may lead to the immediate destruction of the device.

The investigation now turns to MCT devices with larger numbers of cells. As already mentioned, large MCT's may suffer from inhomogeneous current distributions during turn-off. However, the cause which leads to the inhomogeneities may be different for smaller and larger cell arrays. We will first concentrate on "small" arrays with some hundred cells. Such arrays are shown in Fig. 2. A turn-off transient of an array with 400 cells of 15 μm pitch (the cells are the same as those used for the devices referred to in Fig. 7) is shown in Fig. 8. The same experimental conditions were used to obtain the results shown in Figs. 7 and 8. Figure 8 displays the anode turn-off current for the highest dV_a/dt for the square cell array. Although it contains approximately seven times the number of cells of the triple cell line devices discussed so far, the increase in turn-off current is very small. Turn-off transients for the small array for currents larger than 110 mA resemble to those shown in Fig. 7b.

If the number of cells is increased to several thousands (Fig. 9 shows some results for a square array with 2,000 cells of 20 μm pitch), the initial increase of dV_a/dt with I_a is again observed; however, turn-off failures now occur as soon as the maximum of dV_a/dt is exceeded. A minor increase in anode current may then trigger the failure. For the MCT array with 2,000 cells (active area: 0.8 mm^2) the coincidence of maximum anode voltage (1,350 V) together with a sharp anode current peak in Fig. 9a is an indication of avalanche induced charge generation in the depletion region. It is believed that the size of the device is still small enough to inhibit the build-up of a strongly focussed current filament, although an inhomogeneous current distribution is obviously at the origin of the turn-off failure. It may be for this reason that the device is able to recover from the failure and finally turns off. Turn-off failures of similar devices were observed with a succession of up to 5 avalanches without immediate destruction of the MCT (Fig. 9b).

Turning to MCT's with still larger active areas (Fig. 10 shows a turn-off transient for a device with 21,000 cells of 20 μm pitch; the active area is 8.4 mm^2), we observe only a single and probably stable current filament in a turn-off failure mode. Such filaments are known to migrate across the device[16]. Although the externally measured anode current has already decreased to comparatively small values, a filament may evolve after the avalanche breakdown. Since the surface occupied by the filament is small compared to the total device area and the hole current from the anode commutates into this small conducting cross section, extremely high local current densities may result. As a consequence, the excess charge generated in the depletion layer is too high to be shunted

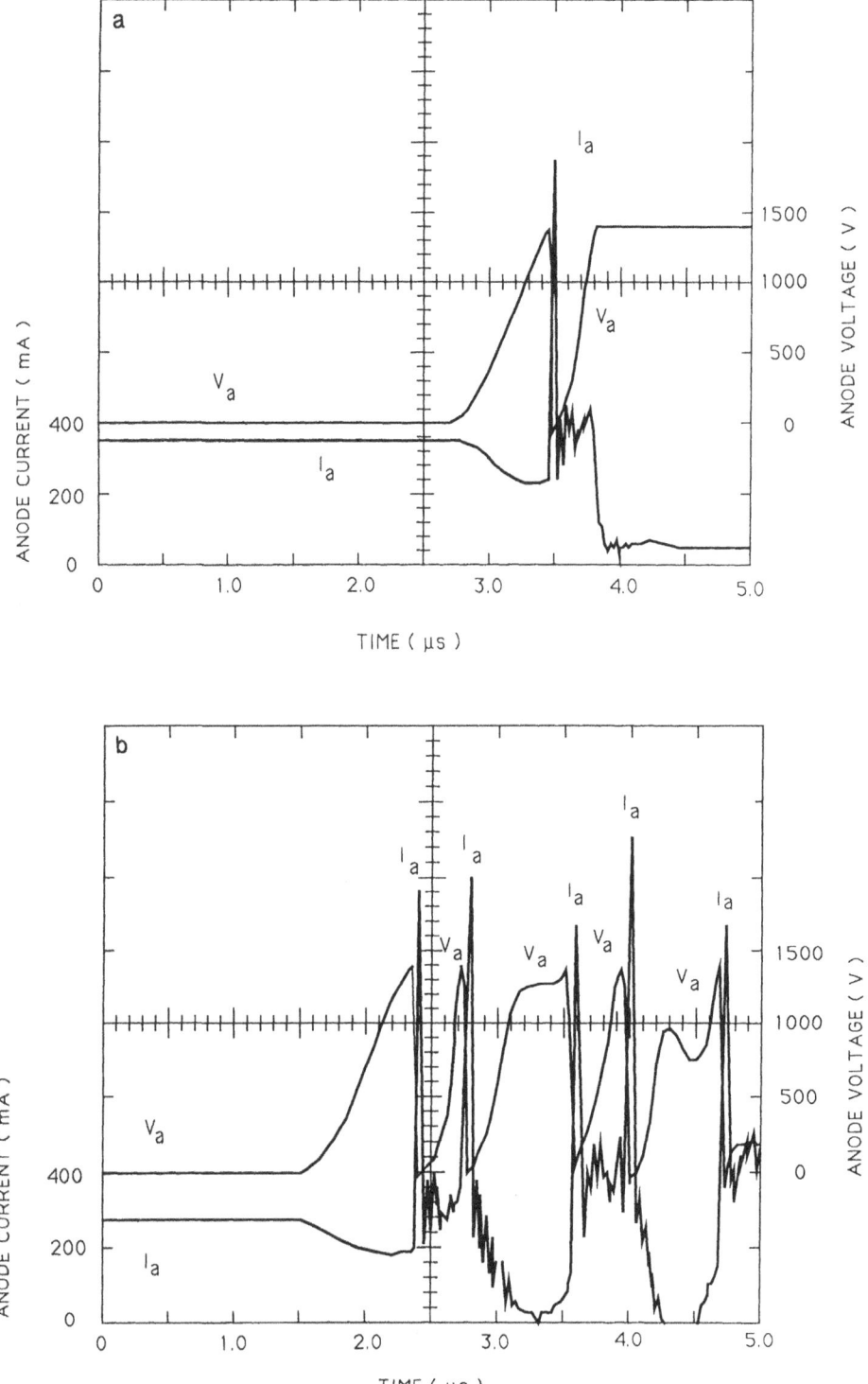

Fig. 9. Snubberless clamped inductive turn-off transient using a 2,000 cell MCT array (active device area: 0.8 mm^2); a) one avalanche breakdown event occurs during turn-off; b) a succession of five avalanche breakdown events occur during turn-off.

by the MOS shorts and the p⁺ emitter. This charge is also responsible for sustaining the emission of the cathode emitter. An operating mode with at the same time full anode current and an anode voltage of 100 V or more is established. The locally generated losses attain such high levels, that the generation of a hot spot is possible. In our test arrangement, a device undergoing a turn-off failure was switched back into the conducting state a few microseconds later. By this means, no permanent damage results.

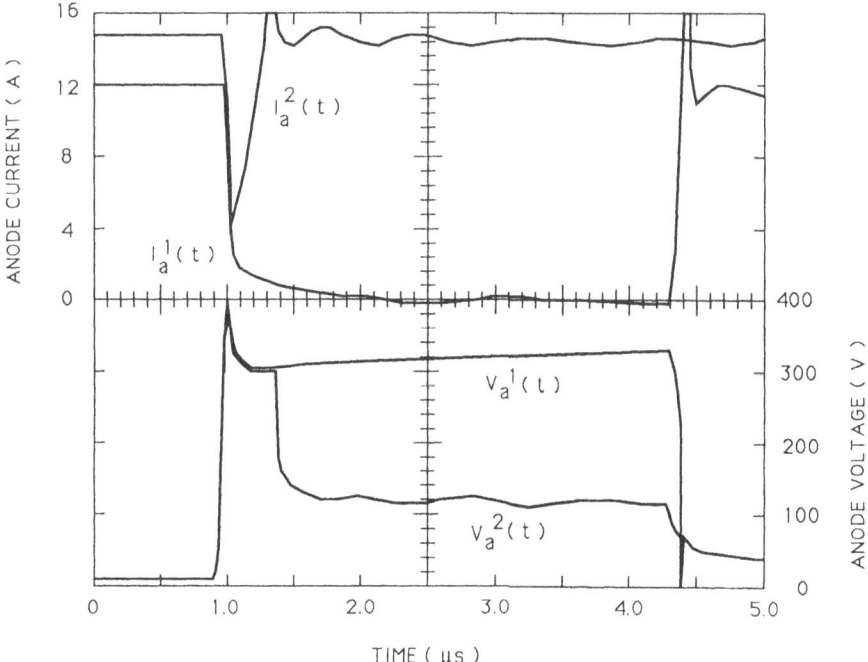

Fig. 10. Snubberless clamped inductive turn-off transients using an MCT cell array with 21,000 cells (pitch: 20 μm; active area: 8.4 mm²); successful turn-off of I_a = 12 A into V_a = 300 V (traces labeled 1) and turn-off failure when trying to turn off I_a = 15.5 A into 300 V (traces labeled 2).

6. LIMITS FOR STABLE MCT TURN-OFF

The experiments, presented in the last chapter, suggest that a turn-off stability criterion can be formulated for devices with a single or only few cells: once the anode current to be turned off increases beyond the value at which maximum dV_a/dt results, a high probability exists for a turn-off failure to occur. Extending our discussion to a fictitious inhomogeneous current distribution in a large device, a stable turn-off will result as long as a localized region subjected to an increased current density will turn-off faster than its surroundings. Under this condition, a negative feedback exists between regions of different current densities resulting in a levelling out of the inhomogeneity. Once the local current density attains levels where dV_a/dt decreases, this region will turn off later than its surroundings. It will thus collect more and more current from the neighbouring regions to form a current filament. Stable operation of a large MCT beyond the maximum dV_a/dt is not possible. This heuristic reasoning is in accordance with our experimental findings. In addition, a numerical investigation of the onset of current filamentation[17] in an MCT leads to similar values of maximum turn-off current densities as found in the study of small MCT test devices.

In the following, we will investigate in more detail the turn-off performance at the stablity limit for various MCT test devices: single cells with various cell pitches (15 to 30 µm), square cell arrays with the same different unit cells but the same active area and finally square cell arrays with an increasing number of 15 µm pitch cells (cell arrangements with 3x3, 5x5, 7x7, 10x10 and 20x20 cells). Results are also presented for larger MCT devices[18]. With this experiment, we were able to identify one of the causes for inhomogeneous current distributions. The dynamic test set-up described above turned out to be impractical for the characterization of very small MCT's. We have implemented a quasi-static characterization procedure for those devices[18]. In comparing the quasi-static technique to the dynamic test conditions, it is clear that no information concerning current and voltage transients is available in the first case. We have compared both methods with respect to the relative differences in turn-off performance for several device structures shown in Fig. 2. The absolute turn-off performance limit, using the quasi-static technique, was typically three times as high as that obtained using the dynamic test. However, the relative differences in turn-off performance among different test devices matched one another extremely well. Thus, the simple quasi-static technique is well suited to derive decision criteria for the selection of optimized cell structures and cathode patterns for MCT's.

For single cell MCT's, we found empirically that the turn-off performance is inversely proportional to the cell pitch[18]. It should be noted, that the doping profiles of the channel region were not changed by the cell pitch variation. Therefore, the differences in turn-off performance can be attributed to the variation of the resistance associated with the lateral current path in the p-base underneath the n^+ emitter (see the discussion in Chapter 2). This empirical cell scaling law equally applies to square cell arrays with different cell pitches (Fig. 2). However, the mean turn-off performance of a cell in an array of typically some hundred cells decreases to 10 % of the value for the corresponding single cell. As discussed in the preceeding chapter, the arrays under discussion are probably too small to allow the formation of a current filament. Nevertheless, a current inhomogeneity was expected to be at the origin of the strong decrease in turn-off capability. To this end, a set of square cell arrays with varying numbers of cells as described above was investigated. The smallest array (3x3 cells) did not suffer a decrease in the mean cell performance or, in other words, the turn-off performance was nine times as high as that of the single cell MCT. However, the mean cell performance degraded to 60 % of the single cell level in the 5x5 cell array and to 30 % in the 10x10 array, respectively. In an attempt to understand these results, we calculated an expected turn-off performance for all arrays assuming that only the peripheral cells would turn-off at 100 % of the single cell performance. A near perfect match between expected and observed performance was obtained. Two-dimensional simulations of multiple cell structures revealed that, due to a plasma encroachment along the periphery at the cathode side of the device, peripheral cells conduct an excess current density already during the on-state[19]. During turn-off of a clamped inductive load, current commutates from all inner cells to those located at the periphery. The experimental results suggest that the peripheral cells turn off homogeneously or, in other words, no additional inhomogeneity occurs during turn-off of the peripheral cells.

The cell scaling law discussed above is also valid for the peripheral cells. For this reason, square cell arrays utilizing smaller cells show a better turn-off performance as compared to arrays of the same active area but designed with larger cells. Although a reduction of the cell pitch by a factor of two entails an improvement in turn-off performance of a single cell by a factor of more than three, there would be no realistic chance to fabricate very large MCT chips with thousands of cells and high turn-off current densities simply by miniaturizing the cells. Due to the periphery limitation, the maximum turn-off current would scale with the square root of the active device area rather than linearly. Our experimental results on MCT's with 21,000 cells (20 μm cell pitch; active area: 8.4 mm^2) are fortunately in contrast to this expectation. Figure 11 shows maximum turn-off currents for anode voltages of 500 and 2,000 V at the stability limit. At the lower anode voltage of 500 V (Fig. 11a), the device is able to turn off a current of 15 A; this is equivalent to a current density of 180 A/cm^2. If the periphery limitation would be valid for this large array, a maximum turn-off current density of only 18 A/cm^2 would be expected. Thus, the strong decrease in turn-off performance, attributed to the current crowding at the periphery, is only observed in "small" cell arrays. Our results indicate that the limitation caused by the periphery effect disappears when the cell count of square cell arrays exceeds several hundred cells. A similar observation was made in[20]. The maximum turn-off current of the large MCT array decreases when the clamping voltage is increased to 2 kV. In this case, a maximum current of 6 A corresponding to a current density of 70 A/cm^2 can be handled by the device. As already noted earlier, these results were obtained for snubberless switching of a clamped inductive load with an MCT.

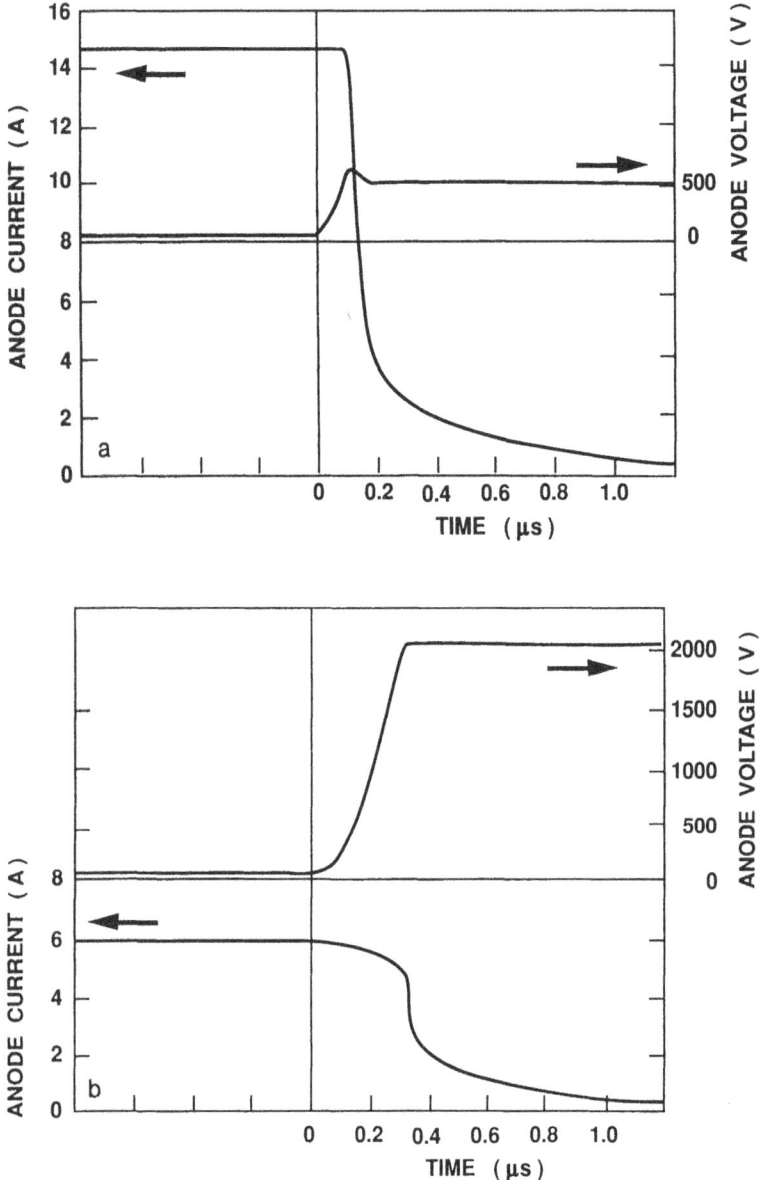

Fig. 11. Snubberless clamped inductive turn-off using a 21,000 cell MCT array (active cathode area: 8.4 mm^2) of a) 15 A into 500 V (180 A/cm^2 vs. 500 V) and b) 6 A into 2,000 V (70 A/cm^2 vs. 2 kV).

7. DISCUSSION

In this chapter, we will analyse the maximum turn-off current density of an ideal MCT (or any other power switch), turning off with a perfectly homogeneous current density distribution across the active area. The data for the MCT turn-off performance will be compared with this absolute performance limit. In addition, we will investigate other probable causes for inhomogeneous current density distributions during turn-off and discuss appropriate means to avoid them.

To estimate the increment of maximum power per area at which "dynamic" avalanche occurs, let us assume that the electron flow from the n+ emitter is completely interrupted. During a fictitious turn-off transient, the anode voltage has attained a high level V_a, which is assumed to be significantly below the static breakdown voltage. Under clamped inductive load conditions, the on-state anode current is supposed to flow until V_a reaches the clamping voltage. The anode current is exclusively carried by holes, injected from the anode p+ emitter. These carriers will move at their saturated drift velocity ($v_d = 8 \cdot 10^6$ cm s^{-1}) across the depletion layer. We assume furthermore that the gradient of the triangular electric field strength E is proportional to the sum of the effective hole doping concentration p_{eff}, supporting the anode current, and to the doping concentration N_d of the n$^-$ base. The breakdown field strength E_{bd} is dependent on the value of N_d; however, for typical doping densities (10^{13} cm^{-3}), it attains $2 \cdot 10^5$ V cm^{-1}. The current density, j_a, is related to the hole density through

$$j_a = q \cdot v_d \cdot p_{eff} \quad . \tag{1}$$

In equation (1), q stands for the elementary charge. Solving (1) for p_{eff} and integrating Poissons equation

$$dE/dx = q/\varepsilon_{si} \cdot (p_{eff} + N_d) \tag{2}$$

across the depletion region yields an expression for the breakdown field strength E_{bd} of the form

$$E_{bd} = q/\varepsilon_{si} \cdot (p_{eff} + N_d) \cdot d_{bd} \tag{3}$$

where ε_{si} is the dielectric constant of silicon and d_{bd} is the width of the depletion layer at the onset of dynamic avalanche, relating E_{bd} to the breakdown voltage $V_{a,bd}$ under dynamic avalanche conditions,

$$E_{bd} = V_{a,bd}/d_{bd}. \tag{4}$$

Solving equation (1) for j_a, using eq. (3) and (4) and eliminating d_{bd}, yields a simple relation for estimating the areal power density P_{aval}, at which dynamic avalanche occurs,

$$P_{aval} = j_a \cdot V_{a,bd} = E_{bd}^2 \cdot \varepsilon_{si} \cdot v_d - q \cdot v_d \cdot N_d \cdot V_{a,bd}. \qquad (5)$$

For small values of N_d, the second term in eq. (5) is quite small; P_{aval} is thus mainly determined by material parameters; for silicon (N_d = $2 * 10^{13}$ cm^{-3}; $V_{a,bd}$ = 1,000 V), a value of 320 kWcm^{-2} is obtained. According to eq. (5), P_{aval} depends on the clamping voltage. In experiments using diodes, similar values for the maximum power densities of approximately 300 kWcm^{-2} were measured. We can now return to Fig. 11 and calculate the power density at the instant of turn-off failure for the large MCT array with 21,000 cells. The simple calculation yields values ranging from 100 to 150 kWcm^{-2}. For the device considered here, there is thus room for for an improvement in turn-off capability by a factor of 2 or 3. If a device attains the areal power density limit expressed in eq. (5), we can expect that the current density is constant across the device area and remains so during the turn-off transient. For the large MCT array (see Fig. 11), this is definitely not the case. Since we can argue that the turn-off failure is initiated by a localized increase of the power density above the limit of dynamic avalanche, we can also estimate the degree of inhomogeneity of current density during turn-off for this particular device. A "current filament cross-section" at the onset of dynamic avalanche breakdown may be determined assuming that the local power density in the filament is 300 kWcm^{-2}. For the experiments shown in Fig. 11, the "filament cross-section" or active turn-off area would amount to one third or one half of the total device area. Obviously, if a homogeneous current distribution could be maintained throughout the turn-off process, anode currents twice as large as those shown in Fig. 11 could be turned off. It should be noted that the use of a snubber circuit can result in an increase of the total switched power. The additional amount of power is continously deposited into the snubber capacitor; however, the maximum power level in the semiconductor power switch is not allowed to exceed P_{aval}. Commonly, the snubber capacitor is used to prevent the switch from reaching this limit.

So far, we have identified the array periphery as a cause for structural perturbations which are responsible for the triggering of inhomogeneous current distributions during turn-off. However, structural perturbations may originate from many causes. Among these, unavoidable fluctuations of technological parameters such as channel doping (threshold voltage, carrier mobility, channel length) or carrier lifetime play an important role. A numerical analysis of the stability criteria revealed that the MCT would form a filament deliberately if a single cell deviates only slightly from all others. However, this situation is quite unlikely to occur in reality in this form; the extent of a specific type of perturbation is expected to be more evenly distributed across the whole cell ensemble. If there is a random distribution of fluctuations which are sufficiently small, a deliberately filamenting MCT must not necessarily be the result. It is believed, that minor fluctuations of many technological parameters are tolerable in the sense, that they will not entail a significant decrease of the maximum turn-off performance. However, a perturbation may very well cause a turn-off failure, if it is large enough and, especially, if it is spatially distributed in a non-random manner.

We have fabricated some of the test structures presented in this paper in two different laboratories. In one case, we used wafer stepper lithography with an excellent alignment accuracy, whereas projection lithography with manual alignment was chosen in the second laboratory. Figure 8 shows the maximum turn-off current of a 400 cell array, fabricated using manual alignment techniques. Employing a wafer stepper, a threefold improvement in anode turn-off current was obtained. This tremendous difference was attributed to the high alignment precision achievable when utilizing a wafer stepper. With the cell structures as shown in Fig. 3, a large misalignment, occuring during the p+ short lithography, may possibly result in a reduction of the MOS channel width available for turn-off by as much as 50 %. We think that such a type of perturbation may be at the origin of the large differences in turn-off performance of the test devices.

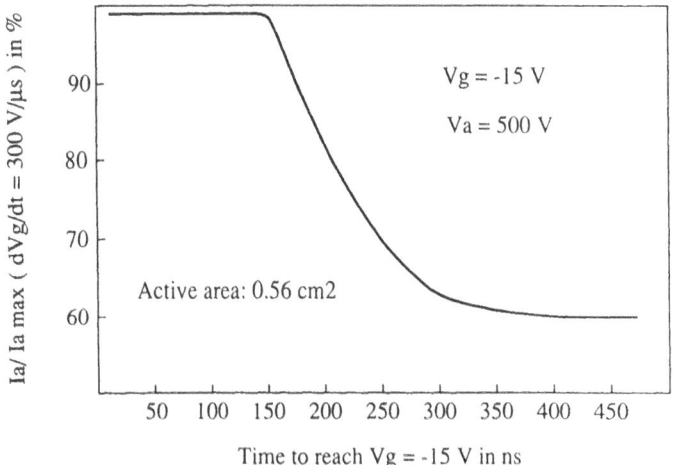

Fig. 12. Relative decrease in maximum anode turn-off current for an MCT array with an active area of 0.56 cm^2 when the rise time of the gate voltage pulse from 0 V to -15 V is increased (snubberless clamped inductive switching); dVg/dt is varied from 30 V/μs to 300 V/μs.

A different type of perturbation may be introduced when the transition of the gate signal from 0 V during the on-state to -12 or -15 V for the turn-off (as used in our experiments) proceeds very slowly. Due to the propagation delay in the polysilicon gate electrode, cells located in different regions of an array may turn off at different moments in time.

Figure 12 shows how the maximum turn-off performance decreases when the risetime of the gate voltage pulse increases. For these experiments, large MCT arrays with an active area of 0.56 cm^2 and more than 100,000 cells were used. With an appropriate layout for the MOS gate metallization, this dynamically induced perturbation becomes only effective for slowly rising gate signals. On the other hand, Fig. 12 shows that an ultrafast gate drive is not required for achieving high turn-off capability of large area MCT's. Using these devices in a set-up for snubberless switching of a clamped inductive load, maximum turn-off currents amounted to 36 A for V_a = 500 V and to 11 A for V_a = 2,850 V, respectively (Fig. 13). It is well understood, that an increase of the maximum gate voltage level results in a higher conductivity of the shorting MOS channel; as a result, higher anode currents can be turned off (Fig. 14). In our future MCT designs, we aim at making more effective use of the available gate voltage swing by reducing the threshold voltage.

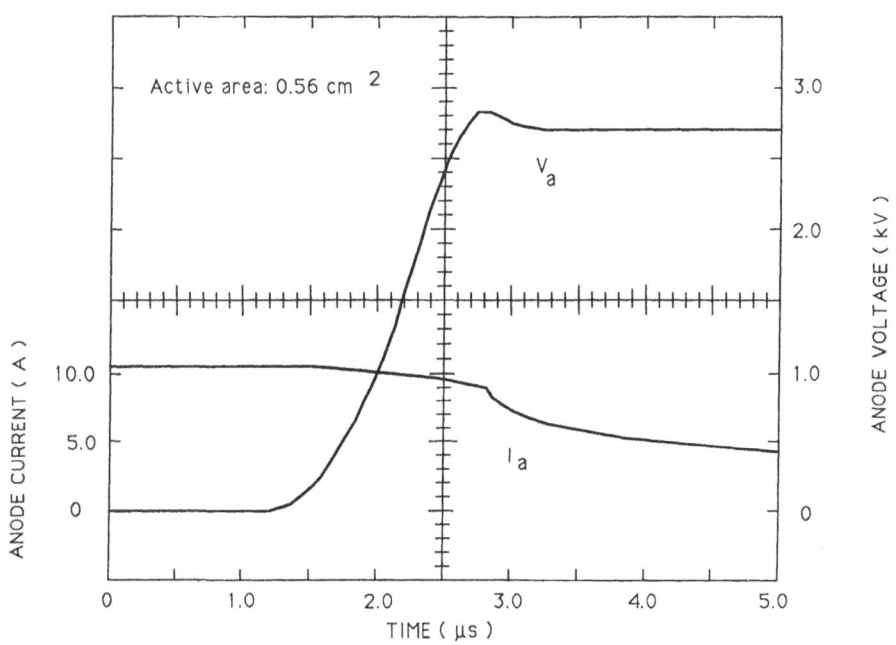

Fig. 13. Snubberless clamped inductive switching using an MCT array with an active cathode area of 0.56 cm^2. An anode current of 10.5 A was turned off into 2,850 V; the cell pitch is 20 μm.

In the last chapter we explained that the peripheral cells in a cell array are subjected to increased current densities in the on-state, giving rise to a perturbation during turn-off. This perturbation is particularly strong for those cells which are located at the corners of an array (see Fig. 2). More ingenious arrangements of cells are helpful in reducing the degree of this structurally induced perturbation[21]. We have found that the addition of permanently shorted cells at strategically important locations in an array is extremely beneficial in order to push the turn-off performance to the limit dictated by the onset of spatially homogeneous dynamic avalanche breakdown. In Fig. 15, results are shown for an array with an active area of 1.6 mm^2 consisting of 4,000 cells with 20 μm pitch. At V_a = 1,500 V, a maximum anode current of 3 A, corresponding to a current density of 200 A/cm^2, was turned off. Since the spatial power density is close to 300 kWcm^{-2}, we suppose that spatially homogeneous turn-off has been achieved using this particular array.

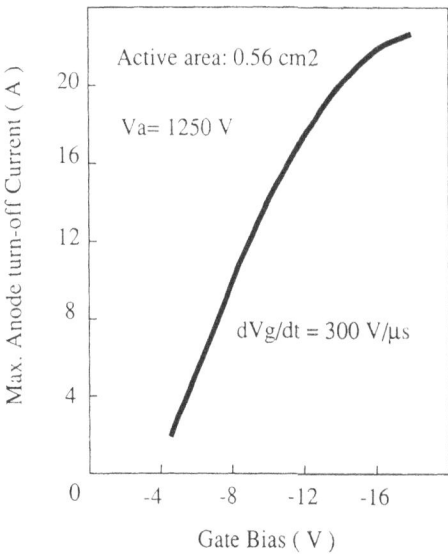

Fig. 14. Dependence of maximum turn-off performance (snubberless clamped inductive switching) of an MCT array on the gate voltage amplitude (active cathode area: 0.56 cm^2; cell pitch: 20 μm; dVg/dt = 300 V/μs).

8. CONCLUSIONS

In this paper, we have presented the state of MCT development for high voltage applications. We have focussed on the critical issue for a four-layer power device, being the maximum turn-off current density, which is known to scale sublinearly with the active device area in GTO's. For several reasons outlined in this paper, it is too early to extrapolate from the present performance level of MCT's to fundamental device limits. However, it was demonstrated that small MCT arrays reach the homogeneous turn-off limit imposed by dynamic avalanche breakdown.

As an outlook to the future MCT development, it may be stated that MCT's have a high potential to become fast, rugged and simple to use power switches. Our development aims at increasing MCT chip sizes without suffering from inhomogeneous current distributions at turn-off, which have been shown to cause the maximum achievable performance level to decrease. The device designer has several tools at hand in order to accomplish this goal. Contrary to power MOSFET's, the MCT is still open for cell miniaturization (minimum dimension DMOS or U-channel MOS, UMOS, cells) and optimization of the DMOS channel region (sidewall spacer technologies). In addition, technological parameter fluctuations can be reduced to a minimum with new, totally selfaligned cell concepts. With respect to the maximum chip size limit imposed by the economics of chip fabrication, the permanent flow of VLSI production knowhow into DMOS technology has made possible the production of large quantities of IGBT chips with up to 2 cm² of active area at very high yield. Further increases in DMOS chip size are expected to appear in the future. Although today, wafer scale MCT's in DMOS technology are beyond the possibilities of semiconductor technologies (except if intelligent repair technologies are used), paralleling of MCT chips in an isolated module has been demonstrated to be a feasible approach to realize MCT's with increased current capability[22]. We have shown in this paper that the MCT is also an attractive candidate for high voltage applications. Despite the up's and down's throughout its development history, the MCT is expected to infiltrate a broad spectrum of power electronic applications in the coming years.

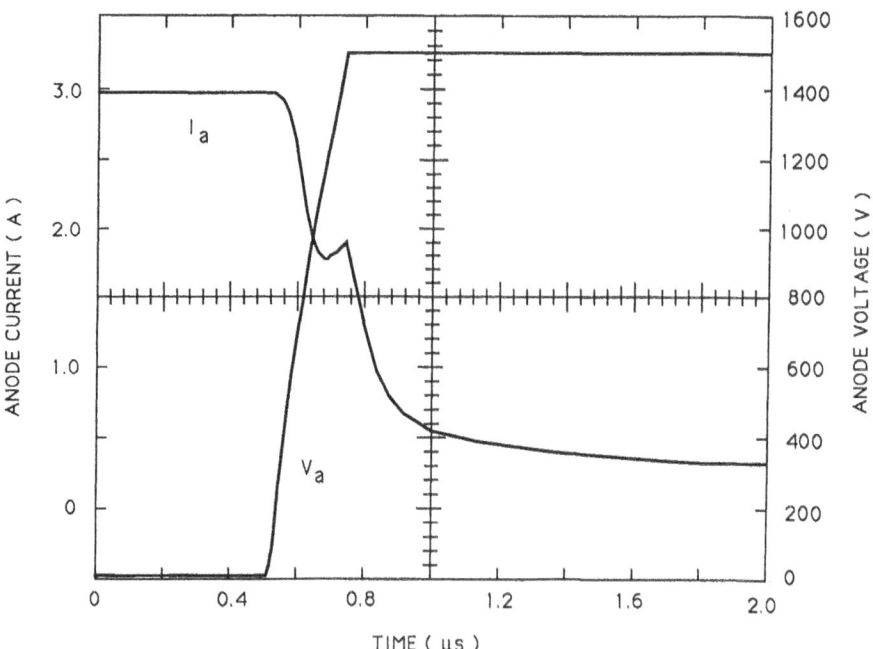

Fig. 15. Snubberless clamped inductive switching using an MCT array with 4,000 cells (active area: 1.6 mm2; cell pitch: 20 μm); $I_a = 3$ A is turned off into 1,500 V (190 A/cm² vs. 1.5 kV).

ACKNOWLEDGEMENTS

During the writing of this manuscript, the author realised how many colleagues at ABB, ETHZ, and CSEM have indirectly contributed to it. In particular, he likes to thank Wolfgang Fichtner and his crew for creating first class simulation tools. André Jaecklin and Kenneth Johanson from ABB research have been very constructive in providing an understanding of important device related phenomena. The tremendous help from R. Vuilleumier and J.-M. Moret from CSEM in Neuchatel during the initial phase of this project is also greatfully acknowledged. Finally, thanks go to Thomas Stockmeier and his crew of the microlab for the manufacture of the devices as well as to Horst Grüning, Hunter Haddon and the electronics crew at ABB research for helping with the electrical characterization.

REFERENCES

1   B. J. Baliga, "Evolution of MOS-bipolar semiconductor technology", in *Proc. IEEE*, vol. 76, pp. 409-418, 1988.

2   C.-T. Sah, "Evolution of the MOS transistor - from conception to VLSI", *Proc. IEEE*, vol. 76, pp. 1280-1326, 1988.

3   B. J. Baliga, M. Chang, P. Shafer and M. W. Smith, "The insulated gate bipolar transistor (IGT) - a new power switching device", *Proc. of the IEEE Ind. Appl. Soc. Meet.*, pp. 794-803, 1983.

4   J. P. Russell, A. M. Goodman, L. A. Goodman and J. M. Neilson, "The COMFET - a new high conductance MOS-gated device", *IEEE Electron Device Lett.*, vol. EDL-4, pp. 63-65, 1983.

5   M. Stoisiek and H. Strack, "MOS GTO - a turn-off thyristor with MOS-controlled emitter shorts", *IEDM Tech. Dig.*, pp. 158-161, 1985.

6   V. Temple, "MOS-controlled thyristors - a new class of power devices", *IEEE Trans. Electron Devices*, vol. ED-33, pp. 1609-1618, 1986.

7   A. Aemmer, F. Bauer, J. Bürgler, W. Fichtner, S. Müller and P. Roggwiller, "Multi-dimensional simulation of MCT structures", *Proc. of the 1990 Int. Symp. on Power Semiconductor Devices & IC's (ISPSD)*, pp. 20-25, 1990.

8   N. M. Darwish and M. A. Shibib, "Lateral MOS-gated power devices - a unified view", *IEEE Trans. Electron Devices*, vol. ED-38, pp. 1600-1604, 1991.

9   F. Bauer, E. Halder, K. Hofmann, H. Haddon, P. Roggwiller, T. Stockmeier, J. Bürgler, W. Fichtner, S. Müller, M. Westermann, J.-M. Moret and R. Vuilleumier, "Design aspects of MOS-controlled thyristor elements: technology, simulation, and experimental results", *IEEE Trans. Electron Devices*, vol. ED-38, pp. 1605-1611, 1991.

10   M. S. Shekar, B. J. Baliga, M. Nandakumar, S. Tandon and A. Reisman, "High-voltage current saturation in emitter switched thyristors", *IEEE Electron Dev. Lett.*, vol. EDL-12, pp. 387-389, 1991.

11   F. Bauer, H. Hollenbeck, T. Stockmeier and W. Fichtner, "Current-handling and switching performance of MOS controlled thyristor (MCT) structures", *IEEE Electron Dev. Lett.*, vol. EDL-12, pp. 297-299, 1991.

12   Q. Huang, G. A. J. Amaratunga, E. M. Sankara Narayanan and W. I. Milne, "Analysis of n-channel MOS-controlled thyristors", *IEEE Trans. Electron Devices*, vol. ED-38, pp. 1612-1618, 1991.

13   R. Stengl and U. Gösele, "Variation of lateral doping - a new concept to avoid high junctions", *IEDM Tech. Dig.*, pp. 154-157, 1985.

14   M. Stoisiek, M. Beyer, W. Kiffe, H.-J. Schultz, H.Schmid, H. Schwarzbauer, R. Stengl, P. Türkes and D. Theis, "A large area MOS-GTO with wafer-repair technique", *IEDM Tech. Dig.*, pp. 666-669, 1987.

15   P. R. Palmer and C. M. Johnson, "Characterizing the turn-off performance of multi-cathode GTO thyristors using thermal imaging", *IEEE Trans. Power Electronics*, vol. PEL-5, pp. 357-362, 1990.

16   M. Stoisiek, G. Wachutka and D. Theis, "2D-simulations and experiments of avalanche injection in GTOs", *Proc. of the 1988 Int. Symp. on Power Semiconductor Devices (ISPSD)*, pp. 48-55, 1988.

17   K. Lilja and H. Grüning, "Onset of filamentation in GTO devices", *Proc. of the Power Electronics Specialists Conference, PESC 90*, pp. 10-15, 1990.

18   F. Bauer, H. Haddon, T. Stockmeier, W. Fichtner, R. Vuilleumier and J.-M. Moret, "Optimization of cathode structures for improved performance of MOS controlled thyristors (MCT)", *Proc. of the MADEP 91 conference*, pp. 270-275, 1991.

19   H. Dettmer, H. Lendenmann, J. Bürgler, S. Müller, W. Fichtner and F. Bauer, "Turn-off behavior of structured MCT cells", *Proc. of the MADEP 91 conference*, pp. 258-261, 1991.

20   V. A. K. Temple, S. Arthur and D.L. Watrous, "MCT (MOS controlled thyristor) reliability investigation", *IEDM Techn. Dig.*, pp. 618-621, 1988.

21   H. Lendenmann, H. Dettmer, W. Fichtner, B. J. Baliga, F. Bauer and T. Stockmeier, "Switching behaviour and current handlingperformance of MCT- IGBT cell ensembles", *IEDM Techn. Dig.*, pp. 149-152, 1991.

22   T .M. Jahns, R. W. A. A. De Doncker, J. M. A. Wilson, V.A. K. Temple and D. L. Watrous, "Circuit utilization characteristics of MOS-controlled thyristors", *IEEE Trans. Ind. Appl.*, vol. 27, pp. 589-597, 1991.

DISCUSSION

D. Silber (TU Bremen)

Some years ago, V. Temple has tried to demonstrate that the MCT is a perfect high temperature device. Did you investigate the high temperature behaviour of your devices ?

F. Bauer

So far, we have carried out all measurements at room temperature. However, measurements at high temperature are planned in the near future, not at 300 °C as proposed by V. Temple but say at 125 or 150 °C.

D. Silber

In your talk, I have missed a discussion on the emitter ballasting concept.

F. Bauer

Emitter ballasting is not applied to our structures. The idea is to avoid the penalty of an additional on-state voltage drop due to ballasting. Instead - as indicated in my last slide - we think that adding other cells to the MCT array, which stabilize the plasma distribution, is a better approach.

D. Silber

Current localization will lead not only to overload of your emitters but also to less shunting. The fundamental device concept might be improved if you reduce your packing density in favor of additional emitter shunts. In your present scheme, only the neighbouring emitter shunts can help. Would you consider a concept where you always have total shunting for everything ?

F. Bauer

Most of our structures consist of more than just the basic MCT cells, structures which are intended to work in the sense you are asking for.

P. Palmer (Univ. Cambridge, UK)

Can you comment on the gate current waveform which has a decisive influence on current redistribution in a GTO ? May I ask you also how you think about paralleling of dies, a rather typical concept for large area MOSFETs.

F. Bauer

Some basic work concerning paralleling of MCT dies has been carried out by V. Temple, formerly at GE and now at Harris, and his published results look quite promising. In our laboratory, paralleling of dies has not been tried out yet, but we intend to investigate it in the coming months.

For the gate drive it is important, of course, how fast we apply the gate signal. As you may have noticed, the gate metallization of our $0.5 \, cm^2$ chip uses gate runners, dimensioned in order to provide a fast propagation of the gate signal. The gate signal propagation is limited mainly by the RC time constant of the poly-gate. In our case, the gate voltage transition from zero to minus 15 volt takes place in 50 ns. We have found that for longer times, roughly between 150 and 200 ns, no degradation in the turn-off performance can be observed. Hence, the driving requirements for MCTs are not as stringent as expected.

G. Wachutka (ETH, Zürich)

You have shown a viewgraph of the FBSOA for different number of cells corresponding to different chip sizes. Do you see any scaling law in this FBSOA, e.g. related to the character or the behaviour of the filaments ?

F. Bauer

We have found scaling laws for very small devices, but we believe that these are not due to a current redistribution of freely moving filaments. For the larger devices, we cannot as yet identify any scaling law; there is not yet enough experimental data.

M. Campagna (ABB Corp. Research, Baden)

Do you have an idea how this structure of mixed cells will look like ?

F. Bauer

Beyond ideas there are realizations. In one approach, we add to the p-channel DMOS cells of the MCT array complete IGBT cells. The result might be called a complementary DMOS technology where the IGBT not only provides MOS-controlled turn-on. In addition, these additional components take over hole current otherwise flowing through the MCT, thus easing turn-off for the array of MCT cells. That is one idea but there are more.

R. Sittig (TU Braunschweig)

Since we have to live with a certain defect density in MOS technology, what are your plans to take into account this problem for large area devices ?

F. Bauer

Considering the level of technology in Japan, their production yield for IGBT chips approaches 80 % for chip sizes close to 3 cm^2. We expect to achieve a similar level of performance, given the time and the investments needed. In addition, there are approaches to provide repair technologies for power MOS chips. To my knowledge, these repair approaches are common place in DRAMs today, but they do not yet seem mature for power devices.

Since INTEL, as a leader in microprocessor development, has recently launched a plan, aiming at realizing microprocessors with a silicon chip of 6 cm^2, highly integrated power device chips might approach sizes of up to 10 cm^2 in the future.

HIGH POWER REVERSE CONDUCTING GTO

P. Streit

Asea Brown Boveri Semiconductors AG
Lenzburg, Switzerland

ABSTRACT

Among modern high power Gate Turn-Off thyristors (GTO's), the asymmetric blocking type has received most attention by far in device development and in applications. This is mainly due to its superior trade-off for on-state vs. turn-off losses and to the high efficiency obtained by optimized anode shorts. Such devices fit any application not requiring reverse blocking capability since they are always used in conjunction with an antiparallel freewheeling diode.

Monolithic integration of this diode into the GTO, therefore, suggests itself for reasons like cutting in half the number of high power components, reducing size and weight of the converters, avoiding stray inductance, and saving on cooling equipment. The overall layout of the integrated solution with respect to area partitioning is further simplified by the fact that all major applications, e.g. voltage source inverters, require full load current in both directions and, therefore, a ratio 1:1 in rated currents for GTO and diode. Basically, this allows to satisfy the requirements of many applications with just one kind of high power silicon component without any sacrifice on flexibility in both overall circuit design and control strategy.

The requirements for such a Reverse Conducting GTO (RC-GTO) are discussed, and its practical design and technology will be explained with special emphasis on GTO-to-diode separation, diode optimization and thermal loading considerations. In addition, both similarities and major differences between the new RC-GTO concept and the well-known reverse conducting thyristor will be sketched. Specifications of available devices are shown and, in an outlook on future developments, the question of maximum controllable current using RC-GTO's will be linked to the development prospects of the two individual components.

1. INTRODUCTION

In recent years, there has been an increasing interest in Reverse Conducting GTO thyristors (RC-GTO's), and a number of components have been developed by different suppliers[1-6]. This is mainly due to the inverter scheme favoured today, the voltage source type explained in Fig.1. Focusing first on single quadrant operation, Fig. 1a, a controllable switch is used for turning the voltage source on and off, and a diode is required for continuity of the current through the inductive load. The two components are not yet antiparallel, but if second quadrant operation is added with a controllable current flowing back into the voltage source, Fig. 1b, a conjugate device arrangement or half bridge will result (Fig. 1c). Adding two such half-bridge circuits leads to four antiparallel element pairs that are typical for a full bridge, (Fig. 1d), capable of four quadrant operation.

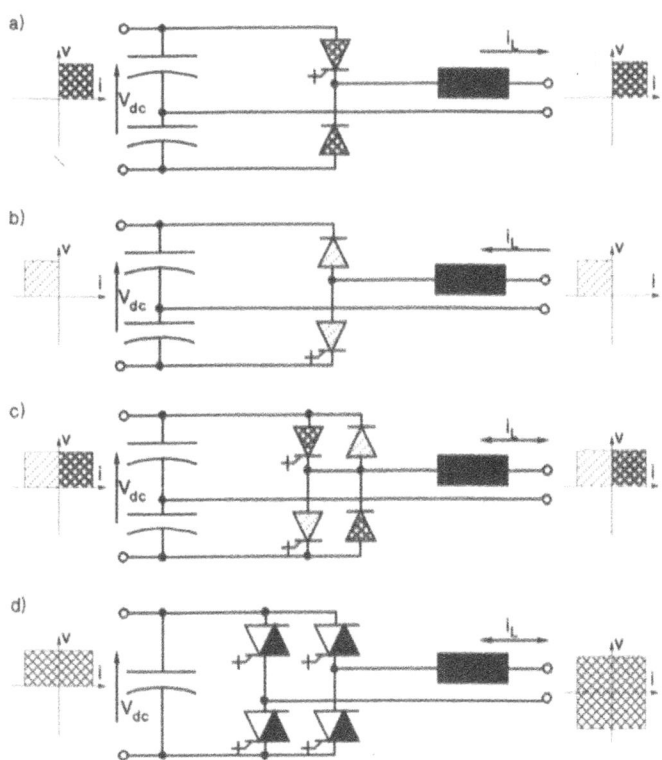

Fig. 1. Schematic representation of switch-mode DC-to-AC inverter configuration; the addition of building blocks a), b) leads to a half bridge c) or a full-bridge d).

Most of the GTO-applications, therefore, require that each GTO is used together with an antiparallel freewheeling diode. Although symmetrically blocking GTO's (Fig.2a) are available from some suppliers, the GTO-type that has received most attention by far is of the asymmetric blocking type required in such circuits. Figure 2b shows that the combination of GTO and antiparallel diode represents: a) a common blocking state, b) a controllable GTO on-state and c) a reverse conducting diode on-state. Offering a device that includes this reverse conducting feature (Fig.2c), therefore, has the following advantages for inverter layout: it cuts in half the number of high power components specified for full current, it saves space, weight and cooling equipment, and last but not least it relieves the user from selecting, evaluating and qualifying two different components, often coming from different suppliers. In addition to these application benefits, the RC-GTO can also be of interest to the device supplier since he has the possibility to offer two functions in just one housing and starting with just one silicon wafer. Savings in material costs will certainly make up for a few extra processing steps, and cost calculations show that the approach looks economically interesting. Evidently, the idea of integration only works under the assumptions that: a) most applications require the same diode, and b) realisation of two reasonably optimal components from the same starting material is feasible.

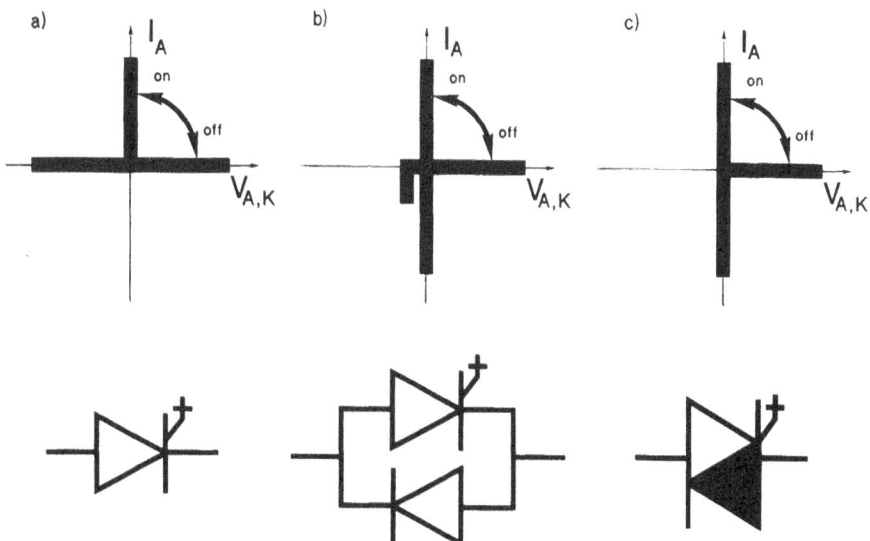

Fig. 2. High power GTO types: a) symmetrically blocking, b) asymmetrically blocking with freewheeling diode c) reverse conducting.

With respect to the first concern, the situation is even more favourable than in case of the conventional reverse conducting thyristor[7]. Figure 3 is a schematic representation of an interesting and quite important difference between the two devices: The conventional thyristor has

a controlled turn-on, however, turn-off has to occur by an externally induced current commutation which, in the reverse conducting component, immediately leads to a conducting diode state within the same
switching cycle (Fig. 3a). Therefore, every switching cycle involves both
devices and all four quadrants. The current, that flows through the
diode, depends on the specific application. In a resonant application,
with a sinusoidal current, thyristor- and diode-currents will be almost
identical, whereas in a DC-chopper, the diode- current will be much
smaller. This is the reason, why reverse conducting thyristors vary in
diode current ratings between full thyristor current and approximately
one third.

The RC-GTO is different in the sense that the GTO-part generally
handles one complete switching cycle, remaining entirely within the
first quadrant and not requiring any diode conductivity (Fig. 3b). As
shown in Fig.1, the antiparallel diode is used in a different switching sequence, which is controlled by the conjugate GTO. However, this second
switching sequence is only shifted in timing and, within one complete
modulation cycle, current waveforms on GTO and diode are almost
alike. This is true for all voltage source switch-mode inverter schemes
and, therefore, all the RC-GTO-applications require the same kind of
diode, rated for full GTO-current.

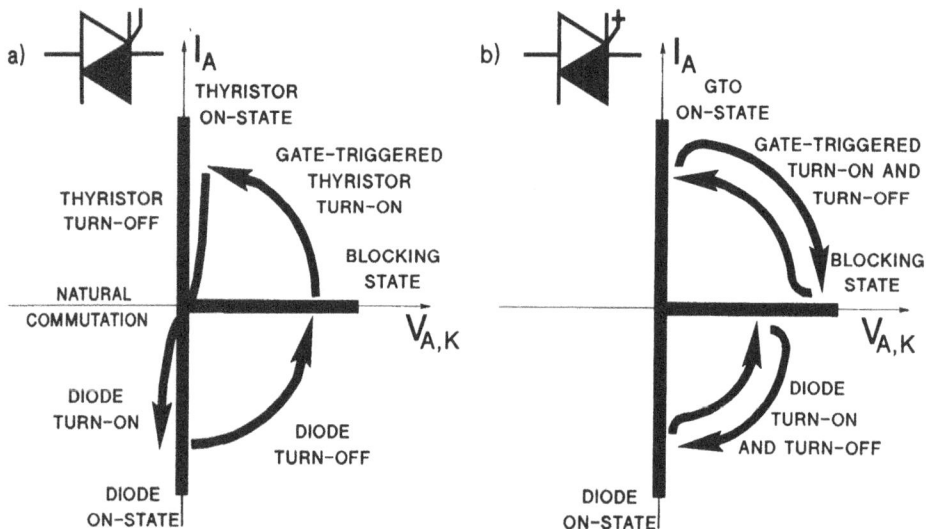

Fig. 3. Principal difference in device operation between a) conventional reverse conducting thyristor and b) reverse conducting GTO.

2. ELECTRICAL REQUIREMENTS

The above general characterization of the application already speci-
fies the main electrical requirements. These are:
a) a common blocking behaviour for both devices,

b) a GTO-specification that corresponds closely to that of an advanced asymmetric device,

c) a fast recovery diode for the same nominal current that can be very specific for GTO freewheeling applications.

Figure 4 shows that characterization of such a diode is different from standard. Normally, a fast diode turn-off is specified using an RC-snubber circuit (Fig. 4a). Characterization is by peak reverse current I_{rr}, reverse recovery time t_{rr} and reverse recovery charge Q_{rr}. One important quality of a standard fast recovery diode is "soft recovery", which prevents detrimental circuit oscillations.

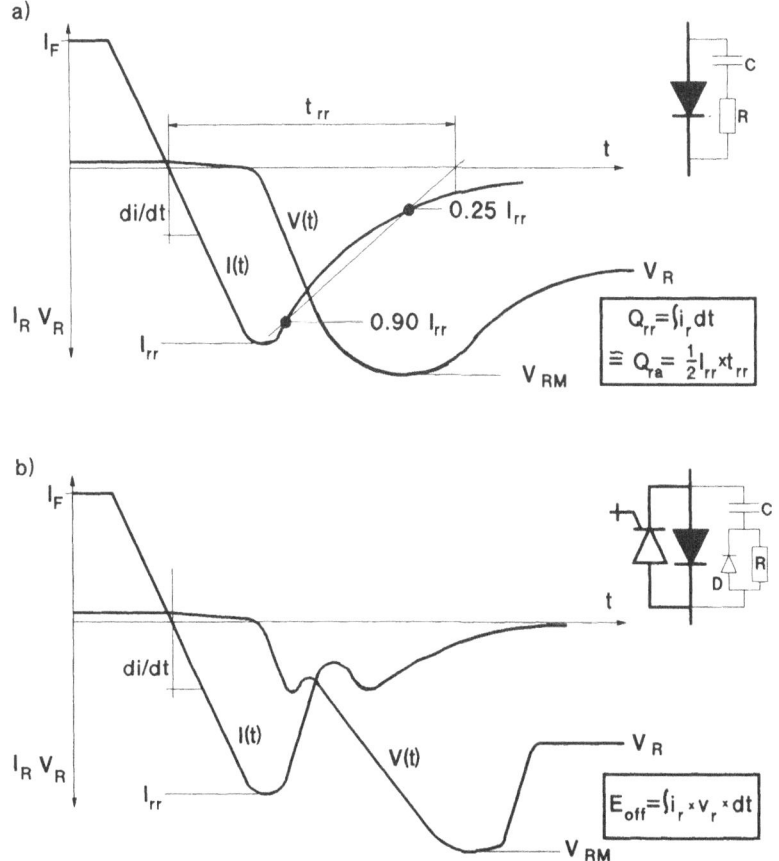

Fig. 4. Difference in characterization of a standard fast-switching diode a) and a dedicated GTO freewheeling diode b), both with the associated snubber circuit.

With the GTO turn-off snubber shown in Fig. 4b, I_{rr} remains essentially unchanged, but the voltage increase is strongly influenced by the circuit, and this also modulates the reverse current. The most important parameter is the turn-off energy E_{off}, determined by integration for one particular snubber configuration. As shown in Fig. 5, peak recovery current I_{rr} is also of some interest to the circuit designer since it influences

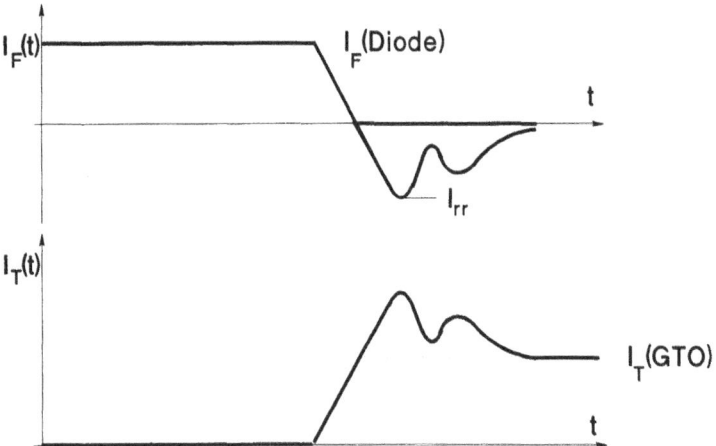

Fig. 5. Influence of reverse recovery diode current I_{rr} on turn-on current overshoot of the conjugate GTO.

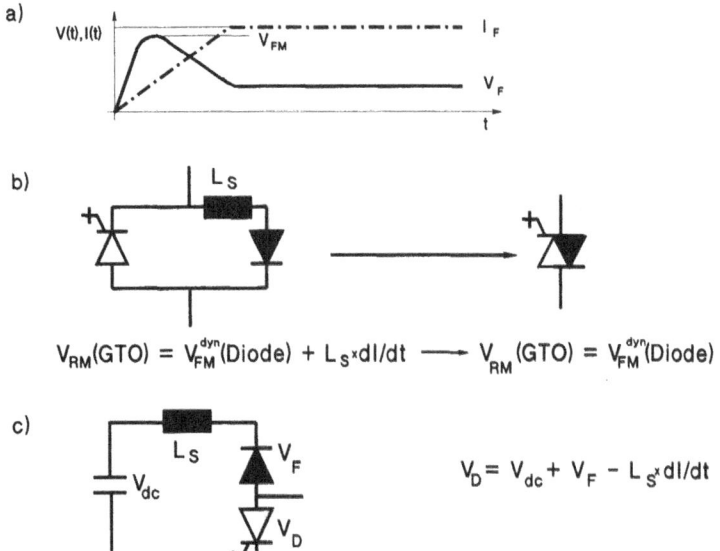

Fig. 6. a) Dynamic diode forward voltage; b) influence of forward voltage on antiparallel GTO-diode combination in comparison to RC-GTO; c) influence of reapplied voltage after turn-off of conjugate GTO.

the current overshoot at turn-on of the conjugate GTO. dV/dt is mainly determined by the turn-off snubber, not by the shape of the diode tail-current. Therefore, "soft recovery" is not an issue.

There is one more dynamic diode parameter that deserves our attention: the dynamic forward voltage at turn-on, shown in Fig.6a. If current is applied with a large dI/dt, the diode produces a transient forward voltage, which can exceed many times the on-state voltage. Schematically, Fig. 6b shows that, in the case of a standard freewheeling diode, this overvoltage, originating from the stray inductance, can drive the antiparallel GTO into reverse avalanche. This is practically impossible in the case of the reverse conducting component because there is no stray inductance. Therefore, the GTO-part of the combined device does not have to be tested for reverse avalanche capability. If, on the other hand, diode overvoltage influences the conjugate GTO's reapplied voltage after turn-off (Fig.6c), then we have to take it into account exactly as in the discrete element case.

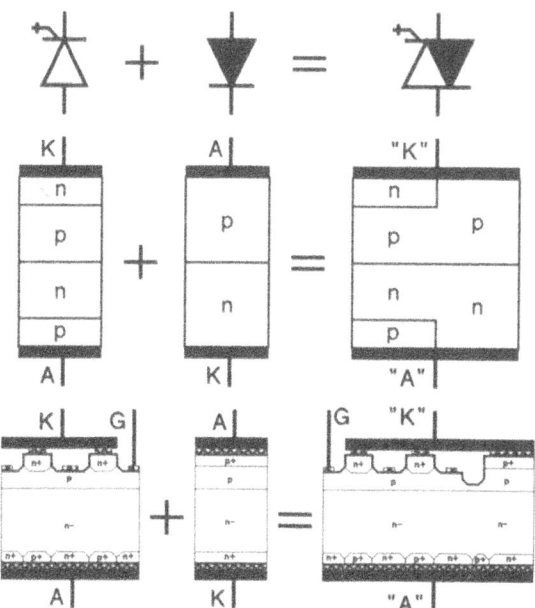

Fig. 7. Schematic representation of the monolithic integration of a GTO and an antiparallel diode.

3. DESIGN CONSIDERATIONS

The upper part of Fig. 7 is a very schematic representation of monolithic integration of an antiparallel diode and a thyristor structure. The

two base layers of the thyristor continue into the diode section, whereas
its two emitter layers have to be masked from the diode area. The main
blocking junction extends through both devices, giving them the required common voltage blocking feature. On a more specific mask level
shown in the lower part of Fig. 7, the anode short structure, which is
typical for the asymmetric GTO, is also shown. It is apparent that this
feature can be used to realize and shape the cathode of the diode.
Furthermore, the low impedance p-base required for the GTO is suitable
as a diode emitter when supplemented by a sintered aluminum contact.
This shows, that some parts of the antiparallel diode can be realized
during standard GTO processing steps, just by properly restructuring the
masks. Other features require extra processing steps. One additional
masking is required for localization of the GTO cathode emitter. Other
additional processing steps will depend on separation technology as discussed in the next chapter.

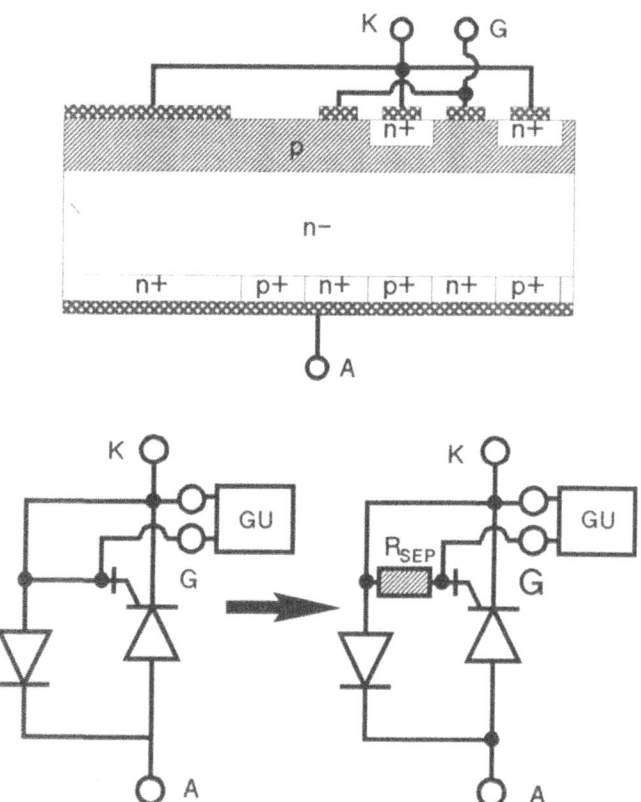

Fig. 8. The gate unit of the GTO requires a separation resistance, R_{sep}, between the p-
base layers of GTO and diode.

4. SEPARATION TECHNOLOGY

The RC-GTO has two specific separation problems. The first one, shown in Fig. 8, is due to the fact that the low impedance p-layer is directly short-circuiting the GTO gate unit. It is very important to find a way of separating these two adjacent p-regions in order to allow for standard gate-unit operation (R_{sep}). The second problem, illustrated in Fig. 9, is a dynamic one: during diode reverse recovery, we have to exclude parasitic GTO turn-on due to possible flow of stored charges from the diode into the GTO p-base. These problems are related since they both originate from the common use of the p-layer by both elements, the GTO and the diode.

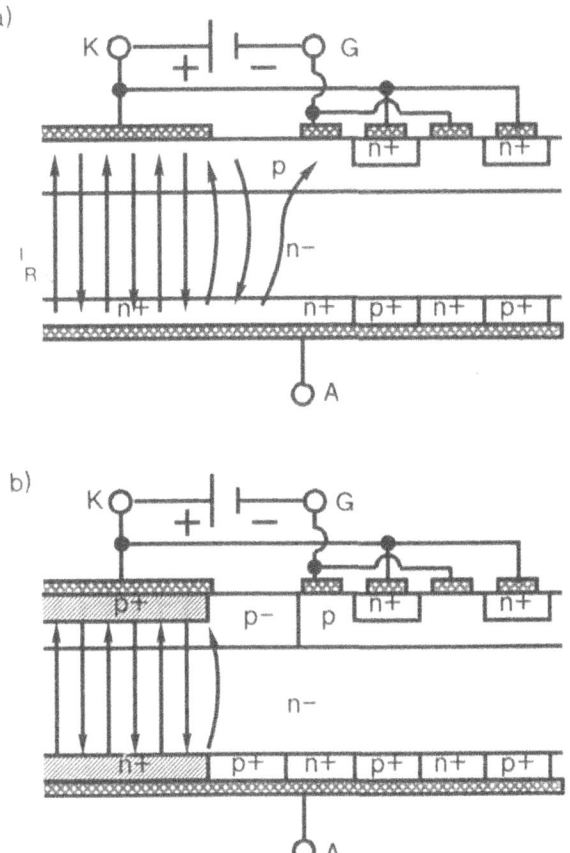

Fig. 9. a) Possibility of dynamic interference leading to parasitic GTO turn-on at fast diode turn-off; b) elimination of this dynamic problem by suitable emitter localization.

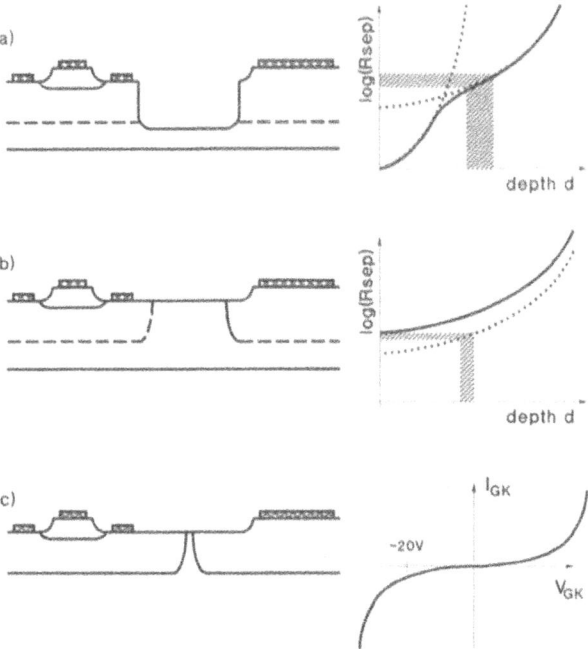

Fig. 10. Schematic representation of three different separation technologies: a) by deep groove etching, shown for the case of a double profile; b) by masking of the high conductivity part of an optimized double profile; c) by separation of the two adjacent p-regions in a narrow zone, using planar technology.

Device separation is the most demanding design aspect of RC-GTO-development. Fig. 10 sketches three different approaches that have been evaluated in some detail:

a) The first one is by deep groove etching into the p-profile. It has been adopted by most suppliers today.

b) The second is also resistive, requiring a double-profile with a masked low-resistance part. The idea is to separate the low-impedance layers by photomasking, but then to link them with a second p-layer of very much lower conductivity, just for blocking.

c) The third possibility shown here is a complete separation of the p-type zones, but only in such a narrow slot, that blocking will not be affected[8]. The approach is somewhat similar to planar device technology. Field penetration to the surface will occur, but since we need only about 20 Volt between gate and diode, a very narrow separation range will do, leading to very small potential differences at the surface. Separation is not resistive in this case, but has some punchthrough-character as shown.

For completeness, we mention that device separation by complete physical separation of the silicon wafer and multiple contouring has also been proposed[5]. However, this approach is only possible in alloyed systems and was therefore not considered since our complete GTO line-up is of the free-floating silicon type[9,10,11], a feature offering great flexibility in realizing the fine emitter structures required for RC-GTO's.

The first two approaches have in common that the electric charges responsible for lateral conductivity are also the ones required for space charge formation during blocking. Therefore, etching too deep or going too low in dopant concentration leads to serious field penetration, whereas, with too little etching, the required separation resistance will not be reached. One solution is to make the separation zone very wide, but this requires surface area of the wafer that will be lost for the active zone.

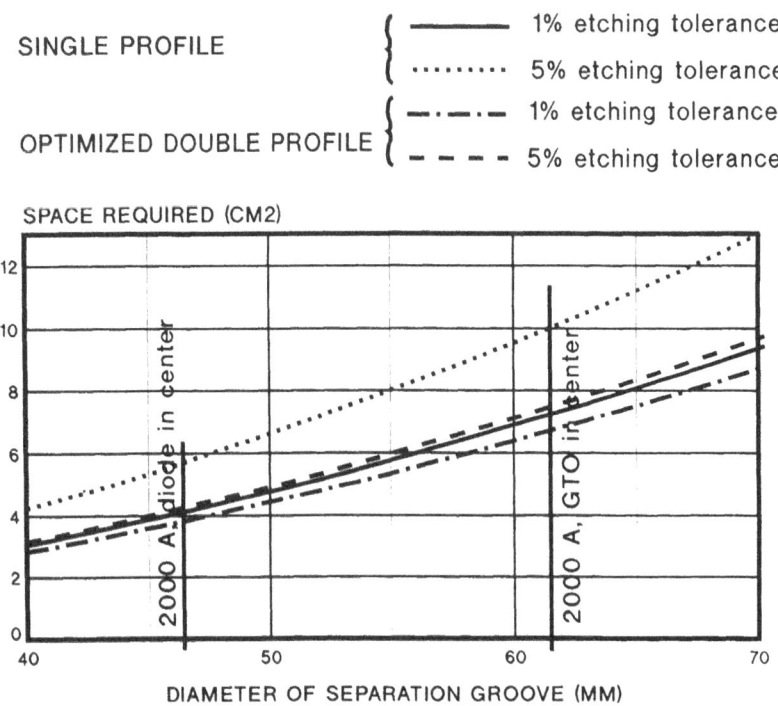

Fig. 11. Space required for resistive device separation, taking into account element size, p-diffusion-profile and etching tolerance.

The gate supply requires a separation of 200 Ω or more. Evidently, the sheet resistivity required depends on device geometry. However, any example will show, that there is a huge factor between the resistivity required for good GTO performance and the one required for separation.

The total charge in the remaining p-layer can be used to estimate at which blocking voltage field penetration will occur. And, with analytical standard profiles, it is possible to evaluate how sensitive this will be on etching depth variations. Then, the area required for resistive separation can easily be estimated.

Figure 11 shows, how it increases nearly quadratically with separation diameter, but also depends on obtainable etching accuracy. The influence of etching tolerance can be significantly reduced by etching into an optimized double profile, and most of the suppliers of groove-etched RC-GTO's are taking advantage of this possibility. However, space required for separating large devices is still very substantial. With this respect it would evidently be preferable to place the smaller area device, namely the diode, into the center, a geometry adopted by one of the suppliers[4]. However, we believe that there are good reasons to have the GTO in the device center rather than squeezed to the edge just because of a rather questionable separation technology. Resistive separation still leads to additional losses in the gate unit, and the design rule for this separation remains very unsatisfactory, mixing device size, lateral geometry, axial diffusion profiles and etching tolerances.

5. NOVEL SEPARATION STRUCTURE

Following a considerable amount of simulation work, experimental verification, yield estimates and cost calculations, the third separation technology, shown in Fig. 10, was finally choosen as a novel approach to the problem[8]. It allows for practically ideal separation, requires less space and is completely independent of device geometry. An additional advantage is its easy adaptation to any device of this kind, making it a flexible approach for future developments too.

Figure 12 shows an Electron Beam Induced Current (EBIC)-picture of the narrow separation zone between the two p-layers. Nominal masking distance is above 100 µm, but due to lateral diffusion final separation is less than 30 µm.

In Fig. 13, the actual design of the device separation in our 2 kV/ 2.5 kV-device is shown. The drawing is expanded in axial direction by a factor of ten, approximately. The outermost segment ring of the GTO-section, the diode section and the negative bevel are also shown.

The separation zone is realized using: a) p-layer separation for the gate unit , b) GTO emitter limitation, with the mesa emitter on the cathode and the short structure on the anode, and c) diode emitter limitation to prevent dynamic device interference, shown on the upper side by a second slot in the low-impedance p-layer, and on the lower side by a p-type ring. We have found, that one of these emitter limitations is generally sufficient to prevent the diode from interfering with the GTO in the sense shown in Fig. 9. None of the problems with parasitic GTO turn-on, as described in the literature[2], have been observed with this new separation design.

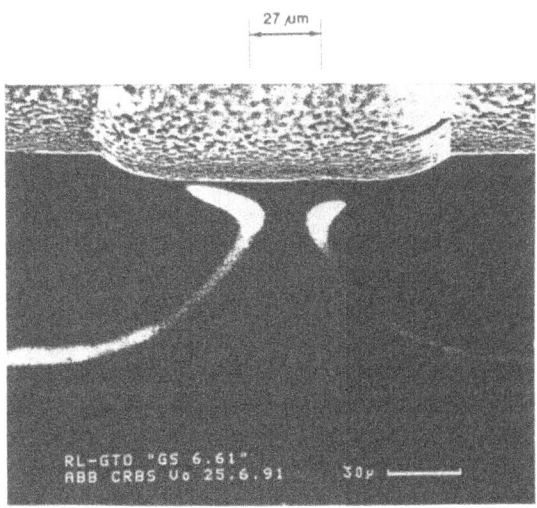

Fig. 12. EBIC picture of p-layer separation according to novel technology.

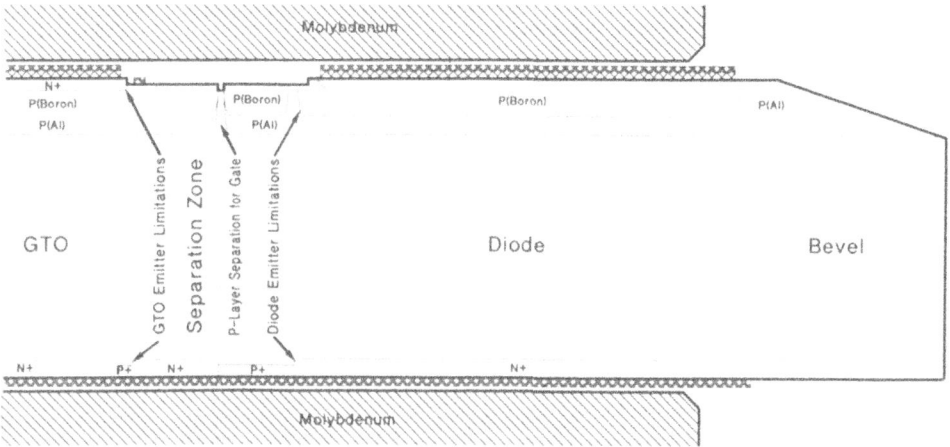

Fig. 13. Cross-section of the RC-GTO available from ABB Semiconductors AG.

6. CARRIER LIFETIME OPTIMIZATION

Figure 14a shows, for the diode of the 2.5 kV RC-GTO, how a data combination of on-state voltage and turn-off loss can be adjusted by suitable choice of the high-level carrier lifetime. After processing, on-state voltage is around 1.2 V and turn-off energy around 7 Ws. In order to comply with the target specs, the turn-off losses have to be reduced below 1 Ws, whereby on-state voltage should remain below 2.5 V. As can be seen, both requirements are simultaneously fulfilled in a reasonably broad parameter range.

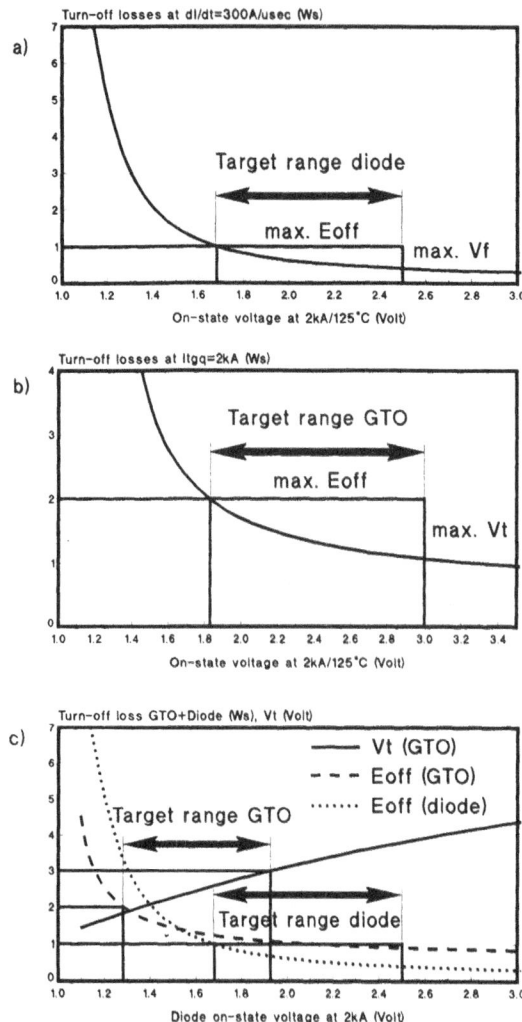

Fig. 14. Trade-off between turn-off losses and on-state voltage of a 2 kA/2.5 kV RC-GTO, influenced by high-level carrier lifetime control: a) diode part; b) GTO-part; c) total losses.

Essentially the same situation holds for the GTO-part as shown in Fig. 14b. Again, turn-off losses are exceeding the limit prior to carrier lifetime adjustment, and electron irradiation is used in order to comply with the target specifications.

In Fig. 14c, the relation between GTO and diode on-state voltage for homogeneous electron irradiation is used to relate both requirements. Unfortunately, there is only very little overlap of the two adjustment ranges. The situation is more transparent in the slightly different representation of Fig. 15: here, both on-state voltages and both turn-off energies are represented as functions of an estimated high-level carrier lifetime. Whereas both parts of the device have an initial carrier lifetime of close to 15 µs, the diode requires a final lifetime at or below 3 µs, whereas the GTO would show its best performance at almost twice this value.

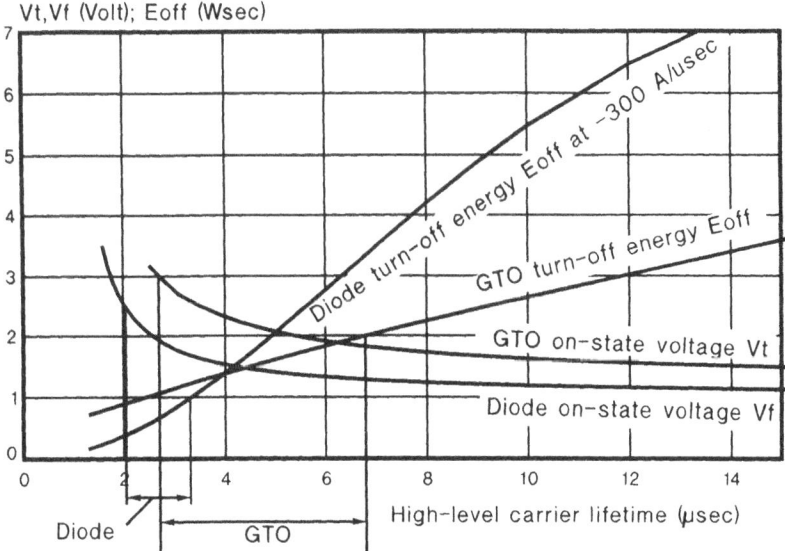

Fig. 15. Representation of device optimization as a function of carrier lifetime.

The immediate solution to this problem appears to be individual lifetime adjustment, using masked electron irradiation. This method works well and is also the most flexible approach. However, it requires two irradiations and, thus, creates extra costs and additional delays in cycle time. Therefore, it would be of interest to shift one of the technology curves in such a way, that a single irradiation can be used to optimize both components. First design adaptations aiming at such a rationalization look promising.

7. THERMAL RESISTANCE CHARACTERIZATION

The definition and characterization of thermal impedance for a multifunctional device deserves some attention. Optimization with respect to cooling is achieved when worst case losses are leading to the same maximum temperature in both elements. However, these worst cases may not necessarily occur simultaneously and, therefore, lateral thermal gradients and corresponding lateral heat transport cannot be excluded. This would lead to a rather complicated equivalent circuit as shown on the righthand side of Fig. 16. After some estimates we rejected it and decided to come back to the simpler scheme shown on the lefthand side of the same figure and already proposed in[7] for the reverse-conducting thyristor. The two impedances shown here and specified in the data sheet are defined for the case of simultaneous and homogeneous heat generation in both parts of the component. They are, therefore, conservative values by definition, and it can be shown that any inhomogeneous heating leading to lateral heat flow can only make things better and not worse. In other words: these impedances are not the values measured experimentally when applying dissipative heat to just one of the components, they are sensibly higher and take into account transimpedances. For a more comprehensive description of the problem we refer to [12].

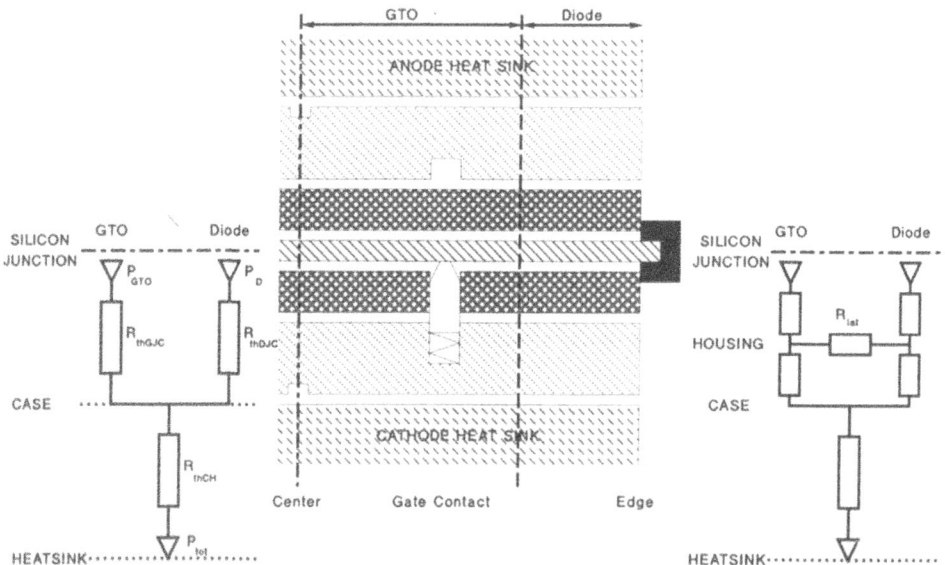

Fig. 16. Heat conductivity in a dual function device and the definition of a suitable equivalent circuit.

8. FUTURE PROSPECTS OF THE RC-GTO-CONCEPT

Representative prototypes of the first RC-GTO designed in this novel separation technology, a 2 kA/2.5 kV-device, are now available, and their performance is characterized by a tentative data sheet. Simultaneously, we have developed a corresponding 2 kA/4.5 kV-device. All the technological problems are solved and some engineering samples have been measured. The 4.5 kV-device has the same size as the 2.5 kV-type. There is no intention to replace the present 2.5 kA-GTO in main-line traction applications since this would require an even larger device.

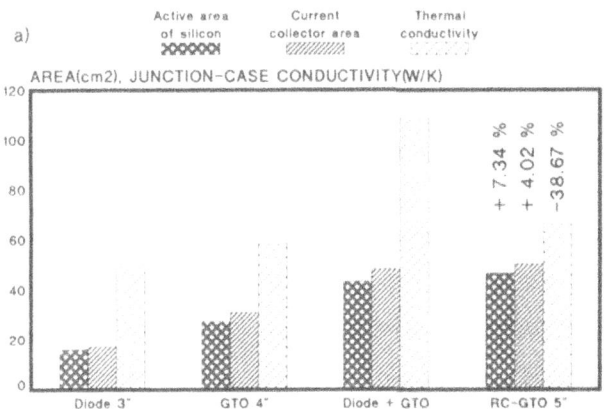

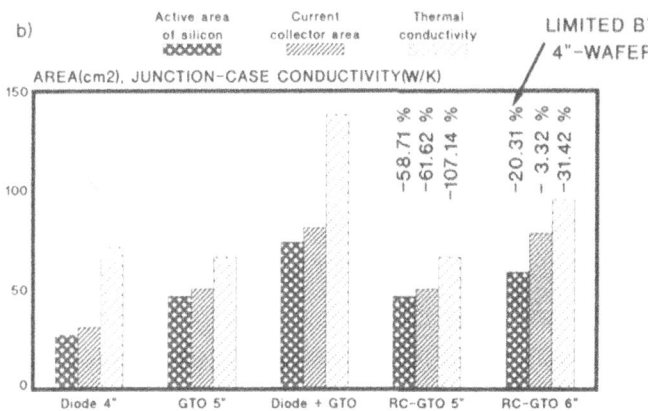

Fig. 17. Size and thermal conductivity comparison of discrete and integrated elements; the percentage values indicate the deviation from the classical solution (Diode + GTO): a) 2 kA/2,5 kV-devices for mass-transit traction applications. b) 2.5 kA/4.5 kV-devices for main line traction applications.

Figure 17 shows the basic reasons for this judgement. Part a) shows a comparison of discrete and reverse conducting elements of the 2 kA/ 2.5 kV-class, with respect to silicon area, current collector area and thermal conductivity. Obviously, the integrated solution presented in this paper is somewhat superior to the discrete element solution with respect to active silicon and to current collector area. The less favourable comparison for the thermal conductivity reflects the conservative character of the thermal data used in the tentative data sheet.

First measurements show that actual values for conductivity are close to the sum of discrete devcie values. For the 2.5 kA/4.5 kV-class, shown in part b) of the figure, this comparison looks much less favourable. A solution in a 5"-housing is very unlikely, and even if we would realize a new 6"-housing, the size would not match the thermal conductivity of the discrete solution. The remaining 20 % reduction in silicon area is now limited by the starting wafer size, not by the housing. A straightforward realization of such a device would, therefore, require that the individual components can handle the power with losses reduced by about 20 %. This is possible, but will require an additional step in direction of low-loss-GTO's and optimized diodes.

9. CONCLUSIONS

In most of present day inverter applications, using asymmetric blocking high power GTO's, the monolithic integration of the required antiparallel freewheeling diode can offer many advantages. The few additional processing steps required for such a device are counterbalanced by material savings, and one optimal GTO-to-diode area ratio should fit for all the main applications.

Electrical separation of the two functions, mainly with respect to the gate of the GTO, deserves special attention. A novel solution to this problem has been presented, which completely eliminates resistive loading of the gate unit and also requires less surface area on the pellet. This flexible solution is easily adapted to new element sizes or technologies.

The prototypes available today are of the 2 kA/2.5 kV type, used e.g. in mass-transit traction applications. The elements still require independent carrier lifetime adjustment, but first design changes aiming at a homogeneous lifetime control look promising. A 2 kA/4.5 kV device, using the same housing has been developed simultaneously and will be available soon. With respect to the highest controllable currents of 2.5 to 3 kA, realization of the reverse conducting types is still questionable with todays device technology, mainly due to cooling limitations.

10. ACKNOWLEDGEMENTS

The author acknowledges efficient support by the whole technical department of ABB Semiconductors AG, and he would like to express his special thanks to Erich Nanser for his great care in preparing and controlling the development batches and to J. Waldmeyer, P. Roggwiller and A. Jaecklin for contributing to the device design and for many fruitful technical discussions.

REFERENCES

1   E. Huang and J.P. Barnes, "GTO with monolithic antiparallel
     diode", *IEEE Proc.* vol.132, no. 6, 1985, pp. 245-247.

2   T. Shinone, M. Asaka, K. Takigami and H. Ohashi, "Isolation
     structure optimization for high power reverse conducting GTO",
     PESC'88 Record (April 1988), pp. 908-914.

3   O. Hashimoto, Y. Takahashi, H. Kirihata, M. Watanabe and O.
     Yamada, "4.5 kV 3000 A high power reverse conducting gate turn-
     off thyristor", *PESC '88 Record* (April 1988), pp. 915-920.

4   M. Ishidoh, M. Yamamoto, T. Nakagawa and F. Tokunoh, "A new
     reverse conducting GTO", *EPE Aachen 1989*, pp. 121-125.

5   S. Yamada, M. Kekura and T. Kawamura, "Reverse-conducting
     gate turn-off thyristor employing a new separation structure",
     Meiden Rev. Series, vol. 89, no. 2, 1990, pp. 13-16.

6   O. Hashimoto, Y. Takahashi, M. Watanabe, O. Yamada and T.
     Fujihira, "2.5 kV 2000 A monolithic reverse conducting gate turn-
     off thyristor", *IEEE Trans. on Ind. Appl.,* vol. IA-26, no. 5, 1990,
     pp. 835-839.

7   P. De Bruyne, J. Vitins and R. Sittig, "Reverse-conducting
     thyristors", *Semiconductor Devices for Power Conditioning,*
     Plenum Press N.Y., 1982, ed. R. Sittig and P. Roggwiller, pp. 151-
     173.

8   A.A. Jaecklin, *United States Patent, 4, 742, 382.*

9   S.D. Prough and J. Knobloch, "Solderless construction of large
     diameter silicon power devices", *Industry Applications Soc.
     Annual Meeting,* pp. 817-821, 1977.

10   A.A. Jaecklin and H. Lawatsch, "Free floating silicon technology
     for high power devices", *Proc. Int. Symp. on Power
     Semiconductor Devices,* 1988, Tokyo, pp. 166-171.

11   A. Schweizer, A.A. Jaecklin and P. Streit, "High power GTOs with
     free-floating silicon technology", *Proceedings of the EPE 1989
     Int. Conference,* Aachen, pp. 749-753.

12   A. Aberg, "Transient temperature response of reverse conducting
     GTO's", to be published in *IEEE Trans. El. Devices.*

DISCUSSION

H. Ch. Skudelny (RWTH Aachen, Aachen D)

Suppose that the silicon area allocated to the GTO and to the diode does not correspond to the character of your load. Can you make use of any flow of heat from one device to the other in that case ?

P. Streit

In traction, simulations of full start and stop phases show that at no time only one element is carrying the entire load. If there is an asymmetry of the load cycle, the hotter element will automatically spread its heat to the cooler one. However, the GTO part can only profit from cooling by the diode part if the temperature of the whole surface is reduced but not if its center stays hot. In contrast, heat flow from the diode, conceived as a narrow ring at the periphery, may reduce maximum temperature. Additional information on these thermal transimpedances are given in a publication by A. Åberg cited in my contribution.

J. Gobrecht (ABB Corp. Research, Baden)

Can you comment why you have not considered alternative or improved packaging concepts in order to overcome the thermal limitations ?

P. Streit

The goal was to design a device based on a housing presently available, i.e. the case of our 3000 A GTO. Since our approach is of the application driven kind, we may consider a different housing if we are sure that this is what our customers want.

D. Silber (TU Bremen, Bremen D)

Can you discuss the possibility of decoupling the p-base diffusions in both devices ? Differences in the p-base depth might lead to further improvement.

P. Streit

The key issue to the integration is how perfect a diode can you produce from a silicon wafer specified for a GTO. So far, we did not discuss differences of the p-base because this would further complicate the technology and we have no evidence ot its advantages. The trench technology, separating GTO and diode, however, would allow for much shallower p-bases and offers the possibility to adapt them. For the 2.5 kV element, the common p-base presents no problem ; for higher voltages, additional investigations will be needed.

REVIEW OF POWER DEVICE CONCEPTS

Roland Sittig

Technical University Braunschweig
Braunschweig, Germany

ABSTRACT

Material characterization, controllable processes and the understanding and computability of the events inside devices constitute the
base for power device production. After these techniques had arrived at
a mature state, devices exhibiting turn-off capability were developed
even for the highest power level.

Following a period of vivid discussions the question about the different application areas and the respectively best suited devices seems to
be settled at present. Some technical arguments causing the separation
of these areas are investigated.

Power MOS-FET's exhibit superior frequency capability and require
low expenditures for triggering. Therefore they are advantageous at low
voltages. Bipolar Transistors exibit less forward voltage drop and are better suited if a higher blocking capability is required. IGBT's additionally
combine this characteristic with a MOS-input. The excellent property of
all of these devices to withstand load short circuit and to not require any
expensive means for protection depends on the base structure and the
total amount of excess charge in the device in its conducting state. Low
on-state losses at the highest voltage level hinder to maintain these
characteristics. In this range GTO-thyristors are applied which have to
be installed with a snubber-circuit. In the future, further progress of
technology can make possible a nearly complete control of carrier
dynamics. Then a strong reduction of switching losses, of protection
circuitry and trigger requirements may be expected.

Power Semiconductor Devices and Circuits, Edited by A.A. Jaecklin
Plenum Press, New York, 1992

1. INTRODUCTION

Let us begin this review by reconsidering the status of power devices around 1970. At that time diodes and thyristors were being produced since more than a decade and had arrived at a quite mature state. First converters for HVDC-transmission based on silicon devices were built and market research predicted an enormous demand for power triacs. The existing know-how was based mainly on empirical experience. It allowed, however, to adopt devices to the requirements of different installations.

Important goals of research activities at that time were to explore material characteristics like crystal defects, doping homogeneity and recombination processes and to investigate physical phenomena inside devices for instance by using infrared emission and absorption techniques. First computer programs were simultaneously developed to cross check these measurements and increasingly also to predict device characteristics. This was the origin of the mighty simulation programs of today which allow to analyse multidimensional, time- and temperature-dependent processes.

2. TECHNOLOGICAL PREREQUISITES

2.1 Silicon Technology

The main problems of device development concerned scattering of characteristics and defects of the starting material, irreproducible processes and the resulting variations in yield. X-ray topographs exhibited "black dots" in the starting material which after processing caused "hot spots" at applied blocking voltage[1]. Resistivity variations of $\rho_{max}/\rho_{min} = 2$ were typical for silicon material above 100 Ωcm and precluded optimization of high voltage devices. And the puzzling behaviour of gold diffusion did not allow to obtain a controllable and homogeneous carrier lifetime.

All of these problems were solved by improving crystal growth, introducing neutron transmutation doping and replacing gold diffusion by irradiation techniques. Moreover, IC-technology offered well controlled processes such that one started to think about applying these to the production of power devices. This way, at the end of the seventies technology was prepared to introduce more sophisticated device concepts.

2.2 Packaging

Progress of power devices always requires a corresponding development of package concepts since heat management, electromagnetic forces or electromagnetic interference directly influence the electrical characteristics of devices. Of great importance in this aspect was the introduction of power modules by Semicron which allowed to greatly simplify device mounting. Direct Copper Bonding of alumina substrates turned out to be a key process to optimize thermal resistances at reduced cost in this technology[2]. The Brown Boveri researchers were happy to demonstrate their ability to carry out this process shortly after first hints about it from colleagues of the GE research center.

A more recent development might gain a similar importance. Schwarzbauer and Kuhnert[3] introduced a new joining technique which can replace alloying or soldering of devices to metal plates. The process is based on a pressure sintering of silver powder at temperatures of about 200 °C. Applying it avoids high temperature treatment as well as any influence on the silicon doping structure and the bond nevertheless can withstand high temperatures and temperature cycling. Due to the low bonding temperature, mechanical stress is so much reduced that even joining of molybdenum plates on both surfaces of a large area MOS-FET could be demonstrated[4].

3. TURN-OFF CAPABILITY

In the course of the seventies the demand for high power devices exhibiting turn-off capability became more and more urgent. These would allow to greatly simplify installations and to reduce cost.

According to the well developed technology at that time, many proposals existed for devices which could meet this requirement. So a somewhat confusing situation arose. Fast Reverse Conducting Thyristors, (RCT), had been introduced, Gate Assisted Turn-off-Thyristors, (GATT), could further reduce turn-off time, Giant Transistors exhibited well suited characteristics at least up to about 1000 V, first Gate Turn-Off (GTO)-thyristors had been demonstrated and data on small samples of more advanced devices appeared in the literature such as: very high power MOS-FET's, Static Induction Transistors (SIT), Static Induction Thyristors (SITh) and Field Controlled Diodes (FCD) or Field Controlled Thyristors, (FCTh).

Meanwhile the assignment of suited devices to different applications can be considered to be settled, Fig. l. Best performance at low power and up to voltages of about 400 V can be obtained by Power MOS-FET's. Above this voltage level up to about 1200 V Bipolar Transistor Modules are predominant, however, they are increasingly being replaced by Isolated Gate Bipolar Transistor (IGBT)-Modules. These devices seem to be the best choice even up to 1600 V or 1800 V. Above 2000 V finally the GTO-thyristor offers the best overall characteristics. More advanced devices will exhibit improved performance in this voltage range. At present, however, their production requires an extra effort that seems to outweigh these advantages. In the following we will discuss some technical arguments that led to the given assignment.

3.1 The GATT-Concept

Devices exhibiting a blocking capability of more than 2000 V have to have a wide base and to attain a low forward voltage drop; good efficiencies of both emitters are required. Transistors, therefore, need not to be considered. One thus had to decide whether the GATT-or GTO-concept would be better suited for applications in this voltage range.

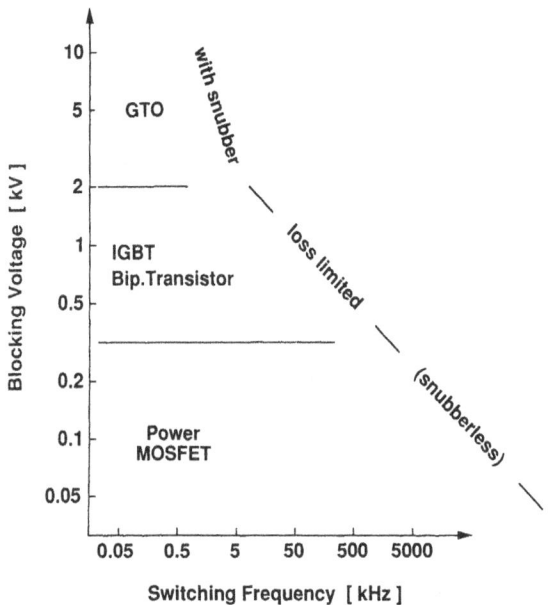

Fig. 1. Schematic arrangement of switchable power devices in a voltage-frequency diagram.

Considering a thyristor structure the GATT-concept consists in applying a negative gate voltage after current commutation up to the next turn-on triggering. At the positive dV/dt the capacitive current induced at the blocking junction is drawn away via the gate electrode, Fig. 2, and thus triggering of the cathode emitter can be avoided. To optimize the structure, cathode emitter-shorts should be omitted since they would short-circuit the gate drive. Furtheron, the effect of the negative gate voltage is limited at the avalanche breakdown voltage of the n-emitter to p-base junction.

With a half emitter width w_k, a sheet resistance of the p-base $\rho_\square$ and a constant current density j at the blocking junction, a voltage distribution according to

$$V(y) = -V_G + j \, \rho_{\square} \left(\frac{w_k^2}{2} - \frac{y^2}{2} \right) \tag{1}$$

is obtained within the p-base. No electron injection takes place when the maximum potential is kept below about 0.3 V:

$$V(y = 0) = -V_G + j \, \rho_{\square} \frac{w_k^2}{2} < 0.3 \text{ V} \quad . \tag{2}$$

With a negative gate voltage of about -10 V applied, a positive voltage rise may take place considerably before all carriers of the stored charge are recombined. Several experiments confirmed that a reduction of turn-off time by about a factor of two can be obtained. The GATT-concept represents a useful and stable mode of operation, however, it cannot turn-off a thyristor. Therefore it had been clear that it would not be used if a GTO solution could be realized.

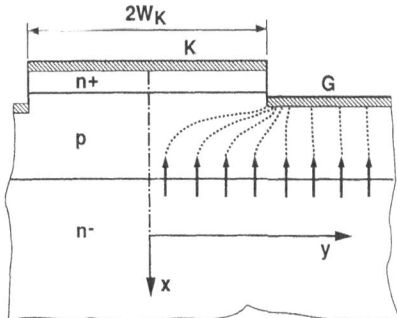

Fig. 2. Flow of capacitive current in a GATT- structure.

3.2 GTO-Concept at Low Turn-Off Gain

At a first glance, the GTO-structure looks identical to the GATT-structure. It is intended, however, to interrupt a thyristor being in its forward conducting state. Therefore, a much higher current density has to be considered. A corresponding narrow width of the emitter segments and low resistivity of the p-base have to be chosen.

As a first approximation the one-dimensional situation shall be discussed. The time dependence of the storage charge Q is given by

$$\frac{\partial Q}{\partial t} = \gamma_n \ j_K - (1-\gamma_p) \ j_A - \frac{Q}{\tau} \tag{3a}$$

where

$$\gamma_n \ (j_K) = \frac{j_n}{j_K}$$

represents the emitter efficiency of the n-emitter and

$$\gamma_p \ (j_A) = \frac{j_p}{j_A}$$

the p-emitter efficiency and τ the carrier lifetime which is assumed to be constant throughout the device.

With $j_A = j_K + j_G$, one obtains

$$\frac{\partial Q}{\partial t} = \left[\left(1 - \frac{j_G}{j_A} \right) \gamma_n + \gamma_p - 1 \right] j_A - \frac{Q}{\tau} \ . \tag{3}$$

In a stationary forward conducting state,

$$\frac{\partial Q}{\partial t} = 0 \quad \text{and} \quad j_G = 0 \ ,$$

eq. (3) results in

$$Q = \tau \ j_A \ (\gamma_n + \gamma_p - 1) \ . \tag{4}$$

This equation is equivalent to the current equation at the middle junction, J_2 in Fig. 3,

$$j = j_n \ (J_2) + j_p \ (J_2) = \gamma_n \ \beta_n \ j + \gamma_p \ \beta_p \ j \ ,$$

yielding

$$\gamma_n \ \beta_n + \gamma_p \ \beta_p = 1 \tag{5}$$

where

$$\beta_n = \frac{j_n \ (J_2)}{j_n \ (J_1)} \quad \text{and} \quad \beta_p = \frac{j_p \ (J_2)}{j_p \ (J_3)}$$

represent the usual transport factors.

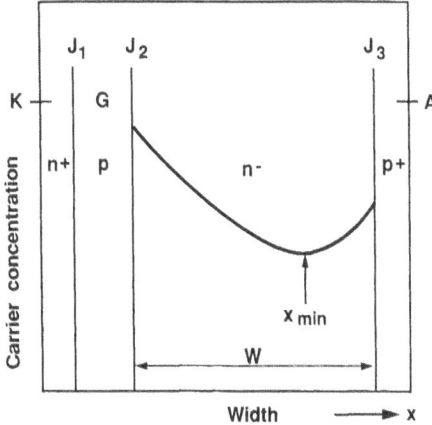

Fig. 3. Thyristor structure and indication of details in the text.

A low forward voltage drop requires a rather large Q, which implies in the on-state

$$\gamma_n \left(j_K = j_A \right) + \gamma_p \left(j_A \right) > 1 . \tag{6}$$

During turn-off, obviously $\partial Q / \partial t < 0$ is required, even for Q approaching zero. Then from eq. (3) it follows that

$$\left[1 - \frac{j_G}{j_A} \right] \gamma_n \left(j_K \right) + \gamma_p \left(j_A \right) - 1 \leq 0 \tag{7}$$

should hold. This means that independent of γ_p a ratio $j_G / j_A > 1$ or a turn-off gain, G, of $G = j_A / j_G < 1$ should always lead to turn-off. From eq. (7) follows that a maximum turn-off gain of

$$G_{max} = \frac{j_A}{j_G} = \frac{\gamma_n \left(j_K \right)}{\gamma_n \left(j_K \right) + \gamma_p \left(j_A \right) - 1} \tag{8}$$

can be expected.

Let us first consider a turn-off process at $G < 1$ with no snubber circuit applied. In this case, the structure has to be designed such that eq. (2) holds at $j = j_G > j_A$. It follows that the cathode current density

$$j_K = j_A - j_G < 0$$

reverses its direction.

Thus during a first phase - the storage time, which is extremely short in this case - the p-base is depleted of excess carriers. As soon as junction J_1 gets depleted, it is driven into avalanche breakdown. The anode current, however, can flow via the gate contact. The carrier distribution within the n-base and the voltage across the device will only slightly be changed. When J_2 gets rid of carriers in a second phase an inductive load will force a sharp voltage rise at least up to the source voltage. This period may be dangerous if "dynamic avalanche multiplication" - as some authors call this phenomenon - is triggered.

The current through the expanding space charge layer is a pure hole current. It contributes a positive space charge density ρ_j according to

$$\rho_j = \frac{j_p}{v_p} \approx \frac{j_p}{v_s} \tag{9}$$

to the space charge of the positive donor ions

$$\rho = q\, N_D^+ + \rho_j \tag{10}$$

where q is the elementary charge, v_p the drift velocity of holes and v_s the scattering limited velocity. As a consequence, the breakdown voltage can be considerably lower at high current densities. Under the assumption of a very short storage phase which does not lead to a redistribution of current density, the maximum j_p will be of the same amount as j_A in the on-state. Taking into account the cathode emitter area as the relevant crosssection, then for instance a j_p of 100 A cm^{-2} would yield

$$\rho_j = 10^{-5} \text{ As cm}^{-3} \approx 6 \bullet 10^{13}\, q \text{ cm}^{-3}$$

This additional space charge density will be of minor influence at doping concentrations as used for breakdown voltages around 1000 V. It would, however, determine the breakdown voltage in devices exhibiting less doping concentration. For high voltage devices, therefore, either a correspondingly reduced maximum current density j_p must be applied during the phase of voltage rise or the dV/dt should be limited by a snubber-circuit.

In conclusion it can be said that operation at low turn-off gain comprises the great advantage of not increasing the maximum current density j during turn-off and thus keeping the threshold for dynamic avalanche at its maximum value. Even snubberless operation seems possible at voltages up to about 1000 V. To take advantage hereof, it has to be assured that the inductive energy which is transferred to the device is kept low enough.

Although there was a strong demand for a high turn-off gain in an early phase of GTO-development, nowadays a low G operation seems to become more attractive[5,6]. Experimental investigations with $I_G = 2\, I_A$, reaching its peak value within a fraction of a microsecond[5], exhibited safe turn-off without a snubber circuit. The medium power GTO withstood a rate of voltage rise of $dV/dt = 15\,000$ V/μs up to 900 V. At present, applications are dominated by high voltage GTO's, however, which are designed to exhibit high turn-off gain. Their characteristics are investigated in the following section.

3.3 GTO Concept at High Turn-off Gain

The demand for high turn-off gain seems to contradict eq. (8)

$$G_{max} = \frac{\gamma_n}{\gamma_n + \gamma_p - 1} \quad .$$

According to this equation γ_n should be chosen as high as possible and γ_p very low. Since a low forward voltage drop requires $\gamma_n + \gamma_p > 1$, eq. (6), γ_p can hardly be chosen less than 0.5 which would yield $G_{max} \approx 2$. Nearly all available high power GTO's exhibit, however, a G of about 4. To understand this deviation, a better description of the turn-off process is needed.

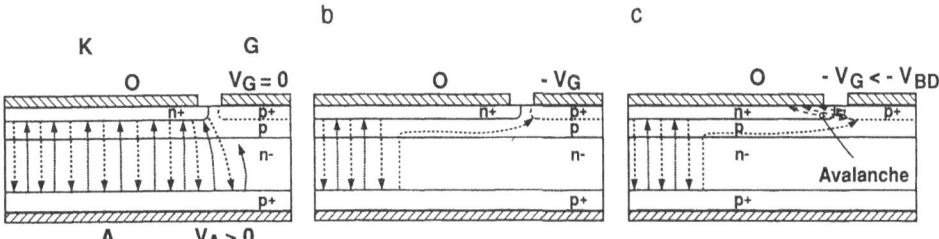

Fig. 4. Current flow in a GTO-structure: a) in the forward conducting state and b), c) during the storage time.

A single cathode segment of a high voltage GTO with an inductive load and a snubber-circuit is considered. According to Fig. 4, a negative gate current will decharge the p-base near the cathode edge. Current is squeezed to the center of the segment with a corresponding increase of current density. One has to be aware that, in the forward conducting state, the vertical electric field in the n-base amounts to about 40 V/cm, causing an electron velocity of only about $4 \cdot 10^4$ cm/s which is more than two orders of magnitude less than the scattering limited velocity $v_s = 1 \cdot 10^7$ cm/s.

Therefore, at the beginning of the turn-off process, a slight increase of forward voltage drop will cause a corresponding increase of current density but induce only a negligible transfer of current to the snubber-circuit. Roughly, the anode current stays constant while current density is steadily increased. Sometimes it is stated that current will be interrupted, when the diameter of the on-area is reduced to about one carrier diffusion length in the p-base, L_p. There is, however, no argument to couple the on-area to the diffusion length. With a reasonable lateral electric field of

$$E_{lat} = \frac{10V}{100\mu m} = 10^3 \frac{V}{cm}$$

and assuming an exponential decay of the lateral minority carrier density $n_y = n_0 \cdot \exp(-y/\lambda)$, the condition $j_{n\ lat} = 0$ would result in a decay length of $\lambda \approx 0.3\ \mu m \ll L_p$. Thus the usual width of a cathode segment and of the gate voltage would allow a hundredfold increase of current density.

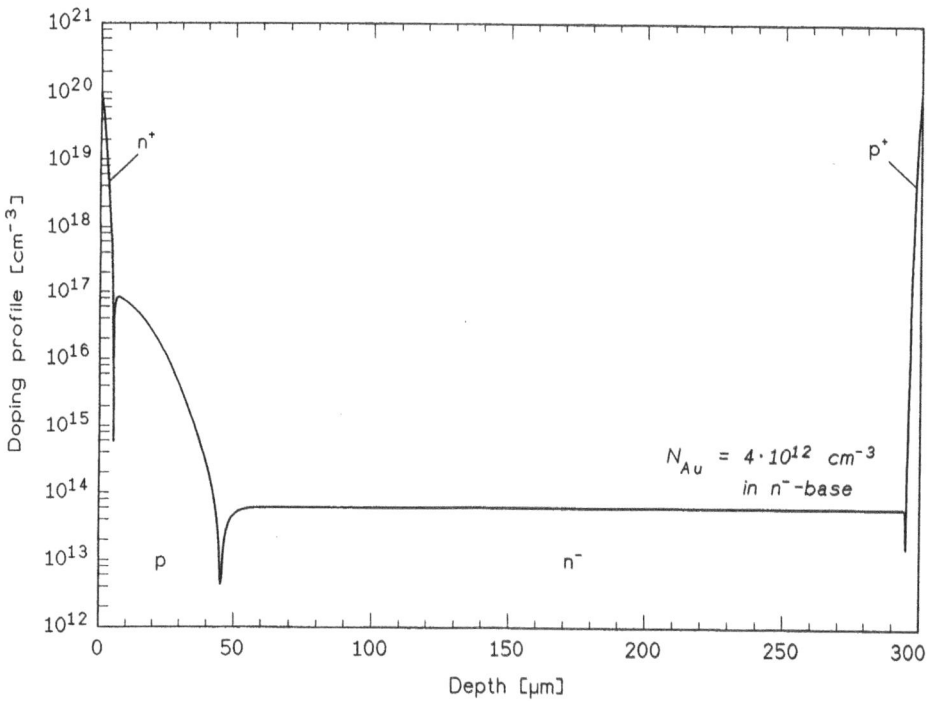

Fig. 5. Thyristor doping profile used for the calculated j-V-characteristics in Fig.6.

One can get an idea of what happens during this phase, assuming that at the center of the segment the current density versus voltage relation will follow the static characteristic of a thyristor. As an example for the structure sketched in Fig. 5, the calculated one-dimensional characteristics at constant temperature are drawn in Fig. 6 for various gate currents.

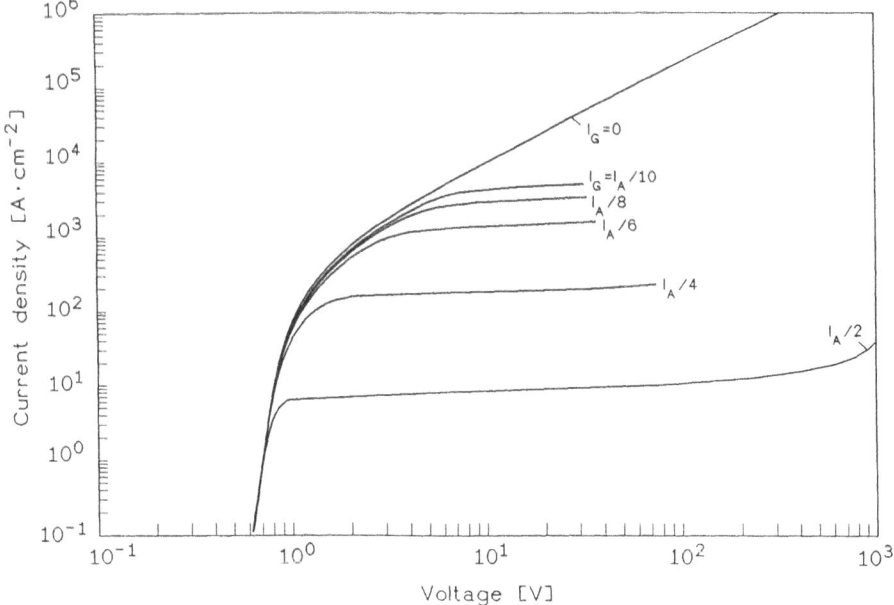

Fig. 6. One-dimensional computation of current density versus forward voltage characteristics with negative gate current as a parameter (room temperature, profile of Fig. 5).

If indeed the current density were raised to 10^5 A cm^{-2} in the presence of a field strength of the order of $4 \cdot 10^3$ V cm^{-1}, then the local power loss would lead to an enormous increase of temperature according to

$$\frac{dT}{dt} = \frac{j\,E}{c} \approx 200\,\frac{^\circ C}{ms} \tag{11}$$

where c is the specific heat of silicon. The corresponding increase of intrinsic carrier density would cause a negative temperature coefficient and thus trigger destruction of the device within microseconds.

The assumption that at the center of a cathode segment the current density versus voltage relation will follow the curve without any negative gate current may be too pessimistic. If instead a constant ratio of gate current to anode current is assumed, the other curves shown in Fig. 6 are obtained. These curves exhibit a considerably reduced current density which only slightly increases with voltage.

Although a dynamic process cannot exactly be described on the base of static characteristics, they allow to get a qualitative understanding: The GTO-structure below a cathode segment is considered as consisting out of several Δy - intervals. The net lateral hole current-flow out of the p-base then represents the respective gate current of that section and the interval will approximately follow the characteristic of the corresponding j_G/j_A-ratio. At the beginning of a constant gate current, the ratio is very high at the outermost interval. It is immediately switched off and its p-base is depleted. As this process proceeds, current density in the remaining on-area steadily rises since anode current is kept constant. Therefore, even at the edge of the on-area, the j_G/j_A-ratio is reduced. Finally, some lateral hole current will flow out of the last interval at the center of the cathode segment. With growing j_A, the current density is limited approximately according to the static characteristic at the corresponding j_G/j_A-ratio and a sharp voltage rise is required to maintain the current through the external inductance. This in turn deviates part of the current to the snubber circuit. Thus the total on-area is further reduced, regardless whether within a single segment or whether some segments are completely turned-off and current density increases at others. From Fig. 6, it can be concluded that areas with the highest local j_G/j_A-ratio are switched-off first. This effect leads to the often observed current redistribution and localized increase of current density[6,7,8]. It can be supported by structural inhomogeneities but would occur even in a perfectly homogeneous structure.

As more and more current flows into the snubber circuit, the j_G/j_A-ratio now is steadily increased until the turn-off condition according to eq. (7) is reached and thyristor action is interrupted. The final size of the on-area up to which this explanation is valid may result from more accurate treatment of the lateral current flow in the transition region. In this mode of operation the maximum cathode current density j_K is steadily increased until its sudden interruption. According to the usual dependence of emitter efficiency on current density, as sketched in Fig. 7, γ_n is reduced during the storage phase and turn-off condition, eq. (7), can be fullfilled at a smaller j_G/j_A-ratio.

If in contrast, a homogeneously distributed negative gate current is assumed, then cathode current density would decrease. This process too should lead to a reduction of γ_n. At the low current density side of the γ-curve, however, there will be no hindrance for a large area device to build up current filaments and in this way keep γ_n at its maximum value. It thus can be expected that current filamentation will always occur at a high turn-off gain.

In conclusion, the turn-off process of high power GTO's, exhibiting a turn-off gain of about 4 and requiring a snubber-circuit may be considered as follows: during the storage time the on-area is considerably reduced at constant anode current. The current density-voltage dependence will be similar to that of a corresponding thyristor. When a further increase of current density requires a sharp voltage rise, current is in part commutated to the snubber circuit. Due to the resulting improvement of the j_G/j_A-ratio the thyristor action is finally interrupted.

Three different failure modes are possible during this process:

- if the gate current is to small, the device will be driven in a stationary mode at high current density and high forward voltage drop, causing a thermal runaway,

- if the thyristor is driven into a state of very high current density, a region of negative temperature coefficient may be reached within microseconds,

- dynamic avalanche breakdown may be triggered during the voltage rise; The increase of current density during the storage phase makes this process much more likely than in the case of low G turn-off.

A remedy concerning the last two failure modes would be some means of current limiting of the thyristor characteristic.

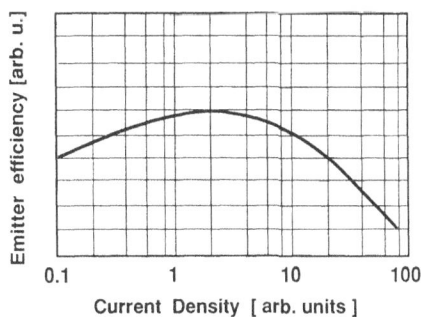

Fig. 7. Schematic dependence of emitter efficiency on current density.

3.4 The Abnormal Thyristor Characteristic

Exactly such a kind of current limiting thyristor characteristic had been discovered by Azuma and Takigami during GTO-development[9]. They further investigated this effect to avoid it, since it deteriorates surge current capability. A numerical investigation yielded that heavy doping effects of the n-emitter have a major influence on its occurence[10].

Recently a more descriptive analysis has been published[11] which allows to better understand the phenomenon. Numerical computations revealed that the voltage increase occurs only in a certain current density region, while the usual characteristic is obtained below and above this region, Fig. 8.

To give an impression of the processes involved in this phenomenon we turn again to Fig. 3 and eq. (5):

$$\gamma_n \, \beta_n + \gamma_p \, \beta_p = 1 \tag{5}$$

We start at a usual forward conducting state and assume $\gamma_n \, (j) \approx \gamma_p \, (j) \leq 1$ near their respective maxima at the chosen current density. If we consider $\beta_n \approx 1$ independent of current density then it follows from eq.

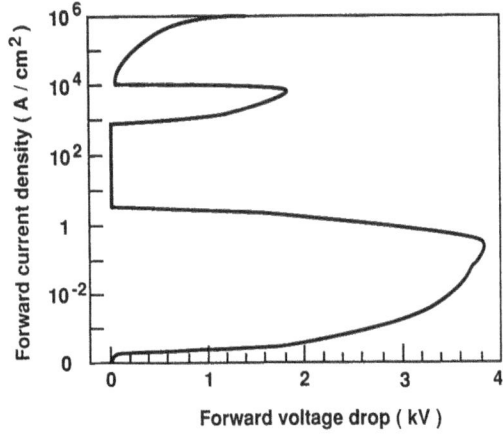

Fig. 8. "Abnormal" current-voltage characteristic of a thyristor structure as calculated by Y. Gerstenmaier[11] (© 1990 IEEE).

(5) that β_p is close to zero, which means that there is a high amount of stored charge recombining in the n-base. At the carrier density minimum x_{min} one has $dn/dx = 0$ which implies that

$$j_n \left(x_{min} \right) = \frac{\mu_n}{\mu_n + \mu_p} \, j$$

and

$$j_p \left(x_{min} \right) = \frac{\mu_p}{\mu_n + \mu_p} \, j \tag{12}$$

where μ_n, μ_p are electron and hole mobility respectively.

An increase in j will reduce γ_n, leading to the condition

$$J_n \ (J_2) \ = \ \gamma_n \ j \ = \ \frac{\mu_n}{\mu_n + \mu_p} \ j = J_n \ (x_{min}) \ . \tag{13}$$

Note that then the carrier density minimum has to be shifted towards J_2. At even higher j, eq. (5) can only be maintained by an increase of β_p. This can be accomplished either by an increase of $n(J_3)$ or by an extension of a space charge layer, SCL, from J_2 into the n-base and a corresponding reduction of the remaining neutral width. The first process can only help for a while since it degrades simultaneously γ_p.

Thus finally the second process takes place which is combined with a sharp increase of forward voltage.

Now in the SCL the space charge, ρ_j, due to the current density has to be taken into account:

$$\rho \ = \ q \ N_D^+ + \rho_j \quad \text{with} \quad \rho_j \ = \ q \ (p-n) \tag{14a}$$

Neglecting diffusion current components in the SCL and using the drift velocities υ_n and υ_p of electrons and holes respectively the current density equations are

$$J_n \ = \ q \ n \ \upsilon_n \ = \ \gamma_n \ \beta_n \ j$$

and

$$J_p \ = \ q \ p \ \upsilon_p \ = \ \gamma_p \ \beta_p \ j \tag{14b}$$

yielding

$$\rho_j \ = \ j \left(\frac{\gamma_p \ \beta_p}{\upsilon_p} - \frac{\gamma_n \ \beta_n}{\upsilon_n} \right) . \tag{14c}$$

When the carrier minimum is shifted to J_2 one obtains

$$J_n \ (J_2) \ = \ \gamma_n \ \beta_n \ j \ = \ \frac{\mu_n}{\mu_n + \mu_p} \ j \ = \ J_n \ (x_{min})$$

giving

$$\gamma_n \ \beta_n \ = \ \frac{\mu_n}{\mu_n + \mu_p} \ = \ \frac{\upsilon_n}{\upsilon_n + \upsilon_p} \ \approx \ \frac{3}{4}$$

and from eq. (5)

$$\gamma_p \ \beta_p \ = \ \frac{\mu_p}{\mu_n + \mu_p} \ = \ \frac{\upsilon_p}{\upsilon_n + \upsilon_p} \ \approx \ \frac{1}{4} \qquad \text{resulting in } \rho_j = 0$$

An increase of j will cause a further decrease of γ_n and γ_p leading to a slight reduction of $\gamma_n\beta_n$ and according to eq. (5) to a corresponding increase of $\gamma_p\beta_p$. As long as υ_p/υ_n remains constant, a positive space charge density $\rho_j > 0$ will be added. With increasing field strength in the SCL, however, υ_p approaches υ_n. Since $\gamma_n\beta_n > \gamma_p\beta_p$ still holds this leads now to a negative space charge density $\rho_j < 0$, which according to eq. (14) increases with j. Thus the total space charge is reduced and increasing current density leads to further extension of the SCL and a corresponding growth of β_p. This finally causes a $dj/dV \to \infty$ at the breakover point of the characteristic. At high carrier densities of about $n \geq 10^{17}$ cm^{-3}, the same ratio υ_p/υ_n can be maintained even at low electric fields due to carrier-carrier scattering. Then the total space charge will completely vanish:

$$\rho = 0 = q\ N_D^+ + \rho_j$$

$$\rho_j = -q\ N_D^+ = j\left(\frac{\gamma_p\ \beta_p}{\upsilon_p} - \frac{\gamma_n\ \beta_n}{\upsilon_n}\right). \tag{15}$$

The breakdown of the SCL is accompanied by an abrupt decrease of voltage which now appears as an ohmic voltage drop across the neutral base. This condition is characterized by:

$$n = p + N_D^+ , \qquad\qquad n \gg N_D^+$$

and a total current density

$$j = q\ n\left(\mu_n + \mu_p\right) E$$

Abbreviating $\mu := \mu_n \approx \mu_p$ one obtains

$$j = 2\ q\ n\ \mu\ E \tag{16}$$

and a recombination current density within the n-base

$$j_r = \frac{Q}{\tau} = \frac{qnW}{\tau} \tag{17}$$

The transport factor β_p is given by

$$\beta_p = \frac{j_p\ (J_2)}{j_p\ (J_3)} = \frac{j_p\ (J_3) - j_r}{j_p\ (J_3)} = 1 - \frac{j_r}{j_p\ (J_3)}$$

$$\beta_p = 1 - \frac{q\ n\ W}{\tau\ \gamma_p\ j} \approx 1 - \frac{W}{2\ \gamma_p\ \tau\ \mu\ E} = 1 - \frac{W}{2\ \gamma_p\ W_\tau} \tag{18}$$

where $W_\tau \equiv \tau \mu E$ corresponds to the mean path of a carrier during the period of one τ. Assuming reasonable values as for instance

$$E = \frac{10V}{100 \; \mu m} = 10^3 \; \frac{V}{cm} \; , \qquad \mu = 10^2 \; \frac{cm^2}{Vs} \quad \text{and} \quad \tau = 10^{-5} \; s \; ,$$

one confirms that $W_\tau = 1$ cm $\gg$ W and β_p can easily approach 1 even at voltages of the order of 10 volts.

The condition $\rho = 0$ is stable. If for instance due to a fluctuation a space charge $\rho > 0$ occurs, then the electric field distribution is changed in a way to increase the electron density and vice versa at a fluctuation to $\rho < 0$.

With $\beta_n \to 1$, $\beta_p \to 1$, and due to a further increasing carrier concentration, one finally arrives at

$$\gamma_n + \gamma_p = 1 \; .$$

Under these conditions, the transport equation for electrons yields:

$$\frac{dn}{dx} = \frac{\gamma_n \; j}{k \; T \; \mu} - \frac{q \; n \; E}{kT} = \frac{j}{k \; T \; \mu} \left(\gamma_n - \frac{1}{2} \right) \tag{19}$$

with k being the Boltzmann constant and T the absolute temperature. Now the current density $j = 2 \; q \; n \; \mu \; E$ can only be further enhanced by intensifying the electric field.

When, at increasing j, $\gamma_n \beta_n$ is reduced to less than

$$\gamma_n \; \beta_n \; < \; \frac{\mu_n}{\mu_n + \mu_p}$$

and when additionally $\gamma_p \beta_p$ can only grow due to the build-up of a space charge region, a current limiting effect will occur. This effect might support safe turn-off of a GTO. If, due to the increasing voltage, current is commutated to the snubber circuit then GTO anode current decreases but now remaining at a nearly constant current density. With a constant gate current being applied, finally turn-off condition, eq. (7), is fullfilled and the device turns off.

To the author's knowledge, the described phenomenon is not used to optimize GTO turn-off. It can be assumed, however, that it contributes to the safe turn-off of most devices commercially available.

4. SHORTCIRCUIT WITHSTAND CAPABILITY

Around 1980, on the basis of the experimental evidence that GTO's can be safely turned-off and exhibit good static characteristics, some device designers supposed that GTO-thyristors would become the most important devices for all voltages above 800 V. Compared to transistors, they show less forward voltage drop and require no continuous base current drive during the on-state. This reasoning proved to be correct above 2000 V only while bipolar transistors were mainly applied up to blocking voltages of 1400 V. The technical arguments inducing this development are that transistors need no snubber circuitry, representing a considerable expense, and that circuit failures can be more easily controlled with transistors.

We will restrict our consideration to the behaviour at a short-circuit of the load. The worst case occurs when the device is in its conducting state. Then the source voltage is suddenly applied to the device, the rate of current rise is only limited by the small stray inductance of the leads.

Under these conditions, a GTO without external means of protection would suffer an enormous current rise, exceeding the maximum turn-off current as well as the maximum allowable surge current destroying the device long before reaching the value of the static current voltage characteristic.

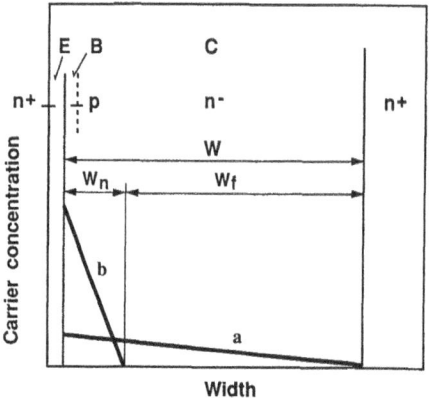

Fig. 9. Schematic carrier distribution in a bipolar transistor:
a) in its forward conducting state   b) at a load short-circuit.

A transistor under the same condition exhibits a completely different behaviour. The charge carriers at a stationary on-condition may be distributed as sketched in Fig. 9, case a. The details in the base layer are neglected. The collector current density amounts to

$$j_0 = 2\,q\,D_n\,\frac{dn}{dx} = 2\,q\,D_n\,\frac{n(0)}{W} \tag{20}$$

and the total stored charge to

$$Q = \frac{1}{2} q \; n(0) \; W \;\cdot \qquad\qquad (21)$$

Due to the sudden increase of the voltage across the device the holes are displaced towards the emitter and concentrated in the base and the neighbouring collector region (case b). As a result, two different regions of the lowly doped collector zone can now be distinguished: a neutral zone, W_n, near the base, highly swamped with excess charge, and a field zone, W_f without any holes.

At the emitter-base junction, the sudden increase of hole concentration appears as an intensified base drive, leading to a corresponding increase of electron injection. Through W_n, this current flows as a diffusion current. If in a first approximation, Q is assumed to remain constant; then n(o) rises up to

$$n(0) = \frac{2 \; Q}{q \; W_n}$$

and the current density to

$$j = 2q \; D_n \cdot \frac{2Q}{q \; W_n^2} = j_0 \left(\frac{W}{W_n}\right)^2 \;. \qquad\qquad (22)$$

Through W_f the same electron current density has to be conducted as a majority carrier current at a corresponding ohmic voltage drop.

At a field strength of about $E_s = 2 \cdot 10^4$ V cm^{-1}, electrons reach the maximum velocity of $v_s = 1 \cdot 10^7$ cm s^{-1}. With a doping concentration, N_D, of the lowly doped collector region, then a current density of

$$j = q \; N_D \; v_s$$

is reached at a voltage drop across W_f of

$$V_f = 2 \cdot 10^4 \; W_f$$

Reasonable numbers of $N_D = 10^{14}$ cm^{-3} and $W_f = 10$ μm for example yield: $j = 160$ A cm^{-2} and $V_f = 20$ V. The voltage, however, increases to a much higher value and enforces a corresponding current density. Then the electron concentration in W_f has to be raised above the doping level. Thus a negative space charge density is built-up:

$$\rho = -q \left(n - N_D\right) \;.$$

A space charge layer spreads from the n$^-$ n$^+$-junction (x=W). The maximum field strength obtained is:

$$|E_m| = \frac{|\rho|}{\varepsilon} W_f = \frac{q}{\varepsilon}\left(n - N_D\right) W_f \ .$$

$$n = \frac{j}{q \ v_s}$$

yields

$$|E_m| = \frac{q}{\varepsilon}\left[\frac{j_0}{q \ v_s}\left(\frac{W}{W_n}\right)^2 - N_D\right] W_f \qquad\qquad (23)$$

and

$$V_f = \frac{1}{2} W_f \ |E_m| \qquad\qquad\qquad\qquad (24)$$

With $j_0 = 50$ A cm, $W = 80$ μm, $(W/W_n)^2 = 10$ and $N = 1 \bullet 10^{14}$ cm^{-3}, one obtains:

$$|E_m| = 1{,}86 \bullet 10^5 \ Vcm^{-1} \ , V_f = 509 \ V$$

$$j = 10 \ j_0 = 500 \ Acm^{-2} \ , \ Q = 1{,}9 \bullet 10^{14} \ qcm^{-2}$$

$$n\left(0, \ at \ j_0\right) = 4{,}8 \bullet 10^{16} \ cm^{-3} \ ; \ n\left(0, \ at \ 10 \ j_0\right) = 1{,}5 \bullet 10^{17} \ cm^{-3}$$

Under these conditions, with the source voltage limited to 500 V or less, the current increase is limited and the device can withstand a short-circuit. It has to be switched-off, however, within some microseconds since otherwise a thermal breakdown at the enormous loss density of $(j \bullet E)_{max} \approx 2 \bullet 10^8$ W cm^{-3} will occur.

Moreover, the example shows that at slightly higher source voltages or amounts of stored charge, E_m reaches a value causing strong avalanche multiplication. The generated hole current represents an additional base drive and can cause an uncontrollable instability. Therefore, it cannot be expected that short-circuit withstand capability may be extended up to higher voltages at similar amounts of stored charge unless additional means to protect the device are provided.

5. DEVICES IN IC - TECHNOLOGY

5.1 Power MOS-FET's

Around 1973, the first confidential communication that Siliconix was developing a power device in IC-technology was spread. Compared to conventional processes, IC-technology was extremely expensive at that time and the data reported of the new device seemed to nearly satisfy thr triggering requirements of a "real" power device. One became conscious, however, of this new possibility and, in the following years, several papers on V- and D-Power MOS-FET's appeared in the literature. Today we need not introduce these devices since they have been well established on the market for more than a decade.

The main advantages of the new trend were the possibility to reduce geometrical structures by more than an order of magnitude, to take advantage of MOS-FET switches and to integrate several functions or devices on the same chip.

The MOS-FET proved to be superior to bipolar transistors due to its high frequency capability and a considerable reduction of trigger circuitry. A typical hindrance of the new technology was the action of unwanted parasitic devices. In power MOS-FET's for instance, the emitter of the bipolar transistor had to become short-circuited and this additional diode had to be taken into account. The strong dependence of the on-resistance on blocking voltage limits its use to application of at most some hundred volts.

5.2 Bipolar Transistors

Devices suited to higher voltages require bipolar injection for a low forward voltage drop. Two different types of devices were developed using IC-technology: A bipolar transistor, the SIRET (Siemens Ring Emitter Transistor), and the IGBT (Isolated Gate Bipolar Transistor). Both exhibited characteristics superior to conventional bipolar transistors, especially concerning safe operating area and maximum switching frequency. Although the SIRET showed less switching losses than conventional IGBT's [12], the IGBT finally got better acceptance due to its MOS-input and the confidence that switching losses of the new device can be further reduced.

Nevertheless, it was interesting to conceive ways how IC-technology could contribute to considerably improved transistor characteristics. In this case, the advantages stem mainly from the reduction of geometrical structures. This allows to narrowly interdigitate base and emitter electrodes so that the complete cross section of the device can be used in forward conduction. Even more important, at a width of the emitter-ring of 5 μm, one can forget about emitter crowding at its edge in the on-state. At turn-off, this emitter can be completely controlled by the base drive, even when avalanche multiplication is triggered at the base-collector junction. Therefore, the safe operating area could be extended to the entire V_{CBO} - I_c rectangle, Fig.10. Numerical device simulation has been extensively used to optimize the various characteristics of this device.

5.3 Isolated Gate Bipolar Transistors (IGBT)

As already mentioned, the IGBT seems to take over more and more of the transistor market. Concerning this device, we have to add some remarks. Again at the beginning, parasitic devices represented a serious hindrance. In this case, the npnp-thyristor caused the well known latch-up failure.

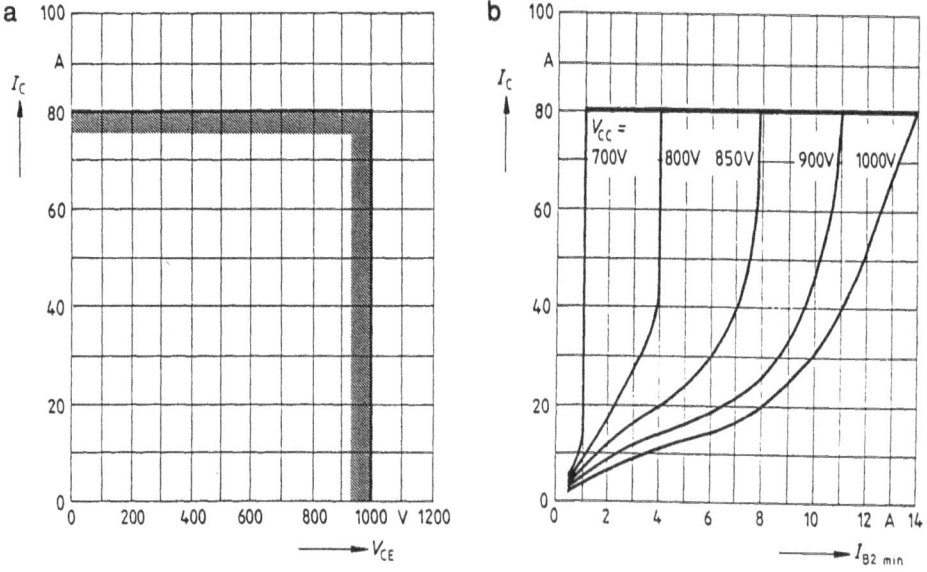

Fig. 10. Safe operating area a) turn-off capability with base drive as a parameter b) of a bipolar transistor in IC-technology.

Just like bipolar transistors, IGBT's too have to show short-circuit withstand capability. At a sudden voltage rise across the IGBT, a similar process will take place as already explained for the transistor. Now the stored electrons have to be considered. They will be accumulated towards the p-emitter. There is no external means to influence this stored charge. It cannot be removed through the MOS-switch and if at high voltages an avalanche multiplication is triggered at the p-well to n^--base junction, the device cannot be controlled any longer. As a consequence, the device should have a rather poor p-emitter or a relatively short carrier lifetime in the n-base. This explains why - for both transistors as well as IGBT's exhibiting short-circuit withstand capability - high carrier injection has to be avoided and a relatively high forward voltage drop has to be taken into account. These problems become increasingly difficult at higher blocking voltages.

Tihany has introduced a very interesting concept for IGBT's[13]. Instead of an epitaxial n-base with n-buffer layer on a p+-substrate, he used the doping of the n-doped starting material for the n-base. To assure easy handling of the wafers, a relatively thick base has to be accepted. This is combined with a very long carrier lifetime and a poor but well controlled p-emitter. Astonishingly enough, these devices need less stored charge than usual structures at the same forward voltage drop. Thus short-circuit withstand capability is correspondingly improved. Moreover, since carrier lifetime does not influence device behaviour, its temperature dependence cannot degrade the characteristics.

5.4 MOS GTO

What can IC-technology contribute to high power devices for blocking voltages above 2000 V? Several results concerning SITh's, FCTh's and MOS-GTO's habe been published but none of these devices is commercially available.

In the following, only the MOS-GTO, known best by the author, is considered, assuming that the pros and cons of the other structures are rather similar. The MOS-GTO originates from the idea that IC-technology allows to densely distribute cathode-emitter-shorts in a GTO-structure and supply all of them with MOS-switches, such that the emitter effiency can be switched from 1 to 0. If a correspondingly distributed structure for turn-on is added, this device would allow to homogeneously turn-on and -off high power via a MOS-input. Additionally, the device should withstand arbitrarily high dI/dt- and dV/dt-values. Especially attractive is that the forward bias of the emitter junction can be used to drive the negative gate current. No external source is required and, moreover, this current has to flow only a very short distance with negligible inductance. Although only a MOS-capacitance has to be charged, the device should switch safely since internal turn-off gain is much less than unity. All of these expectations could more or less be verified on small area MOS-GTO's. What, then, are the problems ?

Concerning technology, it has to be considered that a high power device requires an area of the order of 10 cm^2. At present, however, even well developed MOS-processes create defect densities of about one defect per cm^2. For the production of these devices, therefore, some means to overcome such defects have to be taken into account as for instance in form of a redundancy concept.

Of course, parasitic devices are present in a MOS-GTO too. Fig. 11 shows the schemes of a p-channel and a n-channel structure. Both MOS-switches represent bipolar structures without a base electrode. High injection of the thyristor emitter, E, will supply, however, a sufficient base drive, switch-on the parasitic bipolar transistor and therewith control the emitter efficiency. An isolating layer could completely suppress the degradation of the forward characteristic but requires an additional process.

The main advantage of the structure lies in its ease to turn-off. The lateral current path and the MOS-channel may be considered as an ohmic resistor. And although it is low enough to conduct current densities up to the order of 10^3 A cm^{-2}, only due to the forward biased emitter junction, there still exists a critical current density which cannot be turned-off. Localized maxima of current density may occur, caused by turning-off the GTO-structure or by geometrical inhomogenieties, for instance at the edges of the emitter region. While in a conventional GTO the negative gate current is always focused on the last on-area and gate-to anode-current ratio is increased, in a MOS-GTO the turn-off resistance stays constant and if the critical current density is exceeded the device will be destroyed. Avoiding spots of high current density, therefore, is a major requirement for high power MOS-GTO's.

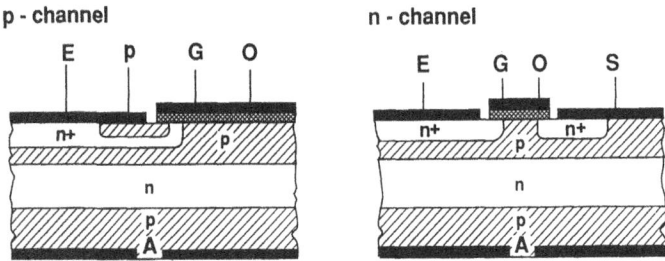

Fig. 11. Structures of p- and n-channel MOS-switches as used in a high power MOS-GTO.

Several ideas have been proposed to overcome these problems, and this author strongly believes that MOS-GTO's or corresponding high power devices in IC-technology will be produced in the future. At present, however, the required effort seems rather high compared to its main advantage of reducing turn-on and turn-off circuitry.

6. OUTLOOK

6.1 Further Improvements

How will progress of high power devices further proceed ? It is obvious that for high voltage devices bipolar injection is required and, in the conducting state, carrier concentration should be rather high. A high amount of stored charge, however, causes problems either during turn-off or in the case of a short-circuit. In the ideal case, anode current density would not rise at any point in the device during turn-off but only decrease. Therefore, one may conclude that a densely interdigitated gate-emitter structure is necessary. Sharply increasing voltages across a device with high amounts of stored charge will lead to enormous power losses. Thus carrier concentration should be reduced before the voltage rises. This may be facilitated when both cathode-emitter and anode-emitter can be controlled by external means. Experimental investigations confirm that turn-off losses can be reduced by more than an order of magnitude with double emitter control compared to single emitter control[14]. Asymmetries of the structure or carrier characteristics require, of course, to adjust the control of the two emitters to each other such that, in a first phase, the device becomes depleted before the final sharp voltage rise and current interruption. When stray inductances of the main circuit cannot be sufficiently reduced then some means have to be offered to take over the inductive energy. Even a futuristic high power device will not exhibit a dramatically increased heat capacitance.

Progress of semiconductor technology and the possibility to numerically simulate the physical phenomena inside devices have allowed to further approach the "ideal switch". The confusing manifold of device ideas around 1980 has settled. At present, a quite clear picture holds about applications and suited devices, but what are the exciting challenges of the future ? On a short term scale, the driving forces stay the same and development will continue along the sketched trend. An important point is understanding and control of plasma dynamics in high voltage devices. Technological aspects will determine how to control the emitters since this is of minor importance for their functioning but relevant for the production costs.

6.2 New Possibilities

Excitement and confusion have moved to another area which will gain importance on a long term scale. High temperature superconductivity may invade power electronics, either requiring devices optimized to much lower temperature or allowing completely new switches based on phenomena related to superconductivity. First ideas and estimates have been published[15]. As a switch, however, simply using the phase transition itself has been proposed. At a first glance, it seems difficult to trigger phase transitions with an accuracy of 1 ms in both directions or to obtain switching frequencies of more than 1 kHz but this is only a weak argument in a completely new field.

Some applications require power devices for high temperature operation. Therefore an investigation of SiC devices seems of considerable interest. At the same time, a strong effort takes place in vacuum microelectronics. New microstructure- or nanostructure-technologies offer possibilities to produce field emitting cathodes. Vacuum devices are suited to operate at high temperature and would exhibit an ultimate electron "mobility". A comparison of these completely different possibilities could be a quite interesting task.

We cannot foresee the development of the next twenty years but electrical energy and power electronics will stay an important aspect of human culture. Development of power devices is expected to offer as exciting challenges as during the last twenty years.

REFERENCES

1   M. Hill, P. van Iseghem, R. Sittig, G. Popp, "Localized homogeneity problems in silicon for high-power thyristors", *Inst. Phys. Conf. Ser.* no. 23 (1975), pp. 522-530.

2   I. Burgess, C. Neugebauer, G. Flanagan, "The direct bonding of metals to ceramics by the gas-metal eutectic method", *J. Electrochem. Soc.* Vol. 122 (1975), pp. 688-690.

3   H. Schwarzbauer, R. Kuhnert, "Novel large area joining technique for improved power device performance", *Conf. Rec. 1989 IEEE-IAS Annual Meeting*, pp. 1348-1350.

4   Stoisiek, H. Schwarzbauer, W. Kiffe, D. Theis, "2000 A/1 mΩ power MOSFET's in wafer repair technique", *IEEE Trans. Electron Devices*, ED 37 (1990), pp. 1397-1401.

5   W. Wirth, "High speed, snubberless operation of GTO's using a new gate drive technique", *IEEE Trans. Industry Applications*, vol. IAS-24 (1988), no.1, pp. 127-131.

6   C. Johnson, P. Palmer, "Improving GTO thyristor reliability by use of linear MOSFET amplifiers and controlled avalanching in the gate drive circuits", *Conf. Proc. of the 4th Int. Conf. on Power Electronics and Variable Speed Drives*, IEEE Conf. Publ. No. 324, London, (1990), pp. 417-423.

7   M. Kurata, M. Azuma, H. Ohashi, K. Takigami, A. Nakagawa, K. Kishi, "Gate turn-off thyristors", in "*Semiconductor Devices for Power Conditioning*", ed. R. Sittig, P. Roggwiller, Plenum Press, (1982), pp. 91-119.

8   T. Yatsuo, S. Kimura, Y. Satou, "Design considerations for large-current GTO's", *IEEE Trans. Electron Devices*, ED 36 (1989), pp. 1196-1202.

9   M. Azuma, K. Takigami, "Anode current limiting effect of high power GTO's", *IEEE Electron Devices Letters*, EDL-1 (1980), pp. 203-206.

10   A. Nakagawa, "Numerical analysis on abnormal thyristor forward
      voltage increase due to heavy doping in gated p-base layer", *Solid
      State Electronics*, 24 (1981), pp. 455-459.

11   Y. Gerstenmaier, F. Pfirsch, "Abnormal thyristor forward be-
      haviour", *1990 IEEE-IAS Annual Meeting*, pp. 1575- 1580.

12   L. Lorenz, "Selection criteria for power semiconductors for
      motor drives", *Proc. of 1990 ISPSD*, Tokyo, pp. 263-269.

13   J. Tihanyi, "MOS-Leistungsschalter", in ETG-Fachbericht
      *"Abschaltbare Elemente der Leistungselektronik und ihre
      Anwendungen"*, Bad Nauheim (1988), pp. 71-78.

14   T. Ogura, A. Nakagawa, K. Takigami, M. Atsuta, Y. Kamei, "6000 V
      double gate GTO's", *Proc. of the Conf. on Solid State Devices and
      Materials*, Tokyo, (1988), pp. 37-40.

15   C. Nowlin, "Power Electronics" in *"Energy Applications of High-
      Temperature Superconductivity"* ed. S. Dale, S.M. Wolf, T.R.
      Schneider, vol. 2 (1990), chapter 4, pp. 1-34, prepared for US
      Dept. of Energy.

DISCUSSION

H. Baltes (ETH, Zürich)
Can you comment on the IGBT diode ?

R. Sittig
In a bipolar transistor, we can remove the carriers by a negative base
drive which is impossible for the IGBT. Since they are stored on the anode side of the device, it is very important for the IGBT to have either a
poor anode emitter or a short carrier lifetime. In this context, the device concept proposed by Tihanyi is very interesting, involving an IGBT
with a thick n⁻-base-layer and a carefully controlled anode emitter-ef-
ficiency. The astonishing thing is that, for the same forward voltage
drop, less carriers are needed for a base width of about 200 μm, than for
a normal width of 30 to 80 μm. There is far too much stored charge in
the conventional IGBT and its short circuit capability can be improved if
one tries to minimize the carrier density during forward conduction.

D. Silber (TU, Bremen)

I have a question about the possibility to protect a device under short circuit while turning it on. We had many discussions about the comparison of the GTO and the transistor in this respect. The point is not only what happens with the transistor under short circuit condition but how can you turn it off before something happens.

R. Sittig

The question is, can you turn the transistor off before it is destroyed during a short circuit. As I have explained, you get some current limitation and if you avoid strong avalanching and you detect the short circuit within the first microseconds, you can turn it off. If you do not turn it off within about 6 microseconds, thermal runaway occurs, the losses being very high in this case. We found that a maximum voltage of about 500-600 V can safely be turned off. Transistors exhibiting higher blocking voltages will not help since the current is there and will cause some dynamic avalanching. Of course, if the stored charge in the conducting state is reduced, you can improve the situation a little bit. In contrast, in the GTO much more carriers are stored and there is no chance to turn it off because the current densities are very high. If we had an emitter efficiency which drops very sharply at a certain injection level, then this could help.

TOOL INTEGRATION FOR POWER DEVICE MODELING INCLUDING

3D ASPECTS

Robert W. Dutton and James D. Plummer

Stanford University,
Stanford, CA, USA

Dedicated to Christiaan Abbas, July 22, 1956 - January 17, 1992

ABSTRACT

The past decade has witnessed an all explosive growth in Integrated Circuits (IC) capabilities, both for high density information processing applications and industrial purposes including sensors, actuators and power devices. The growing possibilities to create "smart" devices of unique capabilities for sensing and controlling diverse electro-mechanical (and optical) systems offer almost unlimited potential. Yet to realize these potentials, heterogeneous constraints must be quantified and used effectively in design. This talk will cover the specific issues of process, device and circuit modeling and the integration issues necessary to enable full exploitation of Technology Computer Aided Design (TCAD), tools for Power Device Modeling. The Process and Device Modeling discussions build on experience with the 2D tools SUPREM and PISCES developed at Stanford and quasi-3D applications for power devices. Both simulation and experimental results are used to illustrate key points. The growing importance of combined integrated circuit and power device applications are illustrated using several approaches to mixed mode simulation. Here the examples are broadened to include both university-based tools such as MEDUSA and SPICE/PISCES as well as commercial circuit/system level simulators. The final aspect of this contribution centers on a framework for tool integration across these various levels of design. Recent progress in establishing a TCAD framework under the

Power Semiconductor Devices and Circuits, Edited by A.A. Jaecklin
Plenum Press, New York, 1992

auspices of the International CAD Framework Initiative (CFI) is reviewed. The paper concludes with an evaluation of progress to date in integrated modeling of power devices and a look into the future of requirements such as full 3D modeling.

1. INTRODUCTION

The evolution of silicon Integrated Circuits (IC) technology over the past two decades has been truly remarkable. In the late sixties and early seventies, the bipolar IC dominated both digital and analog domains. Throughout the eighties, the Metal Oxide Semiconductor (MOS) technology seemed destined to dominate in virtually all applications except high speed analog. As we enter the nineties, a variety of bipolar technologies have re-emerged, both in high density applications such as Bipolar Complementary MOS (BiCMOS) and new high speed devices such as Heterojunction Bipolar Transistor (HBT) structures. Over this same time period, the landscape of technologies for high voltage and power IC applications has changed as well. Pure bipolar thyristor and Gate Turn-Off (GTO) discrete devices gave way to integrated digital/analog structures. In parallel with the emergence of high density MOS came Double Diffused MOS (DMOS), V-Groove MOS (VMOS) and other novel power devices and IC structures. Finally, the further development of merged structures has led to Insulated Gate Bipolar Transistors (IGBT) and other MOS gate-controlled bipolar structures. Clearly, the power device field has kept pace with and benefitted directly from the evolution of the IC industry as a whole.

From a circuits point of view, the power IC industry has paralleled the evolution of the analog IC industry. Starting with small-scale IC's, the level of chip complexity for power circuits tracked the development and complexity of Medium Scale Integration (MSI) bipolar IC's up to about the late seventies. The more recent developments in MOS-based power devices has opened an era of more complex digital circuitry. Hence, through the eighties we have seen a dramatic growth in functional integration involving advanced control logic as well as new gate-controlled power devices.

The above two themes of technology and circuit evolution for power devices provide an important background for this contribution. Namely, there is a long historical connection of power IC's with the bipolar world yet a steady shift towards and influence by MOS technology and circuit concepts as well. In looking at the needs for Technology CAD (TCAD) tools to address the diverse needs involved in modeling power devices, it is essential that we look at the technology and modeling requirements that come from each domain. In addition there are broader systems issues to be considered, for example thermal effects and the influence of packaging.

This paper is organized as follows. First, the early efforts in 1D TCAD are put in the context of DMOS technology development that occurred at Stanford in the seventies. Then recent trends in 2D process and device modeling and other power device structures are discussed, including selected work in quasi 3D simulations. Next the importance of mixed-mode (circuit/device) simulation is discussed along with examples from several groups working in this area. The concepts of TCAD framework are then presented; the growing importance of heterogeneous tool integration is emphasized. Finally, recent trends in 3D TCAD discussed.

2. EVOLUTION OF 1D PROCESS MODELING

Despite the rapid evolution of 2D and even 3D modeling tools for complex device structures, the 1D process simulator SUPREM 3 continues to be used broadly in industry. The reason for this persistent use, even after development of robust 2D tools such as SUPREM 4, goes to the base of problems involved in modeling power devices. Namely, the complexity of modeling bipolar device structures in terms of double-diffused profiles has continued to be a major research challenge, even after more than three decades of investigation. Moreover, the calibration of impurity diffusion effects, especially in the high concentration regime, is complex and more data is needed. Fortunately, a majority of the controlling device physics and critical parameters for power devices occur in the bulk regions as opposed to at the surfaces. Nonetheless, it is instructive to revisit our earliest efforts with DMOS process modeling to understand the underlying problems.

Shortly after the introduction of the DMOS structure[1], the Stanford group was actively developing DMOS high voltage multiplexing switches for ultrasonic imaging applications[2]. As part of that development effort we considered the issues of threshold-voltage control in double diffused MOS structures[3]. Fig. 1 shows the results of theoretical and experimental variations in threshold voltage as a function of the boron predeposition parameter $\sqrt{Dt}$. The models exploited the typical analytical forms used prior to the development of SUPREM. In fact, it was out of this work that it became clear that more accurate process models were needed for devices such as DMOS. The first efforts in 1D process modeling (SUPREM) were discussed in the context of understanding the parameter variations of double diffused devices[4]. Despite encouraging progress in modeling the redistribution effects in double-diffused profiles, the characterization technology still lagged the theory. In an effort to quantify the device effects, the use of the Spreading Resistance Probe (SRP) was further developed and exploited for those early DMOS structures[5]. Fig. 2 shows typical results obtained using the SRP for DMOS de-

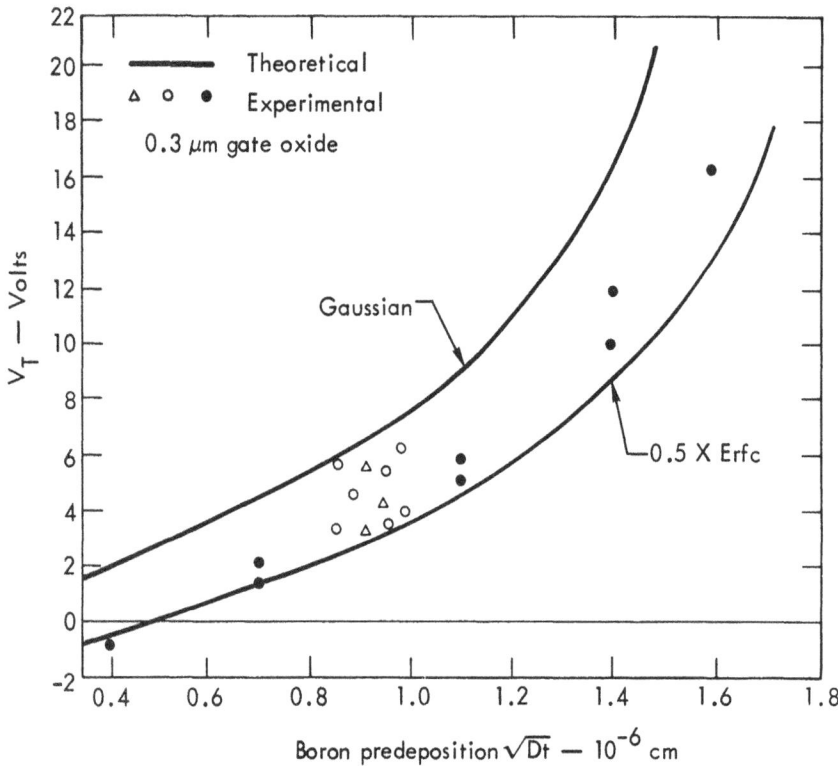

Fig. 1. Comparison of experimental and theoretical values of threshold voltage for DMOS devices with varying boron thermal cycle[3].

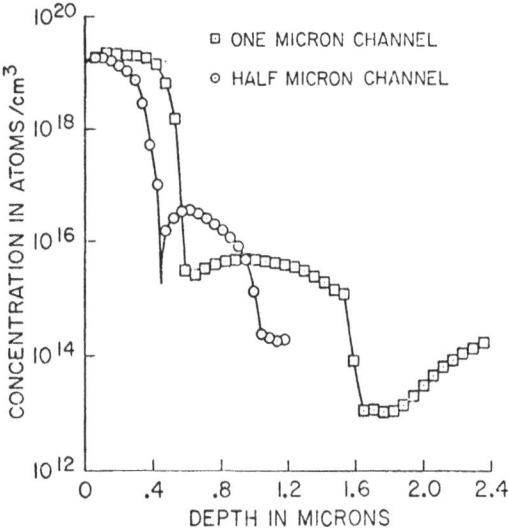

Fig. 2. Measured spreading resistance profiles for channel lengths of 0.5 μm and 1.0 μm[5].

vices with channel length of 1 µm and 0.5 µm. Although many factors, such as probe calibration and surface preparation, can affect results significantly, it was based on the combined results of 1D SUPREM simulations and extensive use of SRP diffusion parameter extraction that subsequent versions of SUPREM evolved and demonstrated utility in the industry. However, as stated in the introduction to this section, the many difficulties of double-diffused bipolar and DMOS structures continue to complicate the development of truly robust and accurate 2D process models. One of the key challenges in the modeling of high concentration profiles is to understand the the role of implant damage on both diffusion and impurity activation. Currently this is one of the most active areas of effort in process modeling and a physically based model is still under development[6].

Having taken a brief journey back to the earliest 1D process modeling efforts with SUPREM and identified an ongoing problem of process characterization and the inherent difficulties of modeling double-diffused impurity profiles, now let us turn to the more recent efforts in 2D modeling. Here we find many significant successes, despite the profile limitations cited above.

3. RECENT TRENDS IN 2D PROCESS AND DEVICE MODELING

It was out of the same DMOS research efforts described above that both SEDAN and PISCES (1D and 2D device analysis) tools evolved. SEDAN was used in a quasi-2D mode to evaluate the DMOS threshold behavior[7]. Based on the obvious 2D effects observed in the DMOS, we in turn developed the PISCES program[8]. Using PISCES, we have subsequently evaluated a variety of power devices including: Field Assisted Turn-Off thyristor (FATO)[9], Conductivity Modulated DMOS (CMDMOS)[10] and IGBT[11]. In this section, we will first demonstrate the advances in 2D power device modeling by way of example. Next we discuss several key device physics effects that are necessary to achieve accurate modeling. Finally we return to the crucial role played by 2D process modeling - especially for new nonplanar devices - which exploits trenches.

Conductivity modulation is one key parameter in the operation of virtually all power devices - it controls the ON-resistance of the structure. As stated in the introduction, the evolution of MOS technology has strongly influenced the development of new power devices. Fig. 3 shows the cross-section of a conductivity modulated DMOS structure that uses trench technology and a novel injector structure to enhance the modulation effects. As can be seen, without the P+ diffusion in the middle of the structure, the device is simply a vertical power MOSFET structure.

The P+ region, shown as buried below the surface, is the source of the minority carriers that are used to modulate the conductivity of the lightly doped epitaxial drift region. In general, this P+ injector can be placed at any depth below the surface or even at the surface. By having the P+ in-

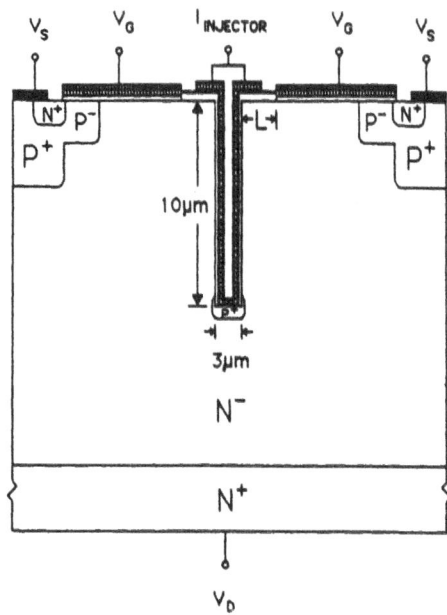

Fig. 3. Basic structure of the CMDMOS device[10].

jector below the surface, it was anticipated that the minority carriers would originate from it and be distributed over a greater (or deeper) portion of the drift region[10]. During ON-state operation, a small hole current is introduced between the P+ and the source in the CMDMOS device. A controlled number of minority carriers are injected into the path of the DMOS electron current to modulate the conductivity of the epitaxial drift region. This causes the various components of the ON-resistance to be reduced significantly. Fig. 4 shows the effects of the current injected at the P+ diffusion on the hole concentrations in the CMDMOS. If a small amount of minority carrier current, $1.0 \cdot 10^{-7}$ A/μm of hole current is injected from the P+ diffusion into the drift region of a CMDMOS device, then, according to Liu's analysis[10], the minority carriers will set up a conductivity modulated area in the drift region. This is indeed confirmed by the PISCES simulation for a CMDMOS device operating under a drain bias of 0.5 Volt and a gate drive of 2.0 Volts, which is illustrated in Fig. 4, showing the simulated hole concentration contours of the device. From the figure it is seen that, contrary to assumptions made in first-order analyses, the holes are not distributed throughout the entire drift region. Instead, the holes have a limited distribution due to the drain bias on the substrate which tends to repel holes toward the upper portion of the drift region. This upper portion is considered to be the "modulated" region since the hole concentration in this region is high enough to significantly change the majority carrier concentration as

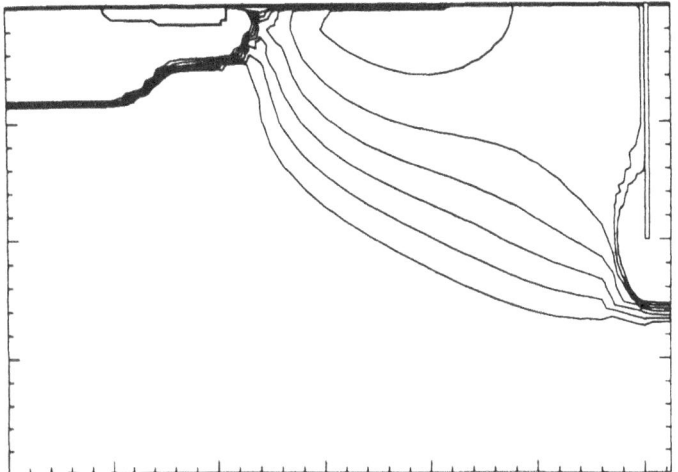

Fig. 4. Hole concentration contours from PISCES simulation for CMDMOS device with buried injector[10].

dictated by charge neutrality requirements. In addition to these static plots of carrier distributions, PISCES has been used to characterize switching events[10]. Owing to the favorable geometric effects of the CMDMOS, minimum switching times on the order of 10 ns are possible.

Another structure that uses the benefits of a trench to achieve improved turn-off characteristics is the FATO[9]. The gated region in the trench serves as a shorting path for the current flow as shown in Fig. 5b. As with the analysis of the CMDMOS, PISCES was used extensively to characterize the ON- and OFF-state charge distributions as well as the transient effects. Fig. 5 shows that the trench structure can be thought of as inducing a second p-type collector for the thyristor's PNP transistor. Current flows through this collector and does not cause the device to latch. This current is analogous to the OFF-current in a GTO device. The area of this induced "turn-off" collector, relative to the area of the primary PNP, will determine its conduction capability and hence the maximum current that can be switched. As might be expected for this MOS gating device, the role of the surface mobility along the trench wall is crucial.

The surface mobility on the sidewalls of dry etched trenches were compared for two processes to that obtained on the wafer surface[9]. It has been determined that electron mobility is substantially reduced whereas mobility for holes is relatively unchanged. Two etching process variations were investigated in this study. Process A used SF_6 and C_2ClF_5 gases in a symmetrical, equal-electrode-area Radio Frequency (RF) reactor with the sample at ground. Anisotropy in etching was induced by polymer deposition on the sidewalls. Process B used a magnetic field enhanced plasma to increase the path length of electrons and hence the

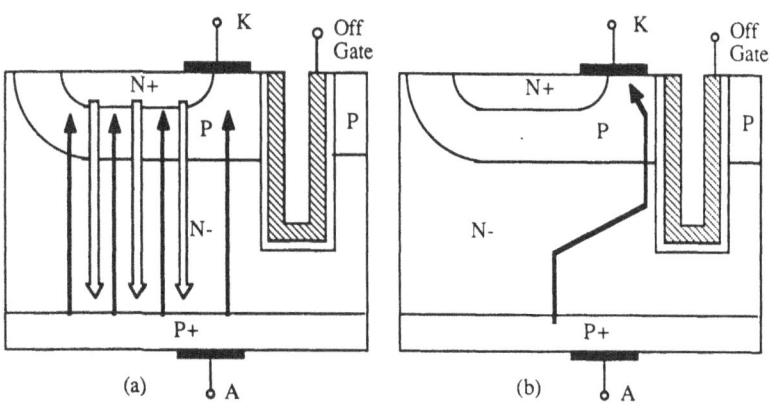

Fig. 5. Current flow paths in the FATO thyristor (a) during the on-state and (b) during the turn-off process. Filled arrows represent hole current; open arrows represent electron current[9].

plasma density, giving a two-times higher etch rate (7000 Å/min). Anisotropy was induced by the formation of an oxide layer on the sidewall that was later removed. There were small but perceptible differences between surface mobilities resulting from the processes as illustrated in Fig. 6[9]. Neither of the etch processes investigated produced n-channel devices with surface quality mobilities. Post-etch treatments such as oxidation and stripping had little effect. It is believed that atomic scale surface roughness is the cause of this mobility reduction. Holes seem to be more immune to these surface roughness effects, thus their mobility is not significantly reduced by the dry etch process.

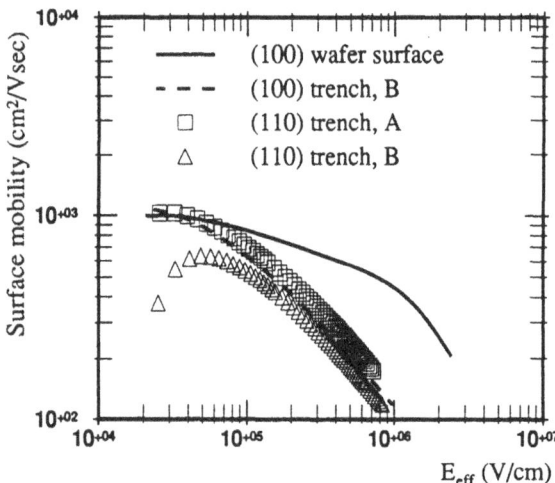

Fig. 6. Electron surface mobility vs. effective normal field for different surface orientations. All trenches underwent a sacrificial oxide post-etch treatment[9].

The above examples have illustrated several trends in the design of power ICs. The effects associated with device ON-resistance and OFF-voltage each bring with them important physical effects as well as possibilities for innovative technology solutions in terms of new fabrication techniques and geometries. Each of these points is now discussed further based on specific examples using PISCES and SUPREM 4.

Aspects of the device physics to be considered here include ON-resistance and OFF-voltage. There are a range of parameters that control ON-resistance. The carrier mobility is influenced by factors such as surface scattering[9] as well as bulk phenomena such as carrier-carrier scattering - both these effects are considered in PISCES.

The implementation of carrier-carrier scattering becomes important when both hole and electron concentrations are high which is an essential factor in lowering the ON-resistance of power devices. Under these conditions, the scattering events become much more probable. Dorkel and Leturcq[12] have proposed a semi-empirical mobility model which includes not only the effects of lattice and ionized impurity scattering but also the carrier-carrier scattering events. The equations that are implemented in PISCES are as follows[13] :

carrier-carrier scattering mobility:

$$\mu_{ccs} = \frac{2 \cdot 10^{17} \, T^{\frac{3}{2}}}{\sqrt{(pn)}} \left\{ \ln\left[1 + 8.28 \cdot 10^8 \, T^2 \, (pn)^{-\frac{1}{3}} \right] \right\}^{-1} , \tag{1}$$

lattice mobility:

$$\mu_L = \mu_{L0} \left(\frac{T}{300} \right)^{-\alpha} , \tag{2}$$

ionized impurity mobility:

$$\mu_I = \frac{AT^{\frac{3}{2}}}{N} \left[\ln\left(1 + \frac{BT^2}{N} \right) - \frac{BT^2}{N + BT^2} \right]^{-1} , \tag{3}$$

where N is the concentration of ionized impurities, and A and B are parameters which depend on the nature of the carriers.

And finally, the mixed-scattering mobility expression which combines the above terms:

$$X = \sqrt{\frac{6 \, \mu_L \left(\mu_I + \mu_{ccs} \right)}{\mu_I \mu_{ccs}}} \qquad \mu = \mu_L \left[\frac{1.025}{1 + \left(\frac{X}{1.68} \right)^{1.43}} - 0.025 \right] . \tag{4}$$

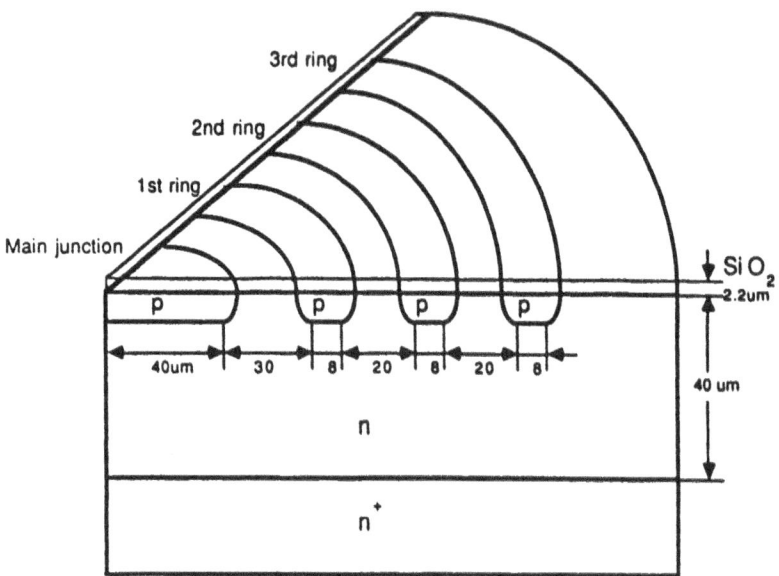

Fig. 7. Floating-field limiting structure with cylindrical symmetry[14].

All parameters used in this implementation are taken from Dorkel and Leturcq.

In addition to the physical parameters such as mobility that affect ON-resistance, geometry plays a key factor as well. For many cases, the 3D influences on device behavior can be emulated using quasi-3D analysis where cylindrical symmetry is used to replicate the behavior in the third-dimension. Such a formulation has been added to PISCES and details of the implementation are described in [14]. Fig. 7 shows a high voltage structure with three guard rings and various geometric parameters as indicated. The device is designed to break down in the punchthrough mode when the depletion region reaches the end of the lightly doped substrate. The doping concentration of the lightly doped n-epitaxial region is $1.5 \cdot 10^{14}$ cm^{-3} with the <100> surface orientation and doping concentration of the n$^+$ region on the backside substrate is $1.3 \cdot 10^{18}$ cm^{-3}.

Figure 8 shows the potential of the reverse-biased floating-field limiting rings. The breakdown voltages are also indicated. Here, the breakdown voltage is determined at the voltage where the current becomes about 100 times larger than the current at the middle voltage range. The breakdown voltage obtained using quasi-three-dimensional simulation with interface charge agrees well with the experimental data. Two-dimensional simulation predicts about a 30-percent higher breakdown voltage than the experimental data. The difference of potential between the main junction and the first ring is crucial in determining the breakdown voltage because this difference is larger than

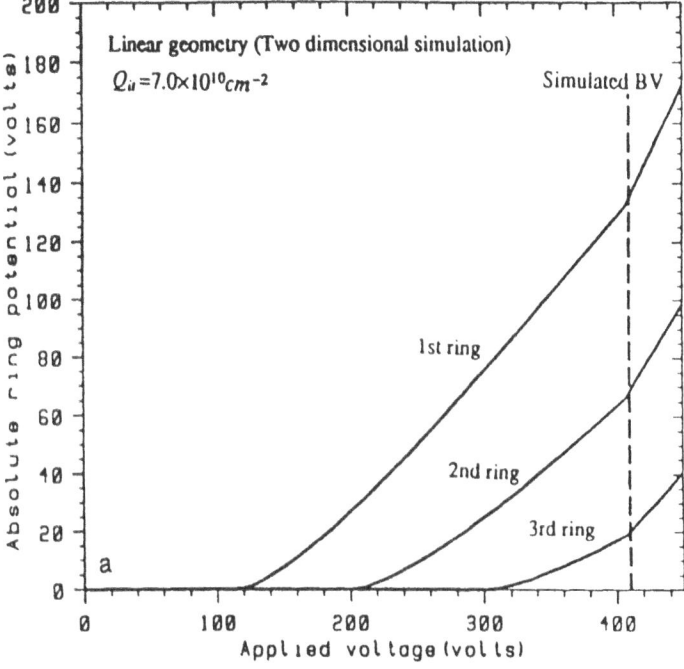

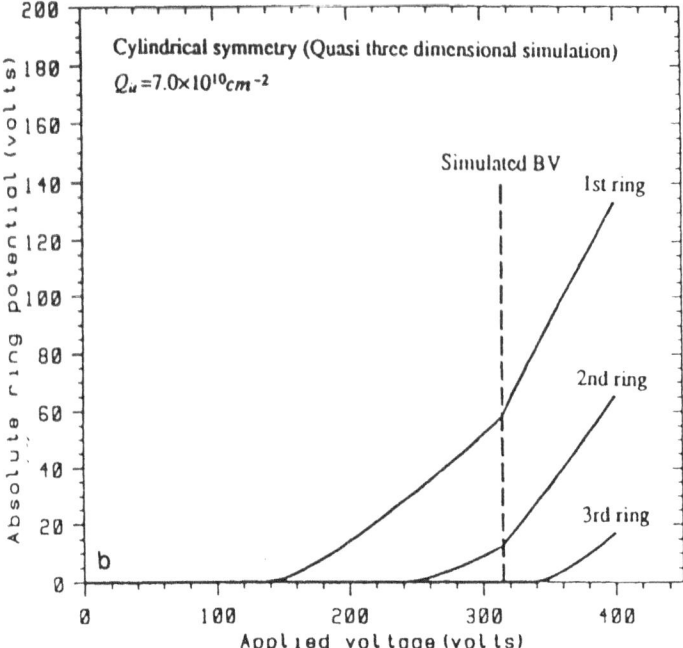

Fig. 8. Absolute ring potential versus applied voltage of the floating-field limiting ring structure with linear geometry with interface charge of $7 \cdot 10^{10}$ cm^{-2} obtained using a) two dimensional simulation,   b) Quasi-three-dimensional simulation[14].

the difference of potential between any of the adjacent rings. Other details of the differences are discussed in [14].

Impact Ionization effects are of critical concern in determining the high-voltage behavior of power devices. Although it is well-known that non-local effects are involved in carrier heating, to date there has been no completely reliable method to treat these effects - with the exception of directly solving for the carrier temperature. In the case of using the local "lucky electron" model, it is essential to base the computations on the term $J \cdot E$ where J is the local current density and E the electric field. Although our formulation of this term was correct[14], the initial implementation was flawed. More recent PISCES versions have corrected that difficulty. As will be discussed in the future trends section, it becomes increasingly important to go beyond the limitations of the drift-diffusion and local model formulations in order to correctly predict the performance of power devices.

As a final theme in evaluating progress in 2D modeling for power devices, let us consider some of the process dependences. The above examples have illustrated the innovative opportunities that come from using advanced processing to improve power device performance. In particular, trench etching has become a key part of both isolation and intrinsic device operation. In contrast to the limitations of present double-diffused and high-concentration diffusion models, advances in trench etching and nonplanar oxidation of complex geometries have generally kept pace with the development of new power device technologies. Figure 9 shows the simulation of simultaneous etching and deposition processes. A two step etching process is considered - the first being isotropic and the second anisotropic - thus creating the "champagne glass structure". In order to correctly simulate the anisotropic effects, the role of polymer deposition is critical. The SPEEDIE simulator accounts for not only such precursor effects but a variety of angle-dependent ion bombardment effects that influence the anisotropic behavior[15].

Figure 9 shows both the simulated results and experimental SEM profiles. Note that, at the lower edge of the horizontal section, a "bump" is created, owing to the buildup of materials due to adsorption/re-emission mechanisms.

Turning to the oxidation of trench and other nonplanar etched surfaces, there are a number of factors that influence the growth of layers over such nonplanar surfaces. In particular, the role of stress on the growth of oxides at both convex and concave surfaces has been the subject of intense interest and investigation[16]. Based on both experimental and theoretical efforts, a more robust numerical implementation has been developed for handling the stress dependence[17,18]. SUPREM IV is now capable of reliable and accurate simulations of complex oxidation conditions used in advanced IC technologies - suitable not only for power device structures but also advanced Dynamic Random Access Memory (DRAM) technologies as well.

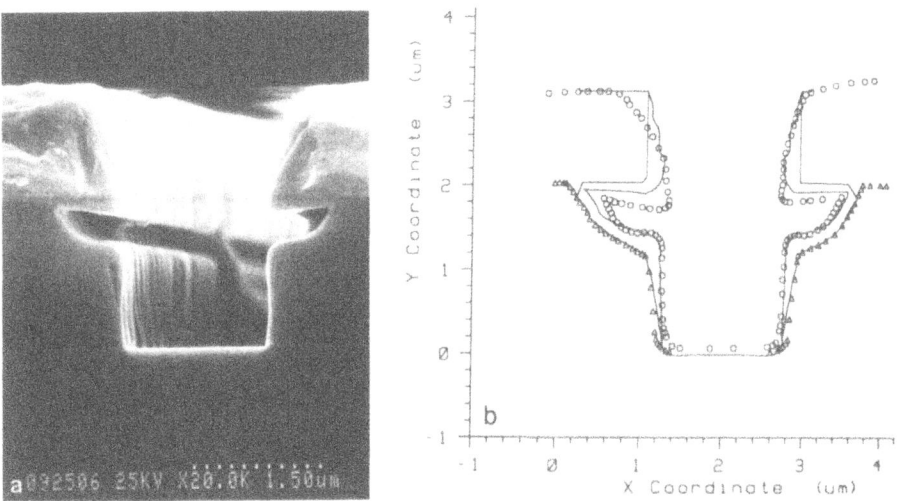

Fig. 9. Simulation and experimental results of simultaneous etching and deposition in a "Champagne Glass" test structure[15]: a) experimental results, b) SPEEDIE simulations

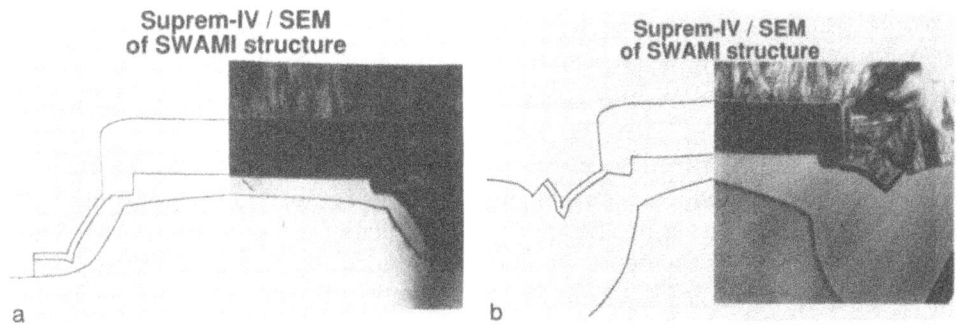

Fig. 10. Comparison of SUPREM-IV simulations to Transmission Electron Microscope (TEM) photographs of a SWAMI isolation process. The initial structure (a) and final structure after oxidation (b) are shown[19].

Based on characterization of properties on the nitride/oxide layers, the improved model coefficients now make it possible to achieve good accuracy on complex structures[19]. For example, Fig. 10 shows the results of SUPREM simulations for an advanced "LOCally Oxidized Silicon" (LOCOS)-process (i.e. the so-called "SWAMI" type). The results show excellent agreement between measurements and simulation, both qualitatively and quantitatively. In addition to these fully-recessed LOCOS structures, SUPREM-IV can handle trenched devices and poly-buffered LOCOS as well.

4. MIXED-MODE SIMULATION FOR POWER IC'S

In addition to the unique features of the device physics and process technology used for power ICs, there is a diversity of circuit constraints that apply to the design of power ICs. At the most basic level, the normal SPICE models used for low power devices quickly fail as the operating conditions go into the high-level injection regime. Moreover, the operating temperature of power devices is higher than that of the surrounding control devices and these effects must be considered. Although there have been various attempts to customize SPICE and its models to account for temperature gradients across the chip (i.e. TSPICE), it is clear that more robust techniques are needed for future power IC designs.

Mixed-mode simulation which couples circuit and device analysis is one attractive answer to the problems stated above. Early work by the Aachen group resulted in the MEDUSA mixed-mode simulator[20]. The first results demonstrated with this approach considered output power devices in an operational amplifier. These devices directly influence transient behaviour of the circuit and by using only conventional device models, the results show major qualitative and quantitative errors in the waveforms[20]. Over the past few years, other approaches and refinements to mixed-mode simulation have emerged. The University of California (UC) Berkeley group has created the CODECS simulator based on SPICE 3 for the circuit analysis portion and the matrix routines are interfaced to a custom device solver[21]. The group at University of South California (USC) has modified PISCES directly and included the circuit matrix terms as part of the device solver[22]. In addition to the device-circuit level of mixed-mode for power devices, other tools are emerging that handle multiple levels of abstraction, specifically for power systems. The SABER simulator provides capabilities to model a variety of nonlinearities besides just the IC's[23]. Although it is beyond the scope of this paper to consider the full spectrum of these mixed-mode simulators in detail, this paper will demonstrate representative examples of results for several interesting cases. The next section turns to the final issue of tool integration which encompasses not only the process and device levels of TCAD but also mixed-mode and other higher level tools that include the electronic circuit and system aspects.

5. INTEGRATING TCAD INTO EDA FRAMEWORKS

The interest and activity in development of CAD frameworks has grown steadily over the last decade. Stimulated in part by the concepts of "structured" IC design, first layout interchange formats were developed (i.e. Caltech Intermediate Form, CIF) followed by formalization of a much broader standard through the Electronic Design Interchange

Format (EDIF) Committee[24]. Over the last few years an international CAD Framework Initiative (CFI) has emerged[25] and initial efforts have focused on the needs related to Electronic Design Automation (EDA). Recently a TCAD working group was formed with the objective to leverage from the broad objectives of CFI and at the same time to develop the essential and unique framework tools needed for TCAD.

To date the interface between TCAD and EDA has traditionally been thought of in terms of lumped values such as circuit model parameters and design rules. EDA framework-based design allows tools to access data from many levels of abstraction (including layout, schematic, netlist, etc.) through well defined interfaces - giving each tool freedom to determine the necessary inputs from all aspects of the design process. Modular interfaces allow framework compliant tools to be easily swapped in and out of the framework, providing rapid transfer of enhanced modeling ability to the user. Extending EDA frameworks to include TCAD is essential for integrating TCAD tools into the design process and to support the mixed-mode simulation tools discussed in the last section.

TCAD frameworks must address both deployment (integrating existing tools) and development (creating new tools). The use of common representations address both areas by minimizing the number of translators needed. It also encourages the creation of shared libraries of functions and services (e.g. I/O routines). By using these library routines within a framework context, developers can focus on the development of the physically-based models that provide the intrinsic benefit of TCAD to EDA. This is especially important in the area of power devices where resources for unique tool development are scarce and efforts must be focused. In the following discussion we review several key aspects of recent progress in developing the TCAD framework - especially the semiconductor wafer representation (SWR) is discussed which is the structural "back-bone" of TCAD .

Semiconductor Wafer Representation[26] - Wafer state describes the structures resulting from the fabrication and simulation of integrated circuits. Examples of wafer state information include topography, dopant profiles, and current density. This wafer state is created and modified via processing steps (e.g. deposition and implantation), and analyzed by device/circuit simulation.

The following describes the representation that the CFI TCAD Framework Group is developing[26,27]. The Semiconductor Wafer Representation (SWR) is based on an object oriented approach - tools query and modify the representation using a functional interface. The representation can be partitioned into two major components - Geometry and Fields. Geometry provides structural and topological information about the wafer while Fields provide information about properties that vary within geometric regions, such as dopant concentration.

The basic unit of <u>Geometry</u> is the cell, a collection of zero-, one-, two-, and threedimensional point sets. Operations on cells include create, destroy, query, and modification. Cells can be created by combining primitive cell - unions of rectangular or triangular cell - or by combining lower-dimensional objects (for example, constructing a face from a set of pair-wise adjacent edges). Because adjacency information must be readily available in TCAD simulators, sets of nonoverlapping cells can be grouped into special sets called "cell complexes". Within complexes, most queries about cell relationships (e.g. adjacency) can be answered quickly and accurately. Queries include classification (e.g. is this point inside this cell complex) and sectioning (e.g. return a 1D slice out of a 2D cell). Modification functions include "inset" (for deposition), and "subtraction" (for etching) .

<u>Fields</u> represent mappings from a domain, usually a cell associated with the geometry, to a range. A specialization of field is "field on mesh". Numerical methods for solving equations that do not possess closed form solutions typically use meshed regions. Facilities are provided for automatically creating and refining several types of commonly used meshes (triangular elements, tensor product). Meshes can also be constructed from lower-dimensional elements in a manner similar to the creation of geometry cells. An important operation is evaluation of a particular field at a point in space (e.g. find the boron concentration at this location). Modifications of fields usually result from application-specific numerical operations, for example, solving the diffusion equations.

Inconsistency between fields and geometry is allowed to enable client applications to define their own set of consistency functions. For example, a string-based etching simulator can modify Geometry but has little knowledge of which algorithm should be used to force the existing "fields on mesh" to follow the new boundary. A subsequent diffusion client, however, could make the old mesh consistent with the new boundary in a way that minimizes discretization error for the diffusion equation.

The following program code shows how a tool simulating the photolithography process would manipulate the wafer representation through the functional interface. Fig. 11 shows snapshots of the wafer between function calls.

```
/* retrieve the existing wafer (Fig. 11a). */
    cell_complex1 = getCellComplex(wafer);
/* deposit_layer uses the geometry component to create a cell based
on the photoresist deposition process. A reference to the new cell is
returned (Fig. 11b). */
    dcell1 = deposit_layer(cell\_complex1);
/* incorporate the cell into the existing wafer (Fig. 11c). */
    wafer.insetCell(dcell1);
/* set the material of the cell to photoresist. */
    dcell1.setMaterial(Resist);
```

The tool developer needs only to provide code for *"deposit_layer"*, since the additional functions necessary to access the wafer boundary and create the new geometry are provided through the programming interface "get" and "insert" methods. Specialized functions such as *"insetCell"* are extremely complex and should be written by computational geometry experts.

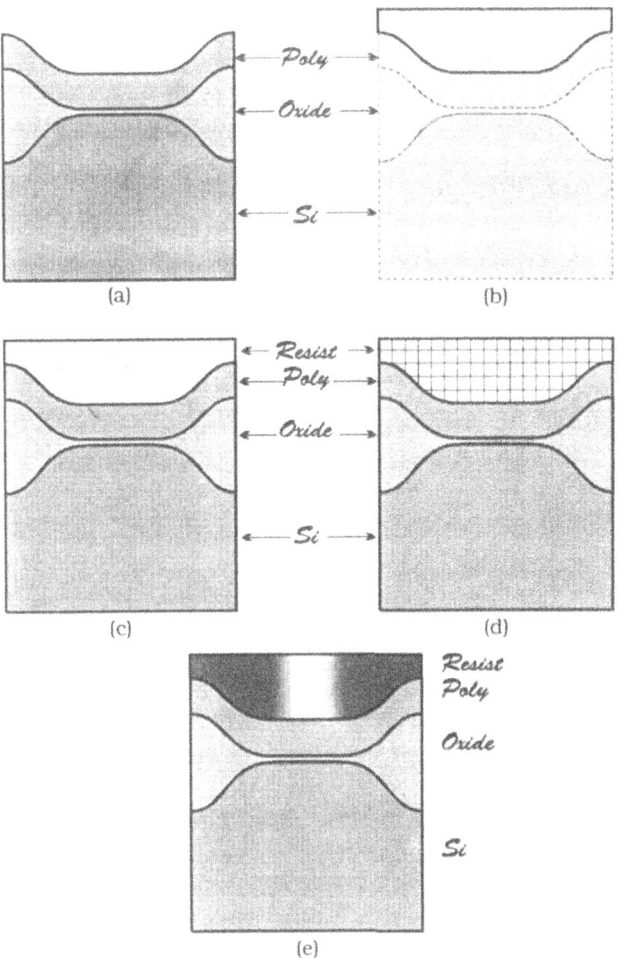

Fig. 11. "Snapshots" during SWR example execution: Before (a) and after (b) calling *"deposit_layer"*, after calling *"setMaterial"* (c), after calling *"createMeshedField"* (d), and after calling *"perform_lithography"*, showing the field values for illumination per square micron (e).

The next code sequence continues processing by having another tool perform photolithography on the deposited photoresist.

> /* createMeshedField creates a new "field on mesh" associated with a particular cell. The cell is automatically meshed if no tensor mesh is present (Fig. 11d). "Hints" describes a data object containing information for the meshing algorithm, such as desired mesh density.
> A reference to the field is returned. */
> exposure = dcell.createMeshedField(TENSOR, hints);
> /* the photolithography client, for example SAMPLE, takes information provided by the cell complex and fields, determines exposure intensity as a function of position, and puts values in the field (Fig. 11e) */
> perform_lithography(exposure, wafer);

Again, the tool developer is only required to provide code for the "perform_lithography" routine. Mesh generation and management of field information are provided through the programming interface. "CreateMeshedField" replaces a long sequence of steps that developers traditionally used to update wafer state. One of the more important aspects of the "createMeshedField" routine is the ability to automatically mesh a cell, an operation that requires knowledge of both computational geometry and numerical analysis. Meshers are difficult to write and should be provided to the typical developer.

As can be seen, the SWR interface makes the manipulation of wafer state transparent to the tool developer. In the past, each model developer typically had his own implementation of the needed operations, instead of reusing modular pieces of code. Because many programs can use the same implementation of the SWR, its providers can concentrate on improved programs for these core services, in turn freeing the users to concentrate on the physics being modeled.

The above discussion of the SWR and the example given in Fig. 11 illustrate some of the underlying issues of the TCAD framework. At the higher levels, there are other system integration challenges, for example control of tool execution and the substantial data base problems that come with applications such as statistical design. Moreover, as suggested by the previous discussion of mixed-mode simulation, there are a growing set of heterogeneous tools with interfaces that will require new methods and paradigms for design using such tools. In the domains of power systems, the range of devices to be controlled is indeed diverse and the complex interactions of not only electrical and thermal effects but also mechanical and even noise sources pose unique challenges for system modeling and tool integration. In the final section of this paper we turn to a more limited but important set of future challenges, consideration of 3D modeling and electron temperature effects.

6. FUTURE TRENDS

In the discussion of current trends in 2D TCAD modeling, we briefly explored two points that are now discussed further - 3D analysis and non thermal equilibrium carrier transport effects. Here we broaden the discussion of these topics in light of recent trends both in the emerging TCAD Framework and High Performance Computing (HPC).

In the area of device analysis, there has been steady progress in development of new 3D tools. However, two limitations have impeded the broad engineering application of such 3D tools - specifically, computation time and user interface related issues. Limitations of computer resources, both in terms of CPU requirements and memory bandwidth, is one major problem. Even with increased CPU performance into the tens-of-MFlop range, the memory size and speed limitations of most Electronic Work Station (EWS) systems support only modest scale 3D computations. Over the past several years the Stanford group has actively pursued 3D device analysis, using a Multiple-Instruction Multiple-Data (MIMD) model for computation-sometimes called "coarse-grain" parallelism. Using a third-generation Intel iPSC/860 we have now demonstrated very impressive results using a 32 processor configuration[28].

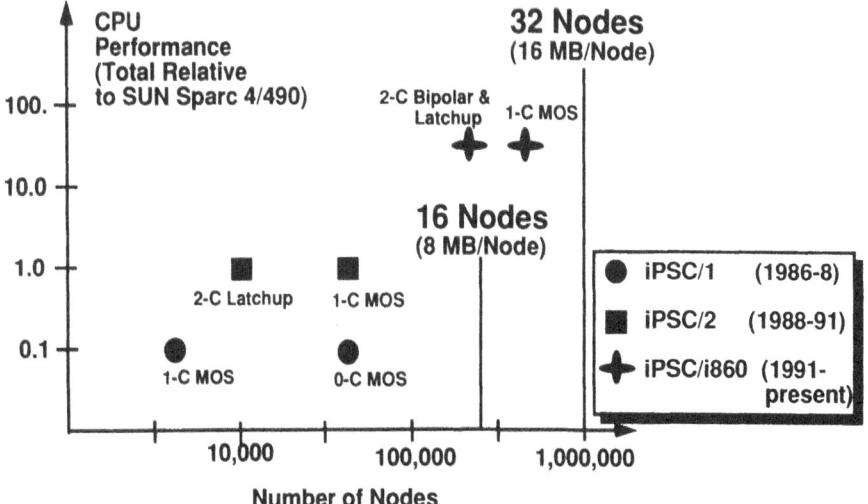

Fig. 12. Log CPU Speedup (units relative to SUN 4/490) versus number of nodes for large 3D problems on three generations of Intel iPSC computers.

Figure 12 shows a plot of Central Processing Unit (CPU) performance compared to a single processor SUN 4/490 versus number of analysis nodes for a variety of 3D MOS, bipolar, and CMOS latchup examples. Also shown on the plot are the memory limitations for the three generations of systems. Because of the substantial improvements in speed of the iPSC/860 system, users have been more willing to input larger problems and the computation time is still reasonable. For example, a 50,000 bipolar analysis takes approximately 2 minutes per bias point[29]. This performance is orders of magnitude improved over earlier generations of hypercubes and is comparable to state-of-the-art super computers. We are now in the process of formulating and testing even larger problems using a prototype Intel system with more than 500 computational nodes. Here, we hope to surpass the performance of present commercially available systems.

This brings us to the second limitation in 3D analysis, the inherent difficulties in formulating the problems and interpreting the results. In contrast to the 2D device technologies cross-sections used extensively as examples in the previous sections, the 3D domain requires a new look at the required data from an IC mask layout perspective. Our recent experience, using the Berkeley SIMPL-IPX[30] tool as a key part of the user interface has been productive and will be discussed. By means of supplemental user panels, interfaces that support not only simplified definition of geometry but also virtual measurement tools for parameter extraction have been demonstrated. Fig. 13 shows a sample of how "virtual" measurement panels are used to both control the simulator execution and also display the results for parametric extraction to support SPICE models. The generalized bipolar structure shown in Fig. 13b is biased according to an interactive electrode selection. An HP 4145 "virtual measurement" panel is used to apply bias conditions, exactly as they will occur using the real equipment (see Fig. 13a). For the bipolar transistor shown, a macro-measurement routine can be used to create a procedure to extract early voltage as shown in Fig. 13c. These "macro routines" both drive PISCES and can also be used to down-load into the actual measurement equipment. The resulting capabilities allow the user to go smoothly and efficiently from layout-based device specification to parameter extraction that supports circuit modeling. Such an automated approach is of growing importance to assist non-experts in the use of TCAD.

The final topic to be discussed in this section is thermal effects in power devices. As stated in the consideration of impact ionization, nonthermal equilibrium effects are now essential in not only power devices but short channel MOS devices for digital applications too. There continues to be a vigorous discussion how to correctly formulate the problem not only for the carrier temperature but also for the lattice and ambient conditions[31]. In the context of the emerging TCAD framework, it is essential to support this type of formulation since it clearly seeks to merge the disparate areas of electrons, phonons and the systems as a whole. In this section of the paper, I will show some recent efforts in improved carrier temperature models.

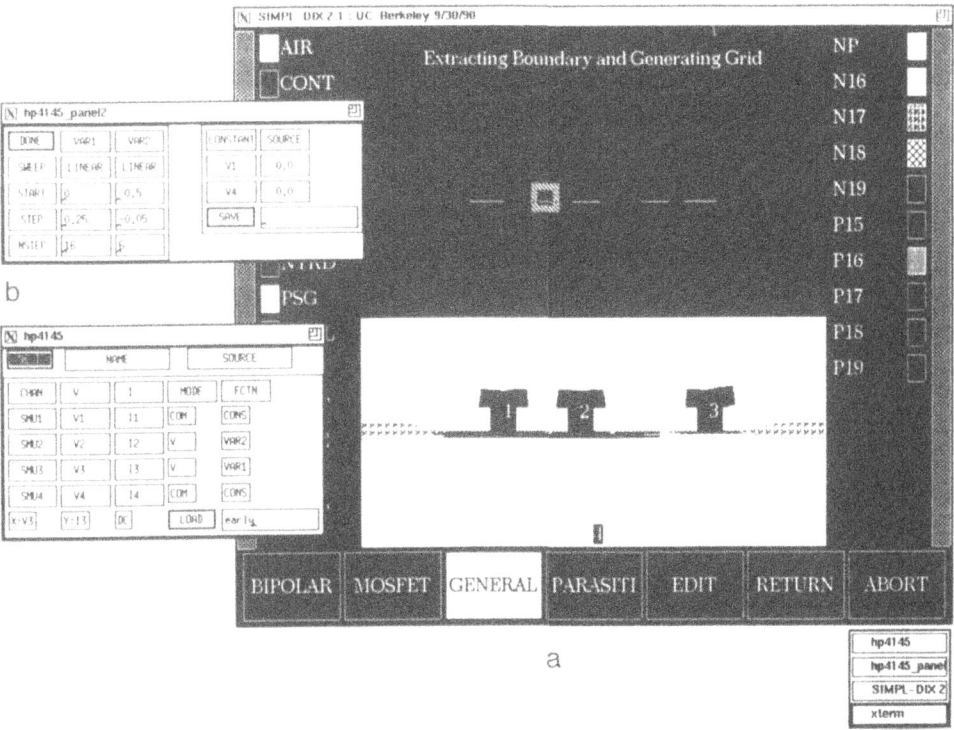

a

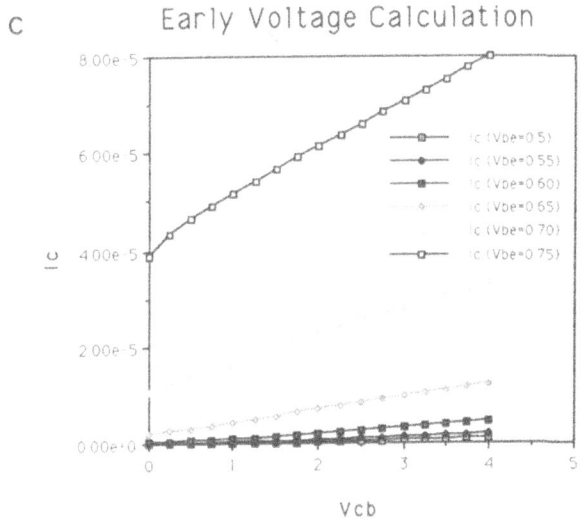

c

Early Voltage Calculation

Steps:

- Use Simpl-dix to construct device profile

- Use "GENERAL" option for Pisces to generate grid

- Use HP4145 front end and load program early_voltage

- Transfer IV.data file to MacIntosh

- Manipulate IV.data until in proper format for graphics package

- Create graph and import into slide program

Fig. 13. Layout and virtual instrument-driven interfaces for TCAD: (a) HP4145 virtual panel interface, (b) bipolar device cross-section and electrode assignment, (c) I-V data and algorithm used for extraction of Early voltage.

Recent work, jointly with University of Illinois, has yielded an improved Energy Transport (ET) formulation[32]. To date the well known problem with Hydrodynamic (HD) type of formulations is the unphysical velocity peaking in transition regions between low and high doping. In the new ET model, the Boltzmann transport equation is re-examined and subtles differences in position of terms and order of integrations have revealed the cause of the peaking anomalies. Moreover, a physical explanation is offered [33]

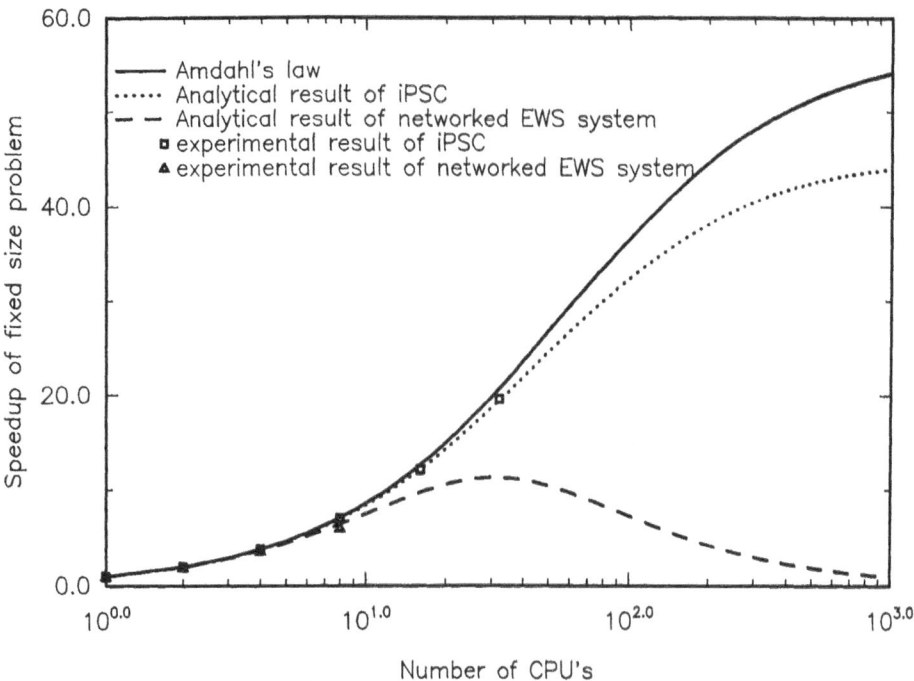

Fig. 14. Plot of CPU speed up factor (referenced to a single CE) versus number of CEs. Data and analytical curves compare performance for the Intel iPSC/860 and a network of EWS[34].

It is generally accepted that, in order to calibrate HD and ET formulations, the use of Monte Carlo (MC) simulators is a major benefit. Yet MC simulation is costly in terms of simulation time. The use of parallel computation offers a major benefit in this area. The use of both a network of EWS and the Intel iPSC/860 have been compared for MC computations in MOS devices[34]. Fig. 14 shows a comparison of speed-up factor versus number of Computational Elements (CEs) for the two approaches. Also shown are the analytical projections, based on equations derived from the experimental data for the two cases. Experiments were performed with up to 32 computational elements (CEs). Clearly, the communications for the hypercube (Intel) system is superior and is sufficient to sustain high performance even with more than 100 CEs. On the other hand, for the network of EWS, the combination of network limitations and software protocols makes it counterproductive to use more than a few dozen CEs. It is expected that improved communications networks will substantially enhance both approaches to MC calculations. Moreover, the opportunities to utilize "windowed" approaches[35] for ET, HD, and MC analysis can provide something like a mixed-mode benefit in understanding the physics with reasonable CPU expenses.

7. CONCLUSIONS

This paper started from an historic view of evolution from 1D TCAD models for power devices, looking carefully at the crucial role of diffusion modeling and parameter extraction based on SRP data. Recent progress in 2D modeling of power devices such as the CMDMOS and FATO was reviewed, based on PISCES simulations. The importance of nonplanar technology for power devices was discussed and advanced process models in this area were illustrated using the SPEEDIE and SUPREM-IV simulators. Next, several of the issues involving mixed-mode simulation were reviewed in light of recent development of both university and commercial tools. The motivation to create a TCAD framework and recent progress in developing a prototype implementation has been discussed. Issues of importance to the power device community focus on both the heterogeneous tool requirements and the ability to effectively leverage from a broader community of IC ECAD/TCAD development effort already in progress. In the area of future trends, the issues of High Performance Computing (HPC) for 3D and requirements for modeling non thermal equilibrium effects were discussed. Results based on use of the Intel iPSC/860 computer with 32 nodes were demonstrated both for 3D modeling and MC analysis.

8. ACKNOWLEDGMENT

The authors acknowledge a number of government and industrial sponsors of this research. We gratefully acknowledge support from DARPA and ARO in the areas of advanced device modeling and HPC. Broad industry support of our TCAD modeling - both the process and device aspects - comes through the SRC. There have been a number of industrial contracts and visitors specifically in the area of power devices. Dr. Christiaan Abbas, from ABB was one of our first industrial visitors and his work led to many key changes in PISCES that leveraged subsequent work. Mr. Akira Yabuta of MEW was another industrial contributor who continued shaping the development of PISCES for quasi-3D power device applications. Other corporate sponsors of this work include: General Electric, General Motors, Matsushita (MEW, MEI, MEC), National Semiconductor, Philips, Kodak, and Samsung. We have been continually stimulated from this diverse group of international contacts and we appreciate ongoing support to continue our TCAD efforts in this exciting area.

REFERENCES

1   H. G. Sigg, et al "D-MOS Transistor for Microwave Applications", *IEEE Trans. Elect. Dev.*, Vol. ED-16, pp. 45-53, Jan. 1972.

2   J. D. Plummer, et al. "An Ultrasonic Imaging System for Real Time Cardiac Imaging," *ISSCC Tech. Digest*, pp. 162-163, Feb. 1974.

3   M. D. Pocha, et al. "Threshold Voltage Controllability in Double-Diffused MOS Transistors," *IEEE Trans. ED*, vol. ED-21, pp. 778-784, Dec. 1974.

4   R. W. Dutton, et al. "Correlation of Fabrication Process and Electrical Device Parameter Variations," *IEDM Tech. Digest*, pp. 609-613, Dec. 1976.

5   D. C. D'Avanzo, et al. "Effects of the Diffused Impurity Profile on the DC Characteristics of VMOS and DMOS Devices," *IEEE J. Solid State Circuits*, Vol. SC-12, pp. 356-362, Aug. 1977.

6   M. Orlowski, "Impurity and Point Defect Redistribution in the Presence of Crystal Defects", *IEDM Technical Digest*, pp. 729-732, Dec. 1990.

7   D. C. D'Avanzo, "Modeling and Characterization of Short-Channel Double Diffused MOS Transistors," Stanford University PhD thesis, March 1980.

8   C. H. Price, "Two-Dimensional Numerical Simulation of Semiconductor Devices," Stanford University PhD thesis, May 1982.

9   C. J. Petti, "Physics and Technology of the Field-Assisted Turn-off Thyristor: A Regenerative Device with Voltage-Controlled Turn-off", Stanford University PhD thesis, Dec. 1989.

10   D. K-Y Liu, "Physics and Technology of Conductivity Modulated Power MOSFETs", Stanford University PhD thesis, September 1989.

11   D. M. Boisvert, "Physics and Technology of High-Speed, Low On-Resistance Power Devices," Stanford University PhD thesis, April 1990.

12   J.M. Dorkel and P. Leturcq, "Carrier Mobilities in Silicon Semi-
      Empirically Related to Temperature, Doping and Injection Level",
      Solid State Electron., vol 24, pp. 821-825, 1981.

13   G. Anderson, et al. "Metamorphosis of PISCES-Application-
      Oriented Transformation of 2D Device Simulation" *TECHCON 1990*,
      San Jose, CA, October 15-18, 1990.

14   A. Yabuta et al. "Numerical Analysis of Breakdown Voltage Using
      Quasi Three Dimensional Device Simulation," *IEEE Trans. Elect.
      Dev.*, Vol. 37, No. 4, April 1990.

15   J.P. McVittie, J.C. Rey, A.J. Bariya, M.M. Islamraja, S. Ravi and K.C.
      Saraswat, "SPEEDIE: Simulation of Profile Evolution During
      Etching and Deposition", *Proceedings SPIE Symp. Adv. Techniques
      for Integrated Circuits Processing*, Vol. 1392, pp. 126-138 (1990).

16   D. B. Kao, "Two-Dimensional Oxidation Effects in Silicon
      Experiments and Theory," Stanford University PhD Thesis, 1986.

17   C.S. Rafferty, "Stress Effects in Silicon Oxidation-Simulation and
      Experiments," Stanford University PhD Thesis, 1989.

18   Y. Oda, et al. "Numerical Techniques on Enhancing Robustness for
      Stress Dependent Oxidation Simulation Using Finite Element
      Method in SUPREM-IV", *IEICE Trans. Elect.*, vol. E75-C, no. 2, pp.
      150-155, Feb. 1992.

19   P. B. Griffin, C. S. Rafferty, "A Viscous Nitride Model for
      Nitride/Oxide Isolation Structures," *International Electron Devices
      Meeting, Technical Digest*, pp. 741-743, Dec. 1990.

20   W.L. Engl et al. "MEDUSA-A simulator for Modular Circuits" *IEEE
      Trans. on CAD*, vol. CAD-l, No. 2, pp. 85-93, April 1982.

21   K. Mayaram and D.O. Pederson, "CODECS: A Mixed-level Device
      and Circuit Simulator", *ICCAD Technical Digest*, pp. 112-115,
      November 1988.

22   J.G. Rollins and J. Choma Jr., "Mixed Mode PISCES-SPICE Coupled
      Circuit and Device Solver", *IEEE Trans. on CAD*, Vol. CAD-7, No. 8,
      pp. 862-67, Aug. 1988.

23   M. Vlach, "Modeling and Simulation with Saber", *Proceedings of The
      Third Annual IEEE ASIC Seminar and Exhibit, Sept. 17-21, 1990.*

24   EDIF Steering Committee, *"EDIF - Electronic design interchange format version 2 0 0"*, Electronic Industries Association, Washington, D. C., 1987.

25   CFI Technical Coordinating Committee, *"CFI Candidate Proposal, CAD Framework Users, Goals and Objectives - Version 0.91, "* CAD Framework Initiative, Inc., Austin, TX, 1991.

26   CFI TCAD Framework Group, Semiconductor Wafer Representation Working Group, *"Semiconductor wafer representation architecture-version 0.1,"* CAD Framework Initiative, Inc., Austin, TX, 1991.

27   CFI TCAD Framework Group, Semiconductor Wafer Representation Working Group, *"Semiconductor wafer representation procedural interface - version 0.09,"* CAD Framework Initiative, Inc., Austin, TX, 1991.

28   K. C. Wu, et al. "New Approaches in a 3-D One-Carrier Device Solver," *IEEE Trans.CAD/ICAS*, Vol. 8, No. 5, May 1989.

29   K. C. Wu, et al. "A STRIDE Towards Practical 3D Device Simulation-Numerical and Visualization Considerations," To be published *IEEE Trans. CAD*, Sept. 1991.

30   E. W. Scheckler, et al. "A Utility-Based Integrated Process Simulation System", *1990 Symposium on VLSI Technology-Digest of Technical Papers*, pp. 97-98, 1990.

31   G. K. Wachutka, "Rigorous Thermodynamic Treatment of Heat Generation and Conduction in Semiconductor Device Modeling," *IEEE Trans. CAD*, Vol. 9, No. 11, Nov. 1990.

32   D. Chen, et al. "A self-consistent Discretization Scheme for Current and Energy Transport Equations," *SISDEP '91*, Zurich, Sept. 1991.

33   D. Chen, et al. "Analysis of Spurious Velocity Overshoot in Hydrodynamic Simulations of Ballistic Diodes," *NUPAD IV (Numerical Process and Device Modeling Workshop) Digest*, pp. 109-114, Seattle, May 31 - June 1, 1992.

34   S. Sugino, et al. "Parallelization of Monte Carlo Analysis on Hypercube Multiprocessors and on a Networked EWS System," *SISDEP '91*, Zurich, Sept. 1991.

35   D. Y. Cheng, et al. "PISCES-MC, A Multi-window, Multi-Method 2D Device Simulator," *IEEE Trans. CAD/ICAS*, Vol. CAD-7, No. 9, Sept. 1988.

DISCUSSION

N. Zommer (ABB-IXYS, San Jose CA)

Is there anything done to standardize the output from a process-simulator to a device-simulator?

R. Dutton

Standards in these directions have come out from the EDIF-committee and the TCAD-working group in Europe, where many industrial groups and universities in many continents participate.

M. Campagna (ABB Corp. Research, Baden)

Is the power industry represented well enough in committees relating to the standardization of software development and simulation tools ?

R. Dutton

I would doubt that enough people have joined from the power side. Issues like benefits from VLSI or trenches for DRAM are extremely important for power devices too. It is important that the people in the power industry see this and that they bring those issues up to the committees.

P. R. Palmer (Cambridge Univ., Cambridge UK)

I would like to have Professor Duttons views on validating the results of such device simulators, particularly in a power device with very high current densities, occuring at the same time as high field regions are rapidly changing in the device.

Could you comment on validation of results, which requires more and more computing power ?

R. Dutton

Regarding power of computing, I do not mean to imply by my hypercube pictures that this is the only way of doing it. We have also done experiments with the same program and algorithms on workstations. With 10-30 of these workstations you may be getting close to half the performance of an equivalent hypercube which is quite promising.

Validation is always a problem, that is why I brought it up. The first point, when we started with 1D process modelling, was the spreading resistance which turned out to be a great life-saver in allowing us to calibrate.

Calibrating temperature is a very interesting thing. There are possibilities to do animation of real experimental work to validate things like temperature. Acoustic measurement is a member of a very exciting new measuring techniques for validating temperatures.

For the dimension control, I do not think that these problems need a lot of work and that they are radically different from the control issues for the DRAM.

The control of the simulator itself is a very interesting issue. I agree that finding out with its help what the physics are, e.g. under dynamic avalanche conditions, will be very exciting.

THE USE OF CAD TOOLS IN POWER DEVICE OPTIMIZATION[1]

W. Fichtner, J. Bürgler, H. Dettmer, N. Hitschfeld,
K. Kells, H. Lendenmann, S. Müller, and M. Westermann

Swiss Federal Institute of Technology
Zürich, Switzerland

ABSTRACT

Modern semiconductor manufacturing technology permits the integration of power integrated circuits with per-chip component densities reaching into the millions. This high-density capability, however, will only be utilized if the design costs can be limited to reasonable levels. Widely accepted for VLSI projects, the use of computer-aided design (CAD) tools has recently become a standard procedure in the development of power technologies and the design of new device structures and circuits. This paper attempts to give an overview of the status quo of CAD tools in power IC design, especially the use of simulation tools and computer-aided layout techniques. We illustrate the possibilities of these tools with examples from a recent MCT design effort.

1. INTRODUCTION

The continuing advances of semiconductor device technology have made it economically feasible to place several thousands to millions of devices on a single piece of silicon. Current developments in semiconductor manufacturing show no sign of a slow-down in this trend towards higher complexity.

---

[1]Extended version of a paper presented at ISPSD'91, Baltimore MD, USA.

Power Semiconductor Devices and Circuits, Edited by A.A. Jaecklin
Plenum Press, New York, 1992

Triggered by these staggering advances in semiconductor technology, the field of power integrated devices and circuits has seen major new developments during the last few years. Power integrated circuits embrace a wide spectrum of different applications ranging from more or less conventional analog circuits that are operated at non-standard voltage levels, to smart power circuits, and to circuits composed of parallel arrangements of similar devices such as power Metal Oxide Semiconductor Field Effect Transistors (MOSFETs) and Insulated-Gate Bipolar Transistors (IGBTs).

Caused mainly by the advent of Bipolar MOS (BiMOS) technologies, analog integrated circuits with voltage levels of 100V or more have emerged for telecommunications and automotive applications. Similar improvements can be witnessed for single power devices such as power MOSFETs, IGBTs, or more modern MOS Controlled Thyristors (MCTs) which have become very-large scale integrated circuits with cell densities exceeding one million.

The design of a power integrated circuit encompasses a large number of individual steps, ranging from the initial specification to architectural considerations and the layout and verification phase. Table 1 gives a schematic representation of the major components in the design of an integrated circuit.

The wide variety of different power electronics systems that contain semiconductor devices does not allow a simple categorization of the various hierarchy levels in Table 1. For example, an analog integrated circuit acting as a Liquid Crystal Display (LCD) driver will require different system and architecture modeling if compared to a Gate Turn-Off Thyristor (GTO) switch in a step-down converter or a power supply. It will therefore come as no surprise that the development of more or less strict system design guidelines has found no equivalent effort in power electronics. If we restrict ourselves to the chip level, however, concepts begin to converge. In this paper, we shall therefore concentrate on the two most important and successful areas of Computer Aided Design (CAD) tools in power Integrated Circuits (IC) design, namely layout generation and simulation.

2. LAYOUT TOOLS FOR POWER IC'S

The diverse group of tools in a modern CAD system relies on a single component, the master technology file, for technology encapsulation. The information in this file includes technology characteristics and process parameters as well as information that governs behavior of graphics devices used by the design tools. Because all design tools use a single source of technology information, consistency is guaranteed.

Table 1. Phase model of power IC development

Development Phase		Task	Result
1	Study	Definition of desired properties	Specification sheet(s)
2	Specification	Development of concept and testing strategy	Chip and test specification
3	System design	Development of modules and testing environment	Functional structure testing frame
4	Implementation	Construction of entire chip layout	Verified layout
5	Integration and test	Wafer fabrication and packaging	Prototype

The technology database permits designers to specify completely the target process on both the mask and the symbolic level. This has several advantages in both the active layout generation phase and the layout verification phase (rule checking).

We have taken several examples from a large MCT project recently completed. In this design, we implemented a variety of different concepts for cell and ensemble combinations into MCTs ranging in size from several microns to more than one cm on the side.

At the start, we optimized the layout of one unit MCT cell based on the process available (for an equivalent CAD effort, see[1]). In the next section, we illustrate the use of process and device simulation tools to optimize the process itself. Fig. 1 shows this layout. It was generated by hand in less than one hour by an experienced designer, verified against the technology database by a design rule check, and subsequently put into the design database.

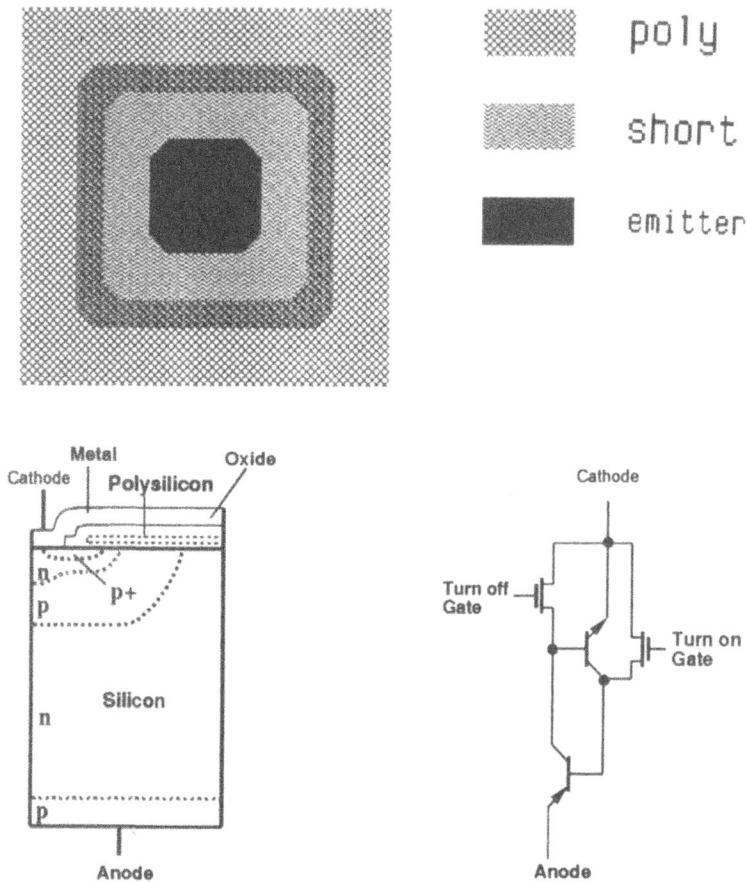

Fig. 1. Layout, schematic cross-section and equivalent circuit of a single MCT cell.

The full power of a modern layout tool becomes obvious once one starts to generate larger cell ensembles. This can be simply achieved by writing "code" in a high-level language that accesses the basic building blocks and arranges them according to the software specification. The possibilities are manifold and can be tuned to any situation that might arise.

The following program creates a circular MCT assembly consisting of a central gate and three 'turns' of evenly spaced single MCT cells (Fig. 2). The number of turns as well as the spacing between the MCT cells are parameters to the procedure. The language is Mainsail and VLSItools VIP (Procedural Layout Language) modules are used for the actual generation of the CIF code[2] (Mainsail is a trademark of Stanford University and VIP is a trademark of VLSI Technology, Incorporated; the use of these tools in this work should not be understood as an endorsement of these tools or a particular company).

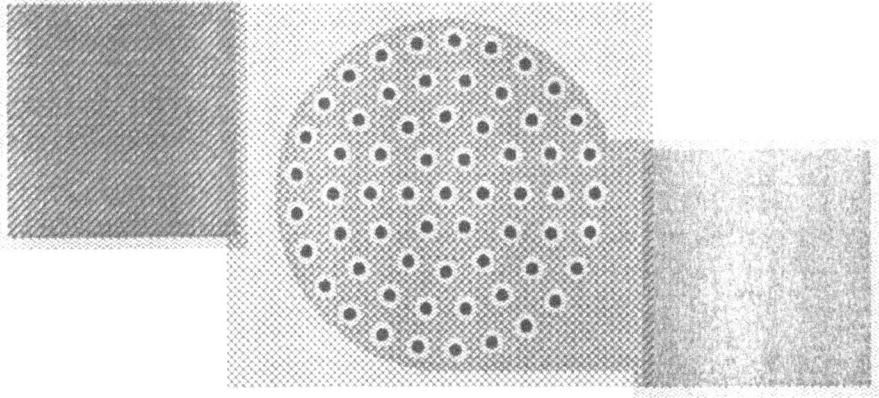

Fig. 2. Circular MCT array, generated through automatic software tools.

```
begin "circular_mct"
# module for generating a circular MCT assembly,
# consisting itself of circular MCT cells.
# 'max_turns' = number of turns
# 'd' = spacing of the cells
sourcefile "(usevip)";
STARTOF(mctrnd);

DEFSTART("circ_mct",[max_turns,d],[<define local variables>]);

#draw the central gate cell
DRAW("i1","vcell:vcell",
     PARAMETERS("[ly]single_mct_cell",ENDARGS));

FOR k:= 1 UPTO max_turns DO BEGIN
  n := <Calculate number of cells per turn)
  FOR i:= 0 UPTO n DO BEGIN
    x,y := <Calculate position coordinates of the cell>
    DRAW(nxy(x,y),"i","vcell:vcell",
             PARAMETERS("[ly]single_mct_cell",ENDARGS));
  END;
END;

ENDDEF;
ENDOF;
```

We have utilized the same approach to generate the real MCT chips in our design. The above code should be understood to be only one of the key sections for drawing the complete MCT ensembles. The pad for the connection of this assembly and the layers covering the entire area of this structure are added in another section of the module. Fig. 3 shows a plot of the final layout.

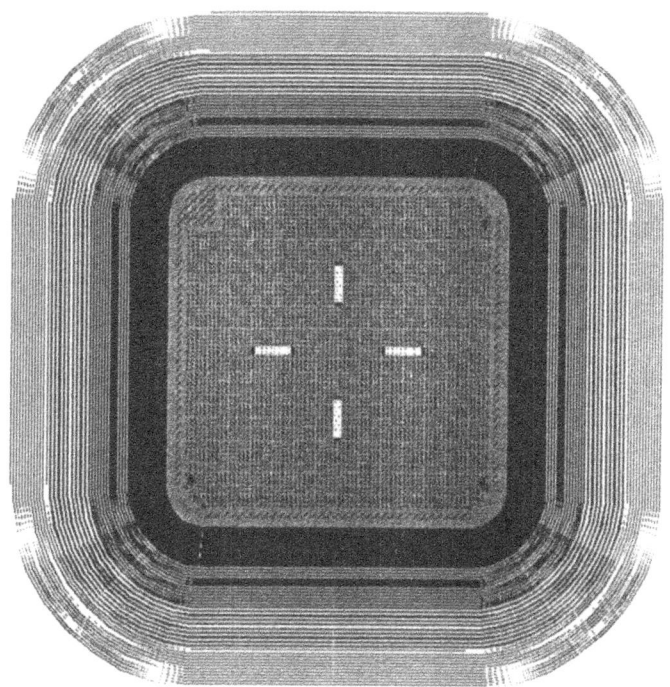

Fig. 3. MCT cell array generated by software.

The advantages of such an approach are manifold. Apart from being correct-by-construction, compiled designs can be easily changed to other cell configurations. Suppose, for example, the technology had changed during the phase and some design rules had to be adapted accordingly. With a minimum of CPU time, the complete layout can be regenerated - again, correct by construction.

3. SIMULATION TOOLS FOR POWER IC'S

Simulation plays an important role in the design of integrated circuits. Using simulation, the designer can determine both the functionality and the performance of a design before committing to the expensive and time-consuming step of fabrication. This is true not only for the more classical integrated circuits, but also for power electronics integrated circuits.

It is important to understand that simulation is more than a more convenience. Through ways that are otherwise impractical or even impossible, a designer can explore function and feasibility of her/his circuit. Influences of manufacturing and the ambient can be analyzed without the need to actually implement the design; testing procedures can be developed and the design can simply be debugged. For example, it becomes possible to study influences of certain fabrication steps without actually fabricating a circuit or device.

In Table 2 we illustrate the simulation and design hierarchy as it applies to power IC layout. For these circuits, three levels of the simulation hierarchy are of particular importance: circuit, device, and process simulation. To use a circuit simulator, the designer enters a design into the computer, usually in the form of a list of components that are connected to one or more nodes. The simulator bases its predictions on models describing the operation of these components.

Table 2. Simulation and design

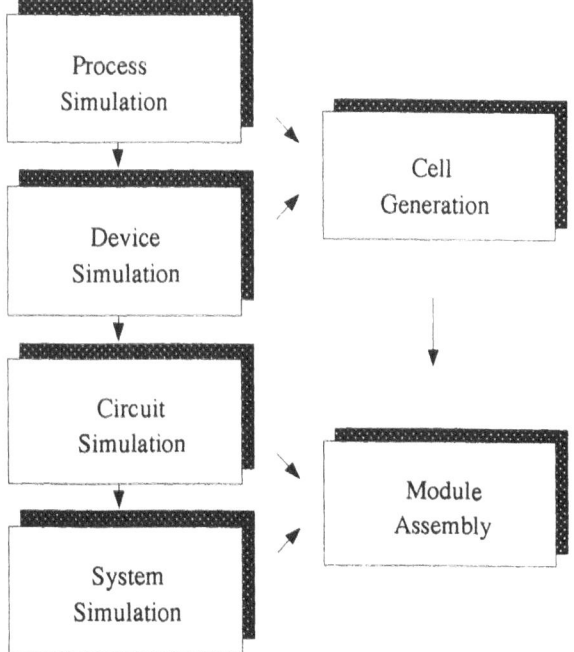

Depending on the hierarchy level under consideration, component models vary widely in complexity. The behavior of a component can in general be expressed in terms of an interconnection of ideal elements.

Usually, only a small repertoire of ideal elements needs to be supported. *Resistive Elements* relate currents to voltages. They include two-terminal elements (i.e. resistors and independent voltage and current sources) and four-terminal controlled sources. *Energy Storage Elements* relate the state of the storage element (capacitor, inductor) to one variable (e.g. voltage for a capacitor). These equations might be linear or nonlinear. Circuit simulation has been recently applied to system-related aspects such as snubber[3] and distortion-sensitivity[4] analyses.

Success and failure of a power circuit simulation hangs and falls with the quality of the models used for the semiconductor devices. While excellent models have been developed over the years for the more classical power bipolar and MOS devices[5,6,7,8], models for "real" generic power devices are still in their infancy. Examples are models for the IGBT[10] and the SITH[11].

One reason for the dissatisfying state of CAD models for power switches can be found in their complex physical modes of operation. It will therefore come as no surprise that power device researchers have taken a more basic approach through numerical process and device simulation.

In process simulation, one tries to simulate all fabrication steps of importance - implantation, diffusion, oxidation, etc. - with a "best effort" physical model, usually at the level of differential equations that are solved numerically in one or multiple dimensions. Over the years, several software packages have been developed that enable excellent process characterization and forecasting (see[12] and references therein). As a representative example obtained with a modern process simulation tool, Fig. 4 shows a surface plot of the net impurity concentration near the surface of a single MCT cell[13,14]. Based on the results of such as simulation, process optimization becomes considerably easier and more straightforward, and narrow margins in critical fabrication steps can be usually avoided.

Using the results from process modeling in the form of geometries and doping distributions, accurate device investigations can be performed, again on a numerical basis in one or more dimensions. Device simulation programs such as PISCES[15,16], BAMBI[17], TONNADDE[18], GENSIM[13] or SECOND[19] are now in widespread use for power device optimization. Other examples can be found[20,21,22]. The capabilities of these programs are manifold, allowing both steady-state and transient conditions with external circuits such as snubbers. Over the past two years, we have developed a series of last generation software packages such as two-dimensional (MESHBUILD[23,24]) and three-dimensional (OMEGA[25,26]) mesh generators, a multi-dimensional device simulator (SIMUL[27,28]) and graphics postprocessors. Compared to older packages just mentioned, these new tools permit simulations of multiple devices structures in one, two and three dimensions such as a two-dimensional thyristor and a one-dimensional free-wheeling diode (in one run). Furthermore, apart from its capabilities for steady-state and transient analyses, lattice and carrier

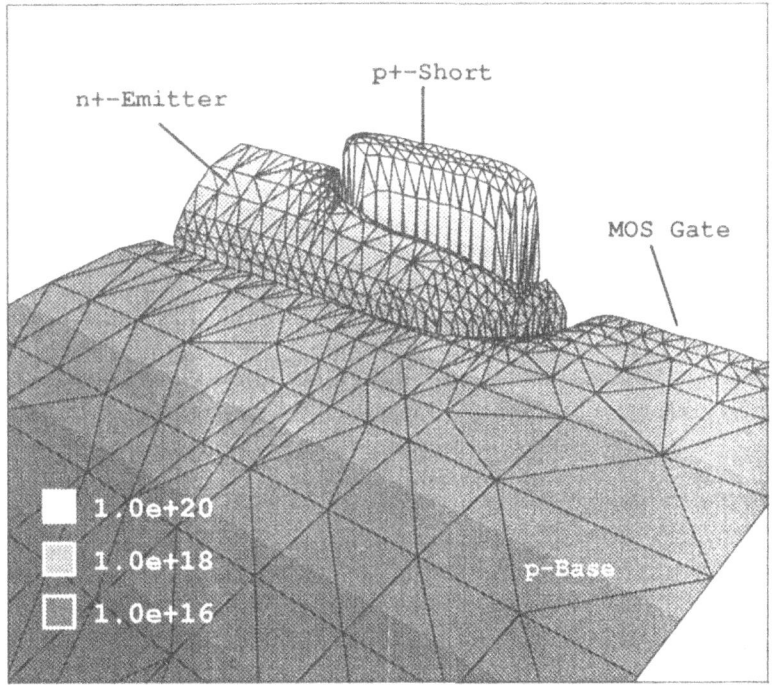

Fig. 4. Doping distribution of the MCT p-channel DMOS switch.

heating phenomena are included. A detailed description of these tools can be found in[27].

The power of these numerical software tools will be illustrated with two examples taken from an investigation of the turn-off behavior of scaled MCT's. In a single "properly designed" MCT cell, turn-off is achieved by switching a gate to the proper voltage levels. For example, if one uses a p-channel DMOS switch, turn-off is triggered by a fast negative gate signal. Figures 5 and 6 show vector plots of the hole current distribution in a half cell at t=0 and t=40ns. While the MCT operates in pure bipolar thyristor mode at t=0, it becomes evident from this plot that some holes entering the n⁺-emitter region are lost in the p-short. Optimizing the fraction of lost holes is absolutely critical for the MCT operation. In reality, a few iterations between process and device simulation have led to an acceptable (and manufacturable) solution.

Switching the gate diverts the holes from the emitter into the p-short. Again, complex trade-offs have to be performed in terms of switch action and emitter behavior.

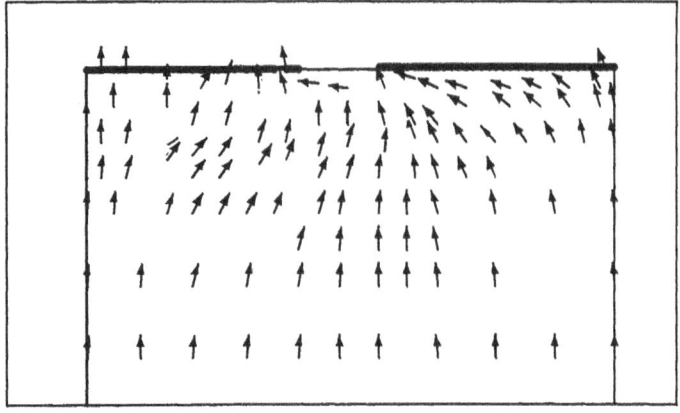

Fig. 5. Hole distribution of a half-cell MCT in steady-state.

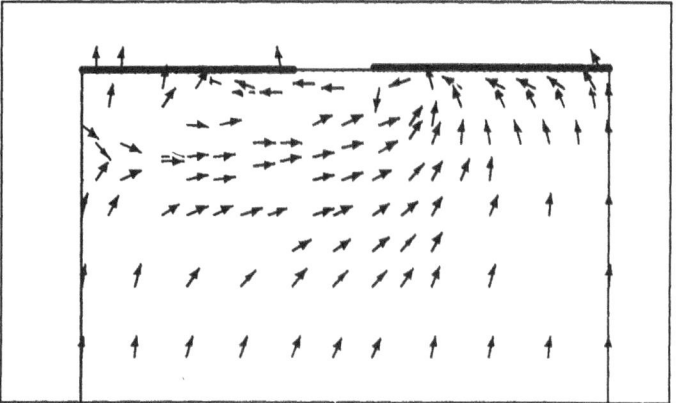

Fig. 6. Hole distribution of a half-cell MCT during turn-off.

The full power of combined process-device-circuit simulation becomes evident if one performs a critical investigation into one of the most important degradation phenomena: current inhomogeneities and filamentation. Only recently, software tools and computer platforms have become powerful enough to tackle these problems at a fundamental level[29,14].

In order to study cell-boundary interactions, we have examined the behaviour of a hypothetical cell ensemble, consisting of three complete MCT cells and one DMOS cell. The cell structure was used to investigate the current distribution near the edge of the chip. In Fig. 7, we display the doping-distribution of the device. (Dark shades mean highly p-doped regions and light ones n-doped regions.)

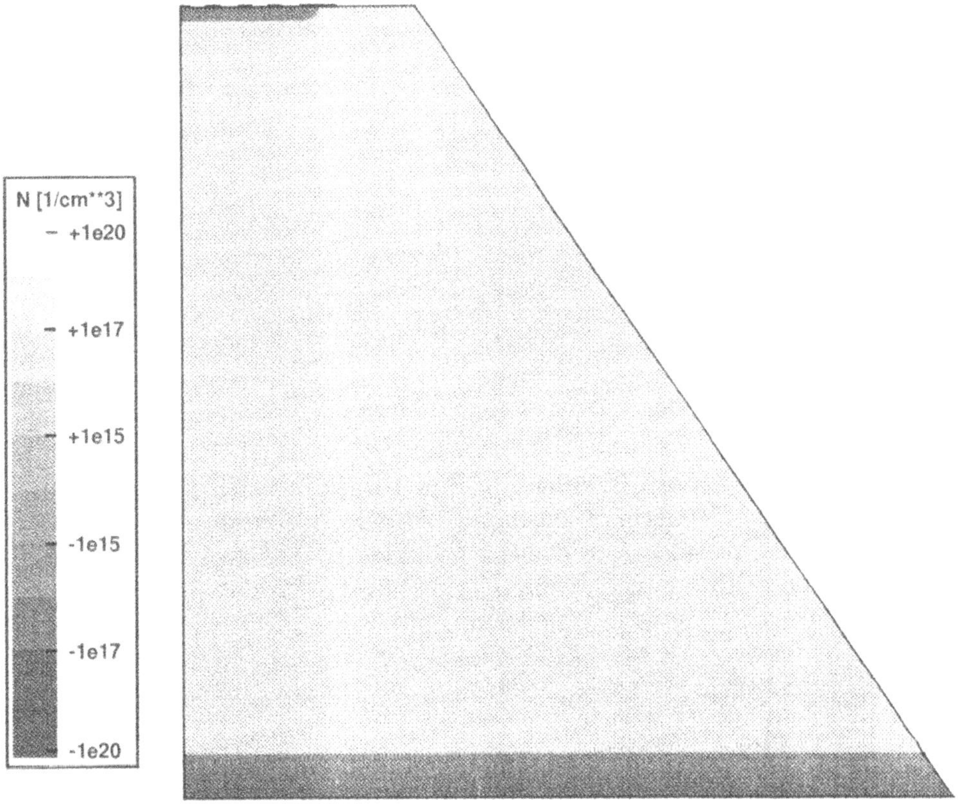

N [1/cm**3]
 − +1e20

 − +1e17

 − +1e15

 − -1e15

 − -1e17

    -1e20

Fig. 7. Doping concentration of the MCT cell structure.

Figures 8 and 9 present enlargements of the structured cathode side of the thyristor. Figure 8 shows three MCT cells to the left and one DMOS cell on the right, i.e. near the edge of the chip. The p$^+$ and n$^+$ regions are short-circuited by a contact. A zoomed view of one single MCT-cell with its MOS-channel is given in Fig. 9.

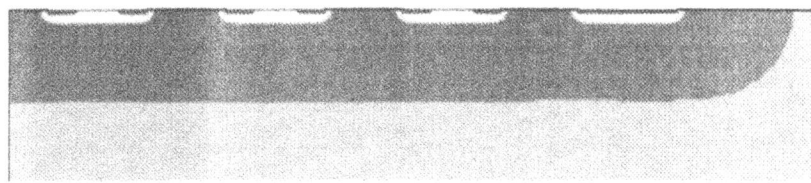

Fig. 8. Zoom-in to the upper left of the MCT cell structure.

Fig. 9. Zoom-in down to single-cell level.

The resulting mesh is shown in Fig. 10 and 11. We generated this mesh in approx. 50 seconds on a SUN 4/290. The mesh contains 5477 nodes in 3937 rectangles and 2473 triangles. It has to be pointed out here that the resulting mesh contains no obtuse angles.

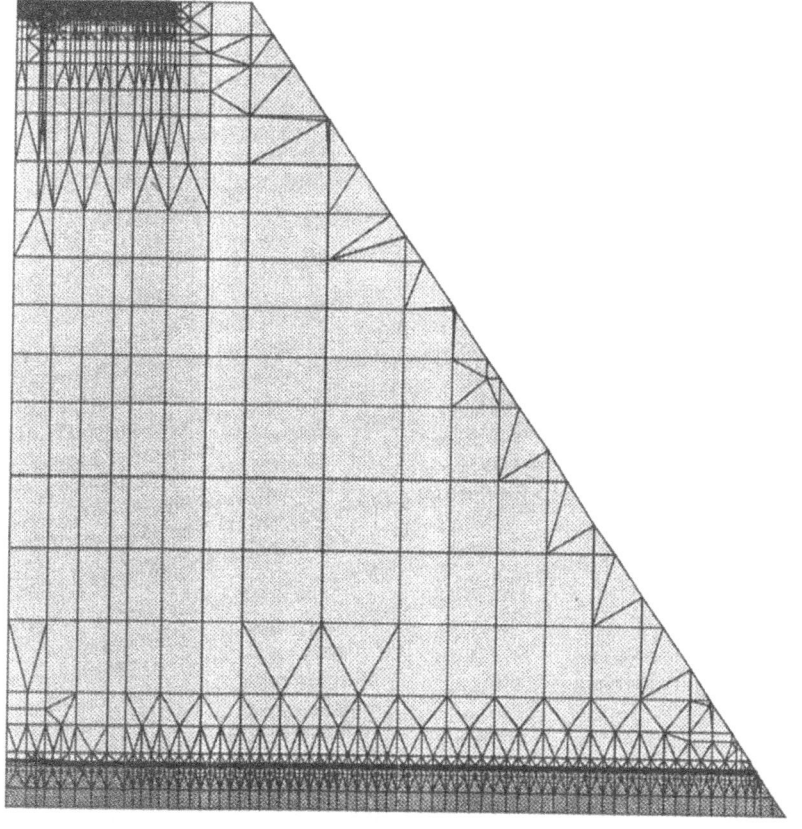

Fig. 10. Mesh generated through MESHBUILD

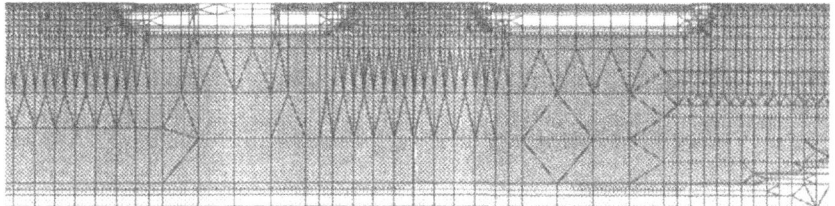

Fig. 11. Zoom-in into the cathode region

For the device simulation, the thyristor was latched by applying 3 V at the anode contact. The distribution of the electrostatic potential is shown in Fig.12. Zoomed views of the electron density and the conduction current density in the cathode region are presented in Fig. 13 and Fig. 14, respectively.

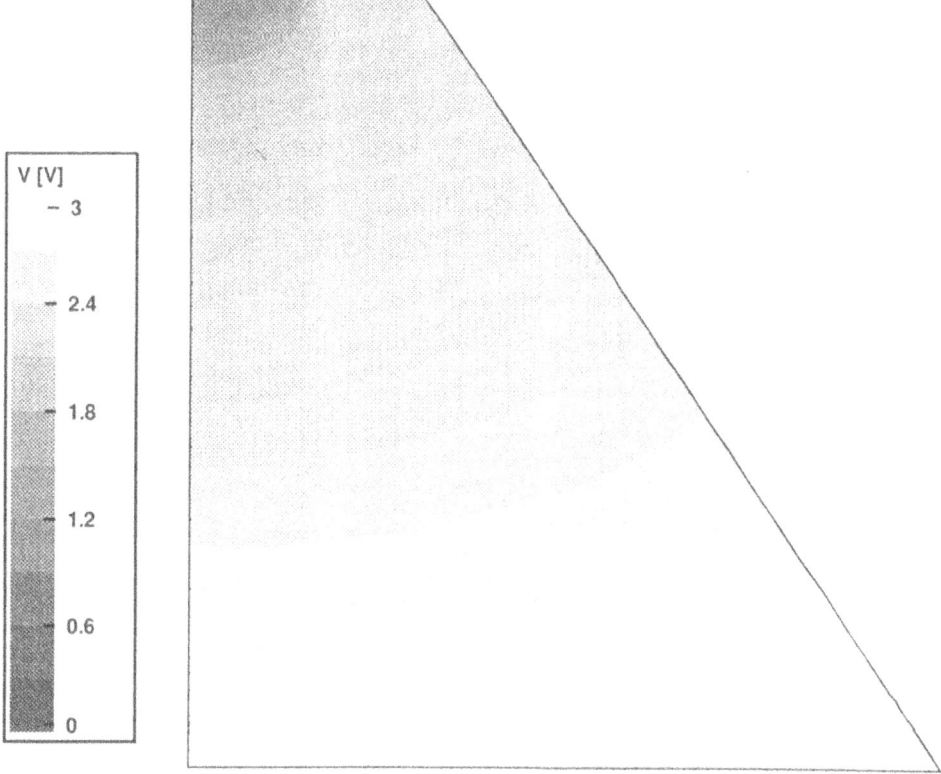

Fig. 12. Electrostatic potential distribution.

Fig. 13. Electron density in the top surface region.

Starting from this initial condition, we simulated a current trace to an anode voltage of around 1.0 V in order to determine the holding voltage, i.e. the minimum voltage that is needed to keep the device latched. The computation was done on a Convex C-220 and took about three hours and fifteen minutes. Some of the resulting currents, flowing at the cathodes of the three MCT cells, are plotted in Fig. 15. One notices that the current, flowing at the rightmost MCT cell, is approximately twice as large compared to that in the two left ones, indicating a severe current inhomogeneity.

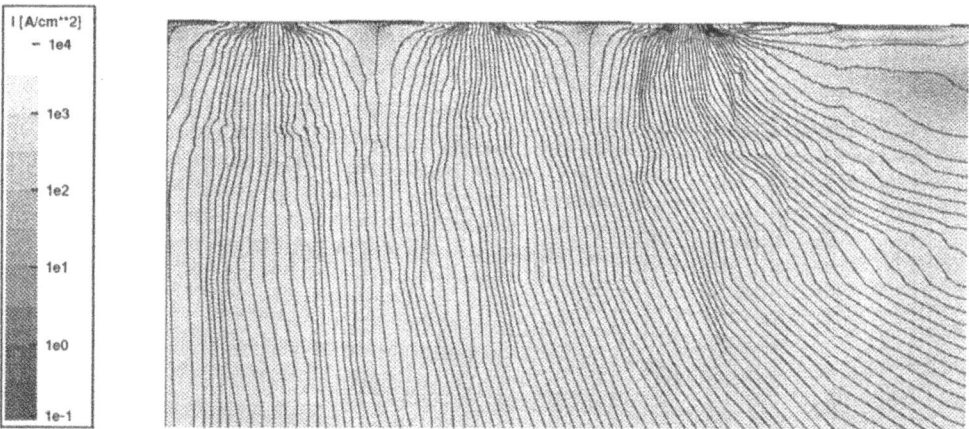

Fig. 14. Conduction current density with overlayed current lines.

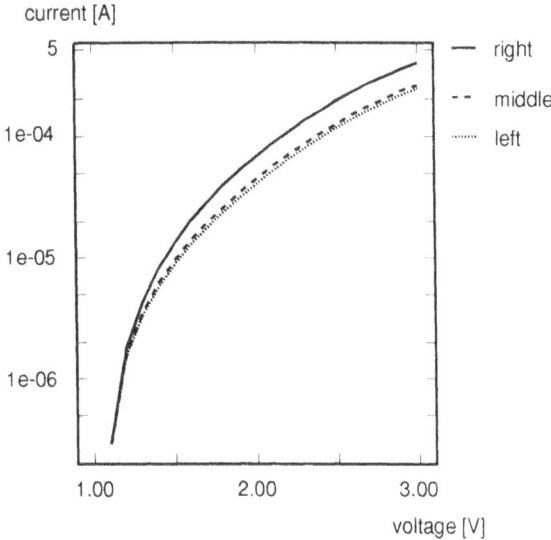

Fig. 15. Cathode currents.

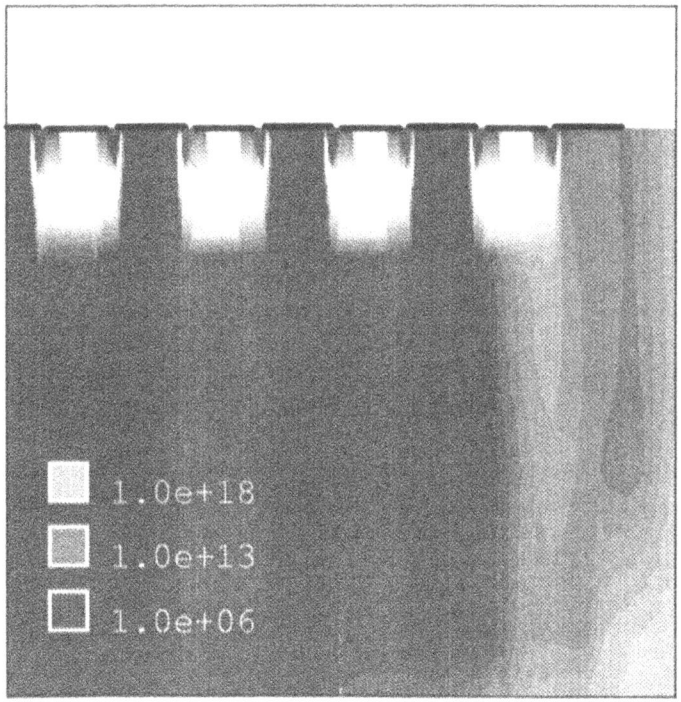

Fig. 16. Electron distribution in a 4-MCT compound ensemble at t = 106 ns.

To critically look at these current flow distributions, we analyzed a structure with four MCT's, embedded in a complete chip plus junction termination, and switched the structure against 1kV, using a inductive load. The simulation time exceeded 5 h on one processor of a Cray XMP-28. Figure 16 shows the major result of the calculation. After applying the gate pulse to turn the device off, current redistribution occurs and the corner device, being flooded by excess holes, is unable to switch off within a reasonable time window.

While the above examples permit good insight into the behavior MCT cells and cell ensembles, it should not be forgotten that devices are in reality three-dimensional. Over the past few years, we have made considerable efforts to develop simulation packages for full three-dimensional semiconductor device modeling[25,30,26,28]. Presently, it is possible to simulate rather complicated and nonplanar device structures in both steady-state and transient mode albeit at considerable computational expense.

The example shown here is the three-dimensional generalization of the cells used above. Fig. 17 presents the doping concentration of a quarter-MCT structure with a 10 x 10 μm^2 surface area and a 500 μm thick bulk region.

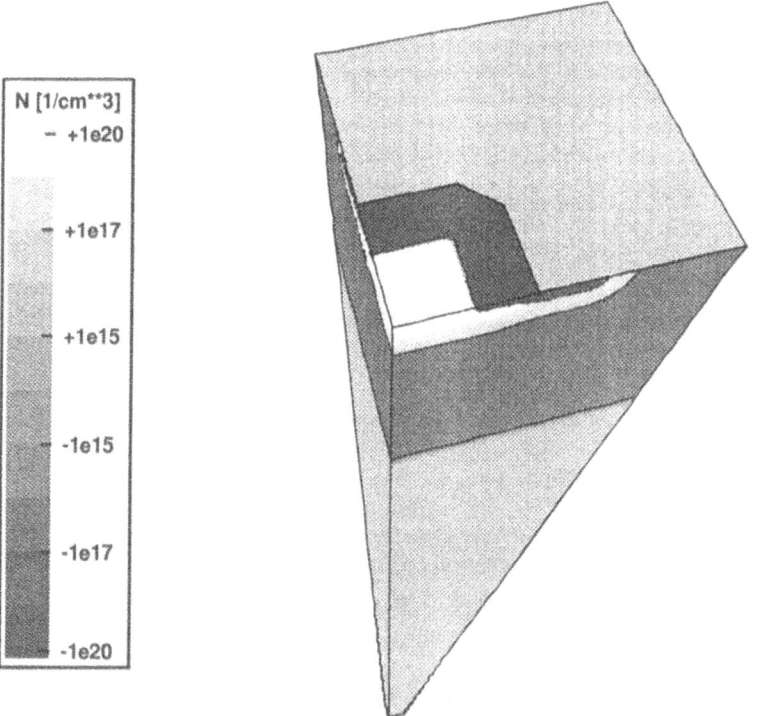

Fig. 17. Doping concentration of the MCT half cell.

Together with structural and geometrical specifications, this doping profile is used as input information to the three-dimensional grid generation process[26]. Fig. 18 shows a top view of the mesh generated. The grid contains 29,584 points and 42,033 elements (17 k bricks, 13.5 k prisms, 7.3 k pyramids and 4.1 k tetrahedra). The generation of this grid required around 1600 seconds on a SUN SPARCstation 1+.

As a representative example, Fig. 19 shows the electron current density distribution in the surface region of the MCT.

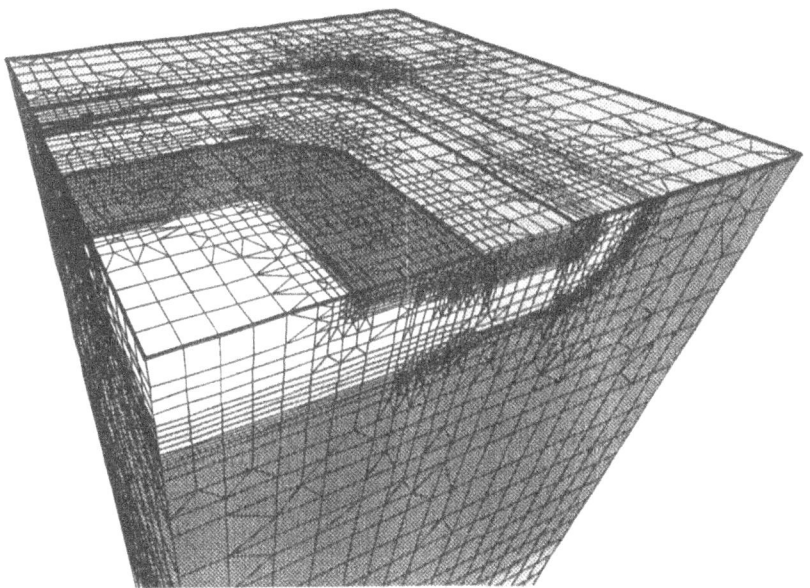

Fig. 18. Final mesh for the MCT cell on top of the impurity concentration.

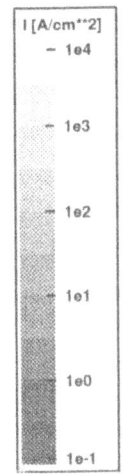
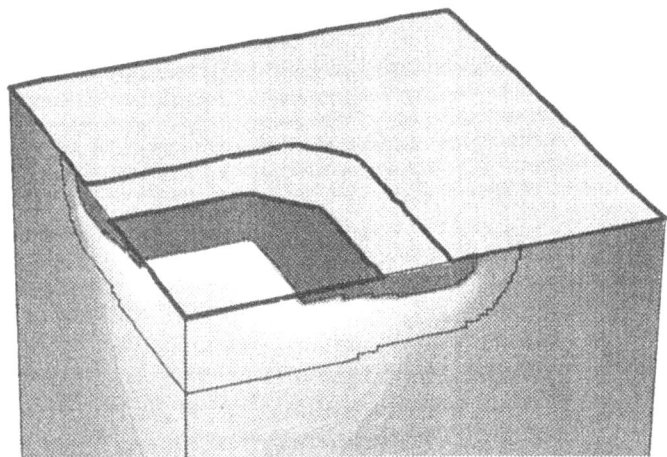

Fig. 19. Electron current density distribution on the surface on the three-dimensional MCT-cell.

4. CONCLUSIONS

In this paper, we have tried to show through examples how modern CAD tools can become a significant factor in the overall design cycle of power integrated circuits. A rather optimistic picture can be painted for the possibilities of modern layout tools. Originating from VLSI requirements, these tools offer unique advantages through automated compilation and verification and fast turn-around.

On the side of power IC simulation, a more critical outlook must be given. While numerical process and device simulation tools have become high quality, indispensible aids for process and device development, circuit simulation and CAD models need additional efforts before they will reach the same level of usability and quality. We have not touched upon other important issues related to power device simulation such as package modeling[31], or simulations for combined control/switching problems. These topics will become more important if higher integration levels at the chip and module level are approaching. In our opinion, combined device/circuit simulation offers the most promising road to a successful design methodology.

It should not be forgotten that in the final analysis - CAD tool or not - all things become a matter of economics. Increasing productivity is important because design time is the critical issue in the fabrication of all products. Expert designers and technologists are a rare commodity and expensive. The better the CAD tools, the more work can be done in the same amount of time with considerably lower risks involved. If one looks at these rather convincing arguments, it would be unfair not to mention that modern CAD tools have also negative aspects. As a first point, one should never forget that modern last generation CAD tools are extremely complicated software products. Their use requires educated users willing to take up a challenge that should not be underestimated. Secondly, modern CAD tools require modern computer hardware platforms. While today's workstations can be purchased at reasonable prices, they are not free, and they have to be maintained. Through their business, power electronics companies are traditionally more conservative than corresponding VLSI units. Nevertheless, they will have to follow the VLSI bandwagon - not only in technology, but also in CAD.

5. ACKNOWLEDGMENT

We would like to thank our colleagues at ABB CRBS in Dättwil, Switzerland, especially Dr. F. Bauer, for their collaboration. The research of H. Dettmer, H. Lendenmann and S. Müller has been supported through KWF Project 1894.1. M. Westermann's work is financed by FSRM Project 90/06 and AT&T. The support of the ETHZ and EPFL supercomputer centers is gratefully acknowledged.

REFERENCES

1   N.J. Elias, "A case study in silicon compilation software
        engineering HVDEV high voltage device layout generator," in
        *Proceedings of the 24th Annual ACM/IEEE Design Automation
        Conference*, pp. 82-88, 1987.

2   *VLSI Technology CAD Manuals*, 1991.

3   T.M. Undeland, A. Petterteig, G. Hauknes, A.K. Adnanes, and S.
        Garberg, "Diode and thyristor turn-off snubbers simulation by
        KREAN and an easy to use design algorithm," in *Proceedings IAS* ,
        pp. 657-654, 1988.

4   Y. Kuroe, H. Haneda, and T. Maruhashi, "A new computer-aided
        method of distortion sensitivity analysis and its elimination
        scheme for power electronics," *IEEE Transactions on Power
        Electronics*, vol. PE-1, pp. 200-209, 1986.

5   I. Getreu, *Modeling the Bipolar Transistor*, Tektronix, 1976.

6   C.H. Xu and D. Schröder, "Modeling and simulation of power
        MOSFETs and power diodes," in *Proceedings of the 18th Annual
        IEEE Power Electronics Specialists Conference*, pp. 76-83, 1988.

7   D.K. Thajur, V.K. Khanna, A. Kumar, N.K. Swami, S.K. Mahajan,
        S.C. Good, and K.L.J. and W.S. Khokle, "Characterization and
        transient analysis of monolithic power Darlington transistors for
        AC drives using CAD as a tool," in *Proceedings IAS*, pp. 504-511,
        1987.

8   C.H. Xu and D. Schröder, "A power bipolar junction transistor
        model describing the static and the dynamic behavior" *Proc.
        PESC'89*, pp. 314-321, Milwaukee WI, 1989.

9   Y. Tsividis, *Operation and modeling of the MOS transistor*,
        McGraw-Hill, 1987.

10   A.R. Hefner, "Analytical modeling of device-circuit interactions for
        the power insulated gate bipolar transistor [IGBT]", in
        Proceedings IAS, pp. 606-614, 1988.

11   D. Metzner and D. Schröder, "A SITH-model for CAE in power
        electronics," *IPEC Technical Digest*, 1990.

12   W. Fichtner, "Process simulation," in *VLSI Technology* [S.M. Sze,
        ed.], pp. 422-465, McGraw-Hill, 1988.

13   A. Aemmer, F. Bauer, J. Bürgler, W. Fichtner, S. Müller, and
      P. Roggwiller, "Multi-dimensional simulation of MCT structures",
      ISPSD Technical Digest, pp. 20-25, Tokyo, 1990.

14   F. Bauer, E. Halder, K. Hofmann, H. Hollenbeck, P. Roggwiller,
      T. Stockmeier J. Bürgler, W. Fichtner, S. Müller, M. Westermann,
      J. M. Moret, and R. Vuillemier, "Design aspects of MOS:
      controlled thyristor elements: technology, simulation and
      experimental results," *IEEE Trans. on Electr. Dev.*, vol. ED-38,
      pp. 1605-1611, 1991.

15   M.R. Pinto, C.S. Rafferty, and R.W. Dutton, "PISCES II: Poisson
      and continuity equation solver," *Tech Rep., Stanford Electronics
      Laboratory, Dept. of Electrical Engineering*, Stanford Univ., Palo
      Alto, 1984.

16   M.R. Pinto, C.S. Rafferty, H.R. Yeager, and R.W. Dutton, "PISCES
      IIB: Suppplementary report," *Tech. Rep., Stanford Electronics
      Laboratory, Dept. of Electrical Engineering*, Stanford Univ., Palo
      Alto, 1985.

17   A.F. Franz, G.A. Franz, S. Selberherr, S. Ringhofer, and
      P. Markowich, "Finite boxes - a generalization of the finite-
      difference method suitable for semiconductor device simulation",
      IEEE Trans. Electr. Dev., vol. ED-30, pp. 1070-1082, 1983.

18   A. Nakagawa and S. Nakamura, "Application of general purpose
      power BIMOS simulator TONNADE-II to double gate lateral
      bipolar MOSFET design", *ISPSD Technical Digest*, pp. 42-47,
      1990.

19   J. Bürgler, P. Conti, G. Heiser, S. Paschedag, and W. Fichtner,
      "Three-dimensional simulation of complex semiconductor device
      structures", in *Proceedings of the 1989 International Symposium
      on VLSI Technology, Systems and Applications*, pp. 106-110,
      1989.

20   E. Masada, M. Tamura, and T. Nakajima, "Simulation of switching
      processes in turn-off thyristors," *IEEE PESC Technical Digest*,
      pp. 91-98, 1988.

21   W.R. VanDell and W.K. Dyle, "SEMINET: further advances in
      integrated device/circuit simulation for power electronics, " in
      *Proceedings of the 18th Annual IEEE Power Electronics Specialist
      Conference*, pp. 69-75, 1988.

22   B.E. Danielsson, "Studies of turn-off effects in power
        semiconductor devices," *Solid State Electronics*, pp. 375-391,
        1985.

23   S. Müller, K. Kells, and W. Fichtner, "Generation of meshes for
        the simulation of power devices" in *VLSI Process and Device
        Modelling*, Kawasaki, pp. 66-67, 1990.

24   S. Müller, K. Kells, and W. Fichtner, "Automatic rectangle-based
        adaptive mesh generation without obtuse angles," *IEEE Trans. on
        CAD*, in press.

25   P. Conti, *Grid Generation for Three-Dimensional Semiconductor
        Device Simulation*, Hartung-Gorre, Konstanz FRG, 1990.

26   N. Hitschfeld, P. Conti, and W. Fichtner, "Grid generation for 3-d
        nonplanar semiconductor device structures," in *Simulation of
        Semiconductor Devices and Processes IV* (W. Fichtner and A.
        Aemmer, eds.), pp. 165-172, Hartung-Gorre, Konstanz, 1991.

27   S. Müller, N. Hitschfeld, C. Pommerell, K. Kells, M. Westermann,
        and W. Fichtner, "Technology-CAD algorithms, implementation
        and results," in *Proceedings of the Second NEC Symposium*,
        SIAM, 1991.

28   K. Kells, S. Müller, W. Fichtner, and G. Wachutka, "Simulating
        temperature effects in multi-dimensional silicon devices with
        generalized boundary conditions," in *Simulation of Semiconductor
        Devices and Processes IV* (W. Fichtner and A. Aemmer, eds.), pp.
        141-148, Hartung-Gorre, Konstanz, 1991.

29   M. Stoisiek, K.-G. Oppermann, and G. Wachtuka, "Turn-off
        behaviour of GTOs with small emitter elements," *Arch. Elektr.
        Übertragung (AEÜ)*, vol. 43, pp. 320-327, 1989.

30   G. Heiser, *Design and Implementation of a Three-Dimensional,
        General Purpose Semiconductor Device Simulator*, Hartung-Gorre,
        Konstanz FRG, 1990.

31   A. Yasukawa and T. Sakamoto, "Simulation for designing
        semiconductor packages with high reliability under thermal
        cycling", in *ISPSD 1988 Technical Digest*, pp. 36-41, Tokyo,
        1988.

DISCUSSION

P. Svedberg (ABB HAFO AB, Järfälla S)

Are you doing any special work to study physical models for power devices, e.g. saturation velocity or temperature dependence?

W. Fichtner

We do a lot of validation e.g. making sure that the process parameters differ not too much from the experiment. We also check the physical models very carefully (e.g. mobility). This is especially important for power devices which have both, bipolar and MOS-mode, simultaneously. The numerics is less of a difficulty, but if the physical models are wrong we will get problems.

R. Dutton (Stanford University, Stanford CA)

In your simulations you showed different contact configurations. Is there any physical basis for the the oscillations in some of the curves?

W. Fichtner

We think this is a grid phenomenon. By applying a finer grid, the fine oscillations disappear.

B.J. Baliga (North Carolina State University, Raleigh NC)

What progress is made towards applying these tools to new materials, e.g. SiC ?

W. Fichtner

Today, I am not aware of any work on SiC. When people have understood that these new exotic material are important enough, it will happen.

R. Dutton

Have you been able to observe and/or simulate dynamic avalanche?

W. Fichtner

In the simulation field, it has been known for quite a while that this phenomenon exists. It has been observed by us, by Siemens and by other people. At this point, we are not capable to study these effects in 3D, in order to get quantitative hints how to avoid it. However, there is good work on 2D and single cells.

MODELING THE LIMITS OF STABLE DEVICE BEHAVIOUR

Klas Lilja

Asea Brown Boveri Corporate Research
Baden, Switzerland

ABSTRACT

The performance of many power semiconductor devices is limited by current filamentation instabilities produced by an inhomogeneous distribution of the current in the device that leads to local over-heating. In bipolar power transistors, in modern Gate Turn-Off thyristors (GTO) as well as in Metal-Oxide-Semiconductor (MOS) Controlled Thyristors (MCT), the safe operating area is drastically reduced by current filamentation problems. In order to design reliable devices for high power applications, adetailed understanding of this phenomenon is required.

In this paper, we will show how the current filamentation instability can be investigated by means of modeling and numerical simulation.

The emphasis of our investigation is on gate turn-off thyristors. We show how and why the current redistribution starts in these devices and how it develops. We give criteria for stability and show the effects of protective snubber circuits.

We also show simulations of current filament formation at thermal second breakdown in bipolar power transistors and discuss its causes.

1. INTRODUCTION

Instability phenomena, which give rise to a development of spatial (or temporal) structures from homogeneous stationary solutions, presently attract the attention in almost all fields of science, from physics to sociology (e.g.[1]). In the field of semiconductor physics, many such phenomena have been found, studied and also utilized (e.g.[2]).

Two basic types of spatial structure formation in semiconductors can be identified by looking at the current density - voltage characteristics, j(V). When this characteristics contains a branch of negative differential conductivity (NDC), σ_{diff},

$$\sigma_{diff} = \frac{dj}{dV} < 0,$$ (1)

the corresponding time-independent states are generally unstable. Depending on the shape of the j(V) characteristics, resembling either an N or an S (Fig. 1), NDC is classified as NNDC or SNDC.

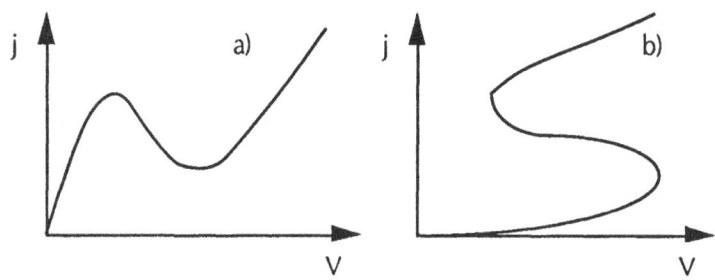

Fig. 1. Current density j versus voltage V for the two types of Negative Differential Conductivity (NDC): a) NNDC, b) SNDC (schematically).

If a device is driven into an unstable NNDC branch, a lateral layer domain formation will occur. Since the NNDC state is unstable, the voltage over the device must be sustained partly by the stable low voltage solution and partly by the stable high voltage solution (Fig. 2a). The best known NNDC phenomenon is the Gunn effect[3].

If a device is driven into an unstable SNDC branch, a vertical layer filament will occur. Since the SNDC state is unstable, the total current must be carried partly by the stable high current solution and partly by the stable low current solution (Fig. 2b). The thyristor is a device with SNDC characteristics.

The latter instability type, leading to the formation of current filaments, is a severe limitation to the Safe Operating Area (SOA) of many high power switching devices, like bipolar transistors and turn-off thyristors (GTO, MCT). This instability leads to a concentration of the current to small spots (current filamentation) and often to device-destruction by local over-heating. We present in this paper an investigation, making intensive use of modern numerical simulation tools, of the onset and development of current filaments. We will study both the causes of the SNDC branch, its stability and under which circumstances the instability becomes critical, i.e. when it prevents proper device operation or even destroys the device. In the focus of the analysis are the modern types of thyristors which can be turned off by a gate signal, e.g. the GTO and the MCT.

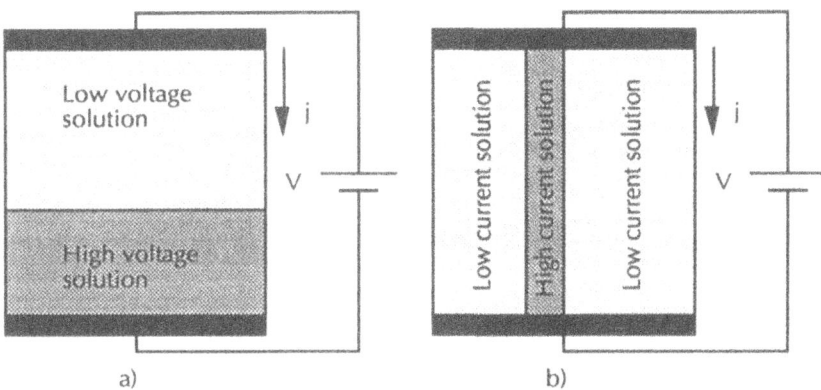

Fig. 2. a) Voltage domains in an NNDC device; b) current domains in an SNDC device leading to a current filament (schematically).

In the thyristor, the device structure itself, consisting of four doping layers, is the cause of the SNDC characteristics. The mechanism is well known and it is utilized in normal device operation. In our investigation of thyristors (Section 4), we go into detailed realistic applications and study when and under which circumstances the filamentation instability becomes critical for the device.

As a further example of how filamentation can be investigated by numerical simulation, we will show a simulation of a Thermal Second Breakdown (TSB) in a bipolar power transistor (Section 3). The charac- teristics of a bipolar transistor is not of SNDC type under normal operat- ing conditions. However, at second breakdown, an NDC branch occurs, caused either by increasing temperature (Thermal Second Breakdown, TSB) or by impact ionization (current induced second breakdown). We will discuss the cause of TSB and show a complete simulation of the de- velopment of a current filament. However, our analysis of bipolar transis- tors will not be as detailed as the treatment of turn-off thyristors.

2. NUMERICAL SIMULATION MODELS AND TOOLS

We use the complete semiconductor transport model in our analysis. The basic model consists of five coupled partial differential equations: Poisson's equation, two continuity equations and two transport equa- tions. If a non-constant temperature must be considered, the set of equations is extended by a heat transport equation. The semiconductor transport model is extensively treated in the literature and can be considered as standard (e.g.[4]). From the numerous applications, we know that it describes most phenomena in the semiconductor quite well.

The foundation of the standard set of equations consists of a mixture between heuristic arguments, derivations from more basic theory and fitting to experimental data. For a discussion of their derivation and limitations see e.g. [4,5]. The six equations of the model are:

Poisson's equation,

$$\varepsilon \nabla^2 \Psi = -q(p - n + C) .$$ (2)

The continuity equations for electrons and holes,

$$\nabla \cdot J_n - q \frac{\partial n}{\partial t} = qR ,$$ (3)

$$\nabla \cdot J_p + q \frac{\partial p}{\partial t} = -qR .$$ (4)

The transport equations for electrons and holes,

$$J_n = q\mu_n nE + qD_n \nabla n + q \frac{D_n n}{2T} \nabla T ,$$ (5)

$$J_p = q\mu_p pE - qD_p \nabla p - q \frac{D_p p}{2T} \nabla T .$$ (6)

The heat conduction equation (for phonons),

$$\rho c \frac{\partial T}{\partial t} - H = \nabla \cdot (\chi \nabla T) .$$ (7)

where

Ψ	is the electrical potential,
n, p	are the electron and hole densities,
T	is the (lattice) temperature,
C	is the doping concentration,
J_n, J_p	are the electron and hole currents,
R	is the recombination - generation excess rate,
$\mu_{n,p}$	are the mobilities,
$D_{n,p}$	are the diffusion constants,
E	is the electric field ($E = -\nabla \psi$),
q	is the elementary charge (absolute value),
ε	is the dielectric constant.
ρ	is the mass density (= 2328 kg/m^3 for Si),
c	is the specific heat (= 703 m^2/s^{2}K for Si),

χ is the heat conductivity,

H   is the heat generation, and

t   is the time.

The transport equations (5) and (6) can be substituted into the continuity equations (3) and (4) in order to obtain the four equations needed for the unknown variables, ψ, n, p and T.

Several models for the physical parameters $\mu_{n,p}$, R, $D_{n,p}$, H and χ have been suggested in the literature; a survey can be found in [4]. In our work, we have mainly used standard models; the details are given in [5]. Note that the heat transport equation and the mechanism of impact ionization are optional features in the simulation program we used. They were included in the simulations only when relevant for the phenomena under investigation (i.e. for the bipolar transistors in contrast to many of the thyristor simulations).

In order to study current filamentation, the system of non-linear partial differential equations (2)-(7) must be solved in at least two spatial dimensions under transient conditions. To this end, we have used an extended version of the well known device simulation program PISCES[6]. In addition to solving the equations above for the semiconductor device, this program also allows for the incorporation of a general external circuit connected to the device electrodes, a feature that we will use below.

3. SIMULATION OF THERMAL SECOND BREAKDOWN IN BIPOLAR TRANSISTORS

Figure 3 shows a typical current-voltage characteristics of the second breakdown phenomenon in a bipolar transistor[7]. Under normal conditions, the transistor characteristics do not contain a negative differential conductivity (NDC) branch. However, when the device is driven into second breakdown, an unstable SNDC branch occurs and, in general, the device is then destroyed by local over-heating.

A number of investigations have been carried out on the subject[8-12]. Two causes have been discussed in the literature, heating (Thermal Second Breakdown, TSB) and impact ionization at the collector substrate boundary (Current mode Second Breakdown, CSB). The CSB occurs in the high current, forward bias region whereas the TSB occurs at lower currents but at high voltages. They are often referred to as forward bias and reverse bias second breakdown.

Various models to describe the phenomenon have been suggested. However, to our knowledge, however, neither the TSB nor the CSB have ever been grasped with a complete physical model containing both a full carrier transport description and a heat transport equation. We will here present a complete simulation of the TSB phenomena, showing the

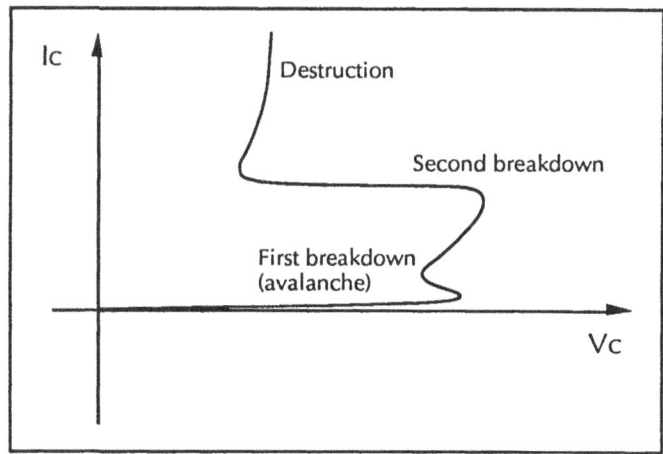

Fig. 3. Characteristic current-voltage waveform during thermal second breakdown.

cause of the SNDC and the formation and development of a filament in a simple case.

3.1 One-Dimensional Simulation of Thermal Second Breakdown

To begin with, we simulate a one-dimensional transistor structure in order to investigate the behaviour in the thermal second breakdown regime. We apply a voltage ramp to the series connection of the device and a resistor with the configuration chosen such that the device is driven into avalanche breakdown while the resistor limits the current to approximately 100 A/cm^2.

We use a standard power-transistor structure for the simulations (Fig.4). The thermal resistance on the contacts were chosen to be 0.4 Kcm^2 W^{-1} (emitter) and 0.8 Kcm^2/W (collector). The high injection carrier lifetime was chosen to be $\tau_n + \tau_p = 1.0$ μs.

The current as a function of the voltage over the device is shown in Fig. 5. The voltage as a function of time is shown in Fig. 6. The variations of potential, temperature and hole concentration across the device are shown in Figs. 7 and 8 for selected successive time-points, as labeled in Fig. 5.

When the voltage reaches the avalanche limit, holes and electrons will be generated at the base-collector junction by impact ionization. These holes supply a base current for the transistor structure and the transistor will start to turn on. This is the "first" breakdown.

When the current is increasing the device starts to heat up. At first, this leads to an increasing resistance of the device, which is due to the negative temperature coefficient of the impact ionization and the mobility.

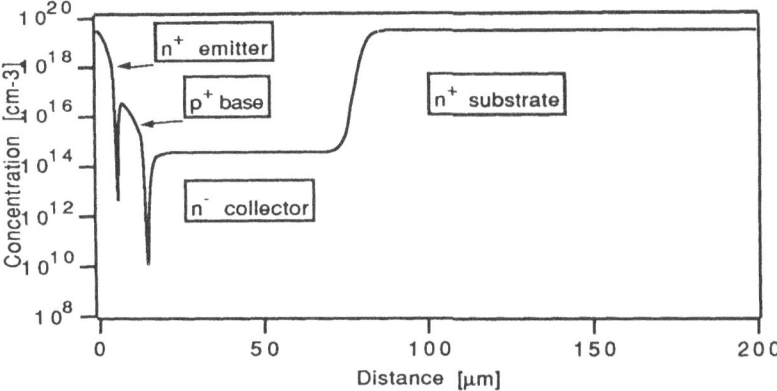

Fig. 4. Doping profile of the transistor used in the simulation.

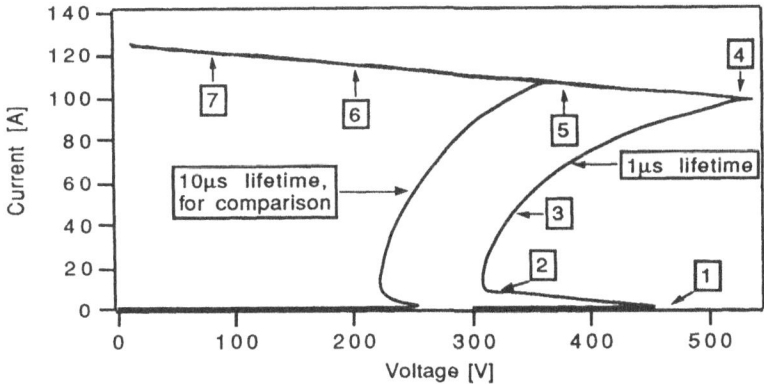

Fig. 5. Current as a function of device voltage for the one-dimensional TSB simulation. The current is normalized to a device area of 1 cm^2. The simulation results for a device with 10 μs lifetime are included for comparison.

However, when the device gets hot enough a second NDC branch occurs. In Fig. 8 we see that, at the turning point of the voltage, label 4 in Fig. 5, the temperature in the transistor collector is around 500 K which is close to the intrinsic temperature of the collector. At this high temperature, we have a considerable amount of thermally generated holes in the n$^-$ collector (Fig. 9) which supply a base current for the transistor. The turning point occurs when the negative temperature coefficient of the hole current generated by impact ionization is surpassed by the positive temperature coefficient of the thermally generated hole current. It is obvious that an NDC branch results because of the positive feedback between transistor current, temperature and thermal generation of holes.

Thus the one-dimensional simulation shows that above a certain critical temperature an unstable NDC branch occurs in a bipolar transistor, caused by thermal generation of minority carriers in the

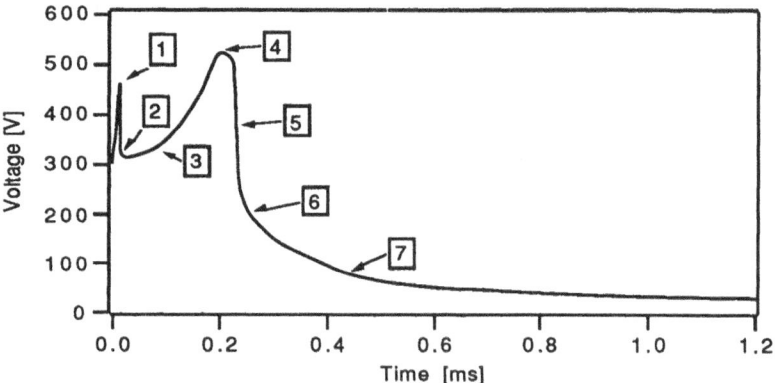

Fig. 6. Device voltage for the one-dimensional TSB simulation.

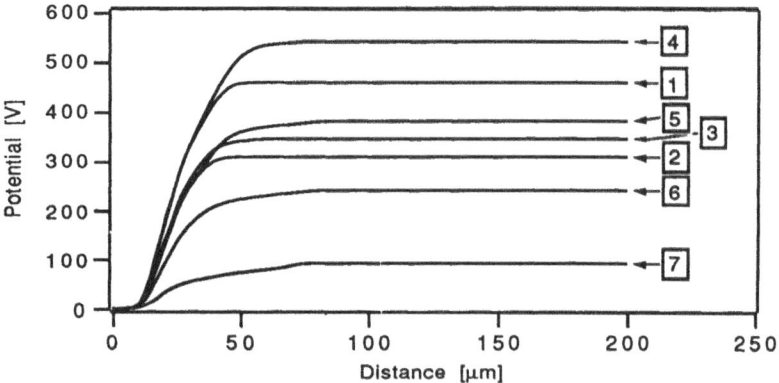

Fig. 7. Potential distribution across the transistor for the time points labeled in Fig. 5 and 6.

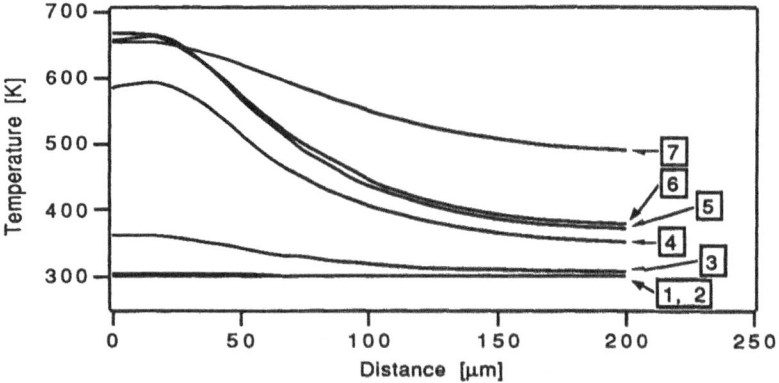

Fig. 8. Temperature distribution across the transistor for the time points labeled in Fig. 5 and 6.

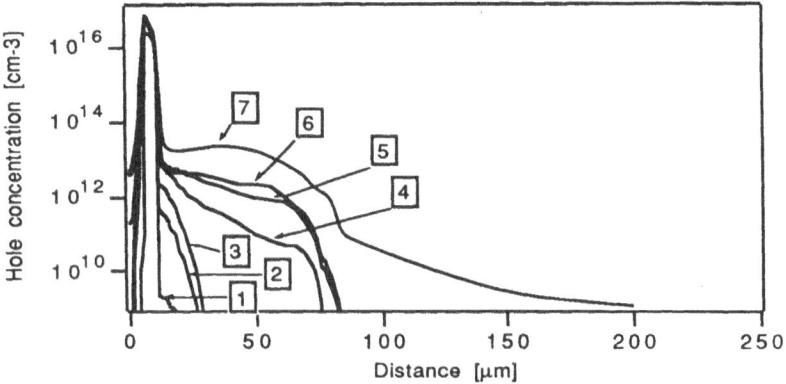

Fig. 9. Hole concentration across the transistor for the time points labeled in Fig. 5 and 6.

collector. The particular form of the current and voltage waveforms during first and second breakdown of a transistor are quite sensitive to a number of parameters, like the carrier lifetime (compare Fig. 5) and the external conditions under which the device is driven into breakdown (see [5]). The general feature of a first avalanche breakdown and a second thermal breakdown are always present, though. Furthermore, we must keep in mind that, since the temperature goes up to almost 700 K (in the next section even 800 K), the accuracy of the physical models may become poor. Hence, when themperature becomes very high the simulations may deviate considerably from a real case.

3.2 Two-Dimensional Simulation of Thermal Second Breakdown; Structure Formation

The positive feedback between current and temperature in the NDC branch and the fact that, when a part of the device entered NDC, the voltage over the device will start to decrease, preventing other parts of the device of entering NDC, makes it is intuitively clear that a current filament can occur at TSB. In [5], an analytical stability analysis, based on the one-dimensional simulations, has been performed. It shows that, with the boundary conditions above (series resistance etc.), a range of inhomogeneous modes will become unstable while the homogeneous mode is still stable, i.e. that a development of spatial structures will occur. Using a simple two-dimensional structure, we can perform a complete simulation of such a structure formation.

We use the same conditions and doping profile as above (Fig. 4). However, we now give the device a lateral width in order to allow for a development of possible structures. We choose a device width of 550 μm and, as we shall see below, this is enough to enable us to see a development of spatial structures. The contacts were extended over the whole

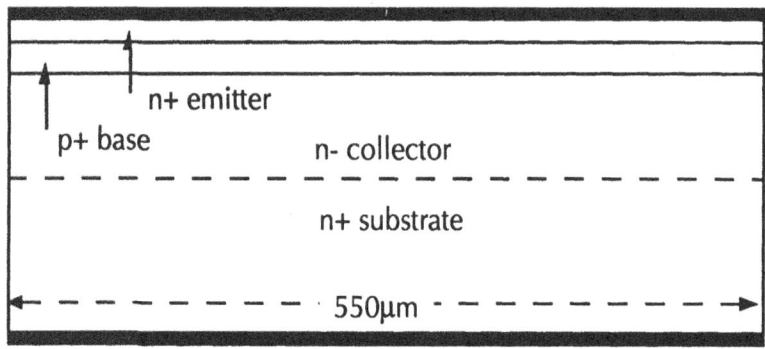

Fig. 10. Device structure for the two-dimensional investigation.

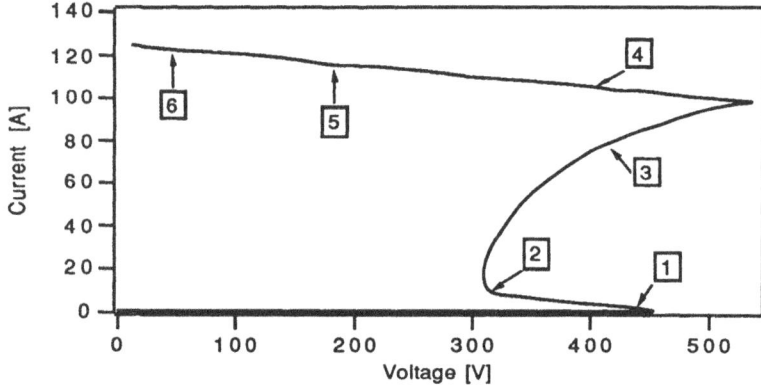

Fig. 11. Current as a function of device voltage for the two-dimensional TSB simulation (identical to the one-dimensional case).

width of the device, i.e. the geometry and doping are homogeneous (Fig. 10). However, we had to make one concession to the mesh-generation: We introduced a small (10 %) perturbation in the collector doping on one side of the structure in order to create a small initial inhomogeneity (see figures below) and thereby control the location where a possible filament would develop.

Figure 11 shows the current-voltage characteristics and Fig. 12 shows the device voltage as a function of time for a two-dimensional simulation as compared to the one-dimensional case from above.

To see if and when a structure formation occurs, we plot for the successive time-points, labeled in Fig. 11 and 12, the magnitude of the current density and the temperature along a lateral cross-section of the device. The points evaluated are in the collector of the transistor, at a distance of 60 μm below the emitter contact. In these plots (Fig. 13 and 14), we clearly see the development of a spatial structure, leading to a current filament. Figures 15 and 16 show the two-dimensional potential and temperature distributions for a number of selected time-points.

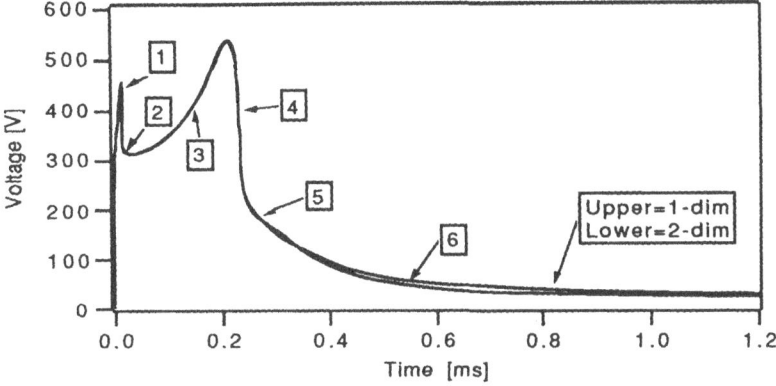

Fig. 12. Device voltage as a function of time for the two-dimensional TSB simulation. For comparison, the one-dimensional curve is also included.

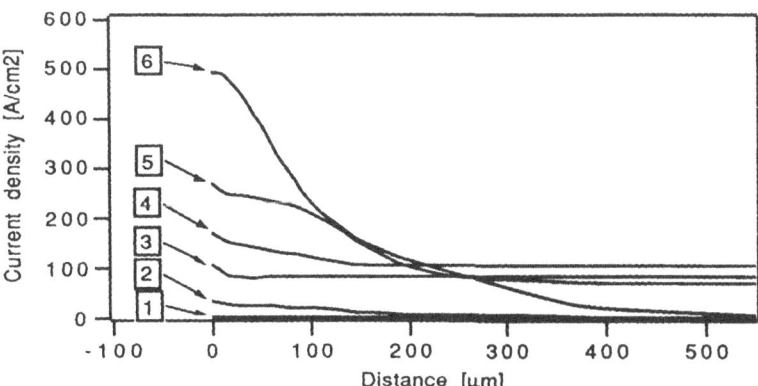

Fig. 13. Magnitude of the current density at a lateral cross-section of the device, 60 μm below the emitter surface, for successive time-points as labeled in Fig. 11 and 12.

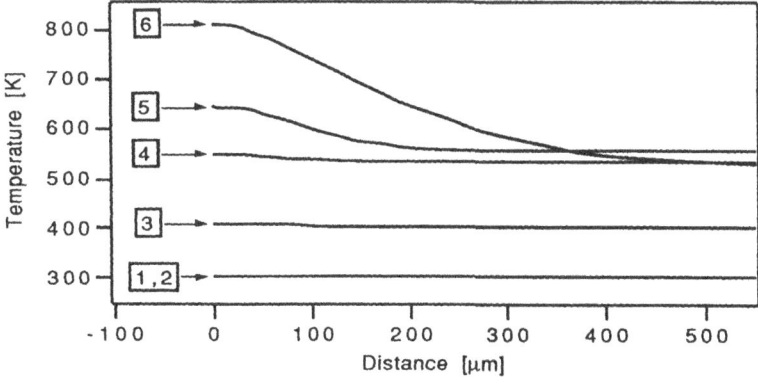

Fig. 14. Lateral temperature distribution in the collector, 60 μm below the emitter surface, for successive time-points as labeled in Fig. 11 and 12.

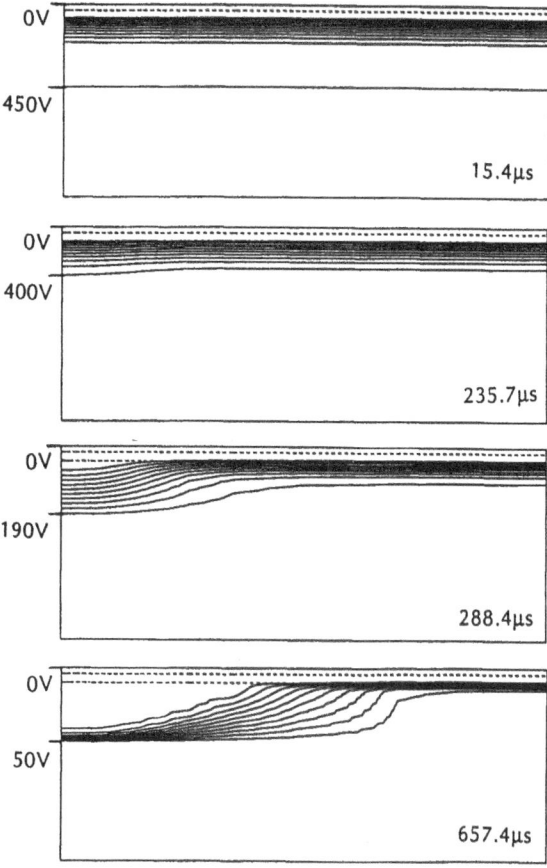

Fig. 15. Equipotential lines in the device at the time-points 1 (15.3 µs), 4 (235.7 µs), 5 (288.4 µs), and 6 (657.4 µs). At time point 6, the spatial structure, called current filament, is fully developed.

We have now seen that the thermal second breakdown is structurally unstable and will lead to the formation of current filaments, and we have presented a complete (two-dimensional) simulation of the occurrence of spatial structures and their development to current filaments. However, comparing our results to a real device, we must keep in mind that a very simple device structure has been used and, as mentioned above, the validity of the physical models is quite uncertain at the very high temperatures occuring in the later part of the simulation. In particular we note that, in real devices, the development from the onset of second breakdown to destruction of the device by hot spots, is a much faster process (µs range) than the filament development in the later part of our model simulation.

Nevertheless, our simulations show that numerical simulation can be used to understand and describe the instability phenomenon occuring at second breakdown. Extending the investigation to more realistic device structures an investigation of the dependence of the device stability on a

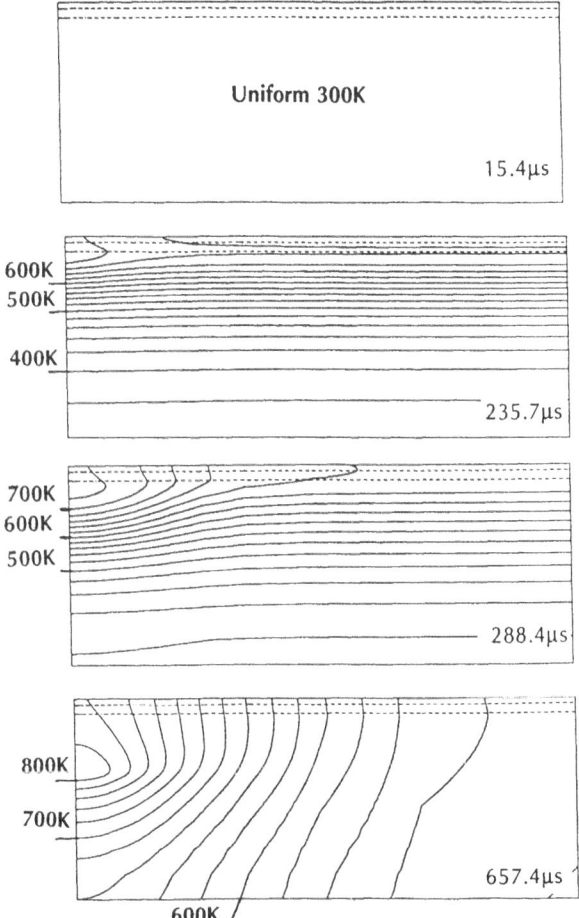

Fig. 16. Temperature distribution in the device at the time-points 1 (15.3 µs), 4 (235.7 µs), 5 (288.4 µs), and 6 (657.4 µs). At time point 6, the spatial structure, called current filament, is fully developed.

number of factors, such as geometry, lifetimes, doping, etc., would be possible.

4. CURRENT FILAMENTATION IN GATE TURN-OFF THYRISTORS

A device with four doping-layers has an S-shaped current-voltage characteristics, even under normal operating conditions. It has a stable high voltage/low current "off" branch and a low voltage/high current "on" branch. They are connected by an unstable NDC branch. Since we know that an S-shaped NDC characteristics can give rise to current filamentation, it is clear that the thyristor must be "handled with care".

Thyristors which can be turned off by a gate pulse (Gate Turn-Off thyristor, GTO[13], and MOS Controlled Thyristor, MCT [14,15]), are particularly sensitive to the current-filamentation problem. Measurements have shown that a very inhomogeneous current distribution develops during the turn-off [15-17].

We have used numerical simulation to get a detailed understanding of the stability of the turn-off process in these devices. We will discuss the influence of the gate drive conditions, of the device structure, and of protection circuitry with respect to the occurrence of current instabilities.

4.1 Methods of Investigation

Characteristically, the cathode of turn-off thyristors must be fine-structured and split into small cathode-cells or "fingers" (Fig. 17). The current inhomogeneities occur when the current is redistributed between these cells during the turn-off process.

In our simulations, a GTO structure as shown in Fig. 17 is used. Also the MCT can be modeled as a GTO cell if, at turn-off, the gate is shorted to the cathode over a resistance. This resistance represents the channel resistance of the integrated MOSFET. Note that in the GTO all cells have a common gate contact, whereas in the MCT the gate of each cell is "separately connected" by its own MOSFET. This has no influence on the onset of the current redistribution but it matters subsequently.

In order to simulate the redistribution, we have applied transient two-dimensional device simulation[1] to different cell configurations. First we consider the case of a perturbed thyristor cell in an infinite ensemble of normal cells (Fig. 18). A possibly deviating current in the perturbed cell will in this case not influence the total current through the device. The voltage, evolving across the device during turn-off, will therefore be the same as the voltage over each single normal cell. This situation is modeled by first simulating the turn-off of a "normal" single thyristor cell and then imposing the voltage evolution on the perturbed cell in a second turn-off simulation (anode and gate voltages for the GTO thyristor and anode voltage only when imitating MCT conditions, respectively). If the current of the perturbed cell remains close to that of the normal cell, then a large multiple cell device will be stable and turn off homogeneously. However, if the current of the perturbed cell starts to deviate strongly from the "normal" current, then the situation is unstable and an inhomogeneous current distribution will develop in a multiple cell device.

When the perturbed cell starts to carry more and more current, the total current and thereby also the voltage evolution across the device will eventually be influenced. Furthermore, interaction between neighbouring cells will become important and needs to be considered. This situation cannot be described by the simulation method above, and to

---

[1] In the simulations of gate turn-off thyristors (Section 4.2 and 4.3), the mechanism of impact ionization and heating were not included (i.e. a uniform temperature was used).

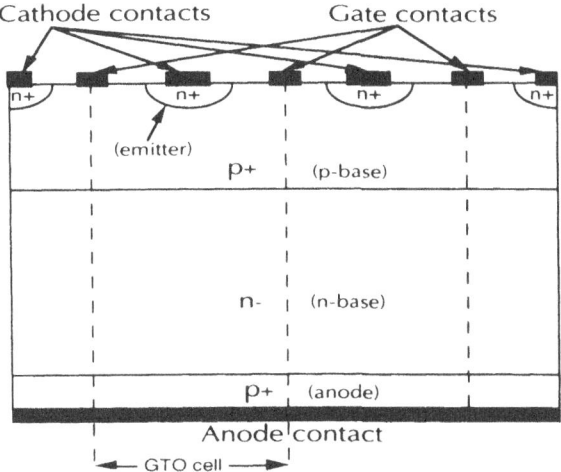

Fig. 17. The Gate Turn-Off (GTO) thyristor.

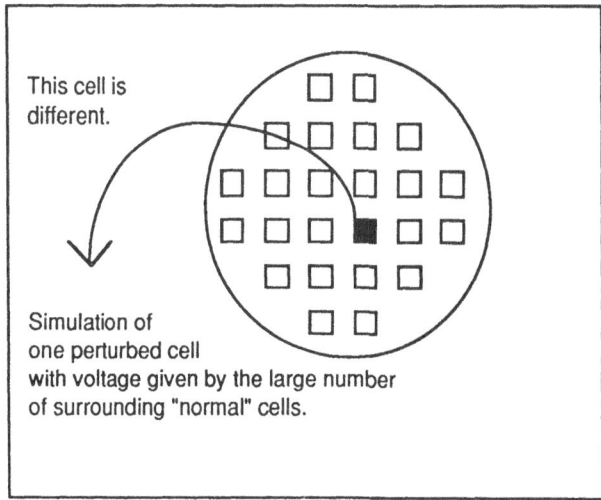

Fig. 18. Simulation-method I

model the continued current redistribution we therefore have to simulate several cells simultaneously. The enormous amount of computer resources required limits us to a very small number of cells. However, we shall see below that, with a somewhat simplified simulation model, we will be able to describe the interaction of any number of cells during a complete (inhomogeneous) turn-off.

Evidently, a real device does not contain one single perturbed cell. The deviation between cells will rather exhibit some unpredictable distribution. This, however, will not influence the validity of our results for the *onset* of an instability. But we have to keep this in mind when we interpret the simulations beyond the onset of the current redistribution.

4.2 Onset of Current Filamentation

In order to describe the general features of the current redistribution, we first discuss the simulation results for a particular device with 5 μm wide cathode n^+ emitters. This device represents a fine-structured GTO and also describes the behaviour of an MCT quite well.

To perturb the cell, we have used slight variations in emitter doping or carrier lifetime. The perturbations chosen lead to a slightly higher on-state current.

To characterize the results on the basis of total currents I, we use an effective turn-off gain, β_{eff}, defined as,

$$\beta_{eff} = \frac{I_{anode,on-state}}{\max\left(I_{gate}\right)}. \tag{8}$$

The significance of β_{eff} is essentially the following: When $\beta_{eff} < 1.0$, the gate current during turn-off can be or is higher than the on-state anode current. This means that the whole device current will flow through the gate electrode during turn-off and that the cathode-current therefore is zero or even reversed, i.e., no electrons are injected into the device from the cathode. If $\beta_{eff} > 1.0$, the gate cannot take all the device current, and the electron injection from the cathode will therefore continue during turn-off.

A normal **single cell** of any of the device-types can be turned off even if $\beta_{eff} \gg 1.0$. However, we shall see below that, within a **large ensemble of parallel cells,** a current filamentation instability will develop whenever the cathode can continue to inject electrons during the turn-off process. This means that, for stability, β_{eff} must be at least less than 1.0, i.e., the gate current during turn-off must be higher than the on-state current of the device.

To turn off the device in the simulations in this section, the gate of each cell is instantaneously (within 1 ns) shorted to the cathode by a small resistance. These are the conditions in an MCT device (MOS Controlled Thyristor) where the gate resistance corresponds to the

MOSFET channel resistance. The gate current is in this case limited by the value of the gate resistance and we use it to control the value of the turn-off gain, β_{eff}. The simulations are done with resistive load.

The turn-off sequence for a single cell with $\beta_{\text{eff}} = 1.1$ is shown in Fig. 19. The storage period is short (around 40 ns, see also Figs. 22 and 23) and the tail period is long due to the high carrier lifetimes ($\tau_n + \tau_p = 12.0 \ \mu s$). In all the figures, we have normalized the currents to a total anode area of $1.0 \ cm^2$, i.e., the unit for the current always refers to a device having a total anode area of $1.0 \ cm^2$, regardless of the number of cells in the device.

When we perturb the thyristor cell (here done by increasing the cathode emitter doping) and investigate the stability as described above, we find that the current in the perturbed cell starts to increase uncontrollably (Fig. 20). Considering the whole device, this means that the current starts to redistribute from the normal cells to the perturbed cell. Let us point out here that, in a single-cell simulation, the perturbed cell turn-off will be identical to that of the normal cell.

With a lower value of β_{eff}, however, a current increase does not occur. Instead, the perturbed cell turns off in conjunction with the normal cells; i.e., it is stable and a current redistribution will not occur during turn-off. Figures 21-23 show a number of simulations with different β_{eff}. From the figures it can be seen that for 150 amp/cm^2 we have stability when β_{eff} is less than about 0.9 ($\beta_{\text{crit}} = 0.9$) and for 350 amp/cm^2, β_{crit} is about 0.76.

The reason for the instability is due to the fact that the electron injection from the cathode of the perturbed cell starts to increase. This increase occurs because *a perturbed cell, lagging behind its normal neighbours in the turn-off sequence, will experience a "forced" increase in anode voltage at the end of the storage period when the voltage of the normal cells starts to rise.*

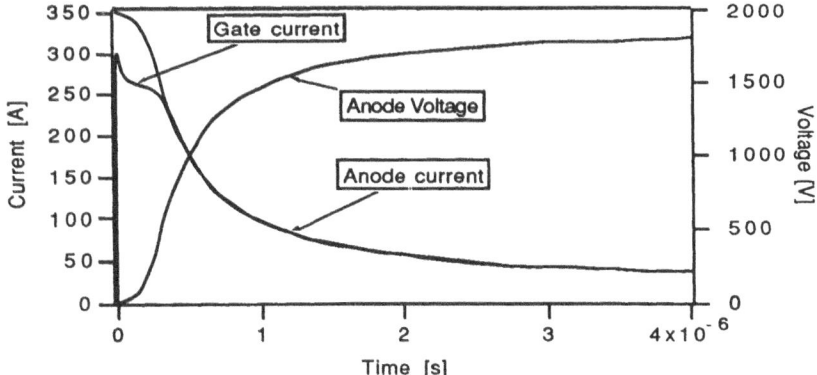

Fig. 19. Current and voltage waveforms during turn-off of an unperturbed cell of device-type 1, with $\beta_{\text{eff}} = 1.1$ and $J_{\text{on}} = 350$ amp/cm^2 (in all figures, the current has been normalized to a total anode area of $1.0 \ cm^2$).

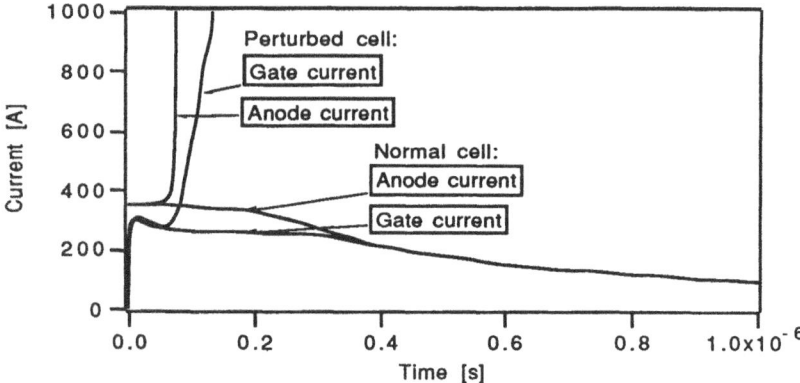

Fig. 20. Comparison between the turn-off sequence of the normal cell from Fig. 19 and a perturbed cell, turned off as described in method 1.

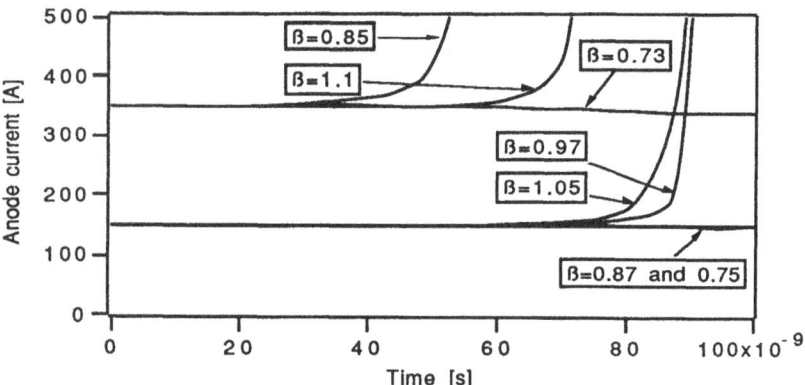

Fig. 21. Anode currents of perturbed cells (method 1) of device-type I. The curve for $\beta_{eff} = 1.1$ is the sequence from Fig. 20.

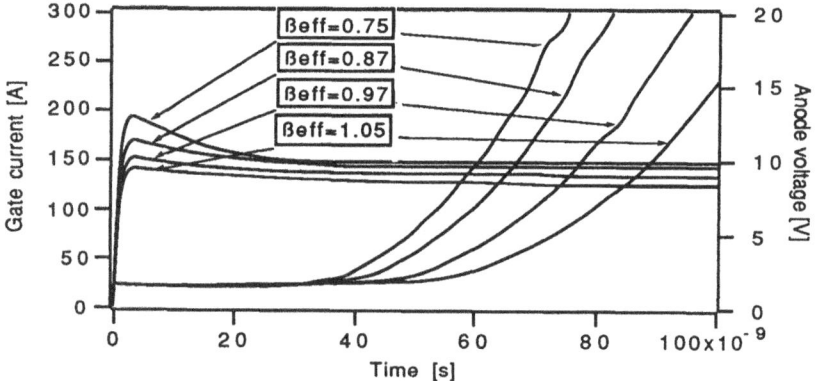

Fig. 22. Gate currents and anode voltages of the normal cell for the sequences from Fig. 21 (150 amp/cm^2). The anode voltage has been calculated with the normal cells and imposed on the perturbed cells.

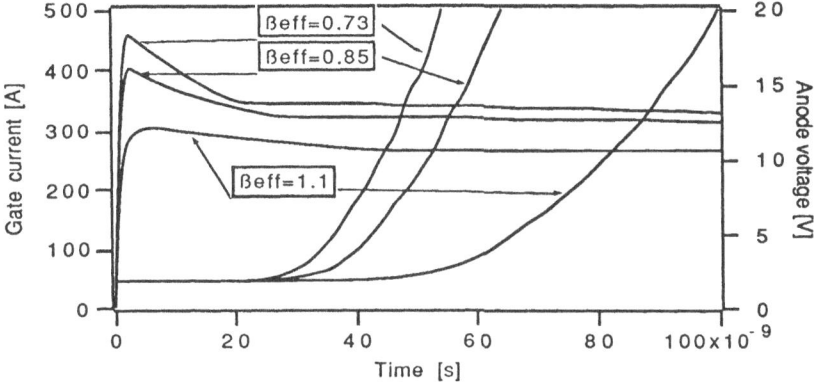

Fig. 23. Gate currents and anode voltages of the normal cell for the sequences from Fig. 21 (350 amp/cm^2).

The amount of electron injection from the cathode, is given by the voltage across the cathode - p-base junction, $V_{p\text{-base}}$, by a "diode characteristic",

$$I_{cathode} = I_0 \left[\exp\left(\frac{qV_{p-base}}{k\,T} \right) - 1 \right]$$

(9)

where k is the Boltzman constant and I_0 the diode saturation current.

The current through the gate is also determined by $V_{p\text{-base}}$, but the dependence is linear and not exponential,

$$I_{gate} = \frac{V_{p-base}}{R_{gate}},$$

(10)

where R_{gate} includes both the external gate resistance and an effective internal resistance of the p-base between the gate contact and the emitter.

Figure 24 shows the cathode emitter current and the gate current as functions of $V_{p\text{-base}}$ (schematic). For a given total device current (anode current), $V_{p\text{-base}}$ is given by the requirement that the sum of cathode and gate currents must be equal to the anode current. If the anode current is low (or, equivalently, the gate resistance is low) there will be almost no electron injection from the cathode and the whole current will flow out through the gate (case 1 in Fig. 24). This is the situation when $\beta_{eff} < \beta_{crit}$.

If the anode current is high, however, $V_{p\text{-base}}$ will be higher than the threshold voltage of the cathode - p-base junction, V_{th}, and there will be a considerable amount of injected electrons ($\beta_{eff} > \beta_{crit}$, case 2 in

Fig. 24). The maximum anode current for which there is no electron injection is approximately given by V_{th}/R_{gate}.

The response to the forced voltage increase in perturbed cell will be an increase in the current. Now, if $V_{p\text{-base}}$ is higher than V_{th} ($\beta_{eff} > \beta_{crit}$), the current increase will almost entirely consist of an increase in the cathode electron injection. This electron injection slows down the turn-off process and delays the cell even more. *In this case, the increase in cathode current cannot be controlled by the gate, and the cell will be unstable.*

However, if $V_{p\text{-base}}$ is below the threshold voltage ($\beta_{eff} < \beta_{crit}$) the (small) increase in the current of the perturbed cell is "handled" by the gate and does not lead to any significant increase of the electron injection. Since there is no electron injection, an increase of the current through this cell will speed up the turn-off process and the delayed cell will catch up with its neighbours. *The perturbed cell is stable.* Without electron injection, the turn-off is similar to that of a transistor which is stable.

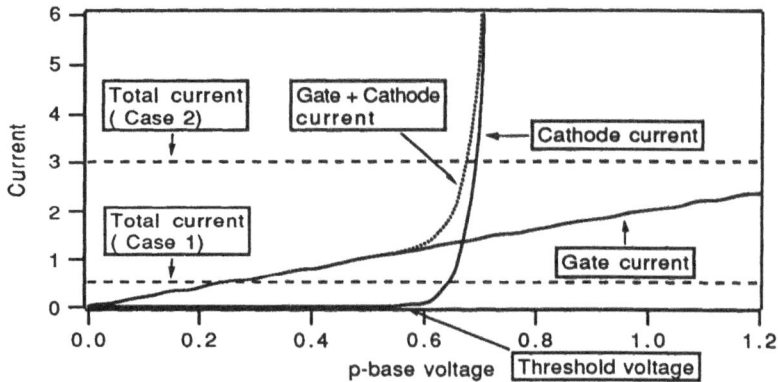

Fig. 24. The figure shows the cathode and gate currents as functions of the p-base voltage (schematic). If the total (anode) current is small the cathode current is negligibly small (case 1). If the total current is high, however, there is considerable injection from the cathode (case 2).

Next, we show a simulation of two neighbouring cells, a normal and a perturbed one, for the unstable case, $\beta_{eff} > \beta_{crit}$ (Figs. 25 and 26). After the end of the storage period, the current in the perturbed cell increases until the cathode-current of the normal cell has fallen to zero. At this instant, the whole (cathode-) current of the device is carried by the perturbed cell. However, since the device consists of only 2 cells, the current is still not too large to prevent the cell from turning off.

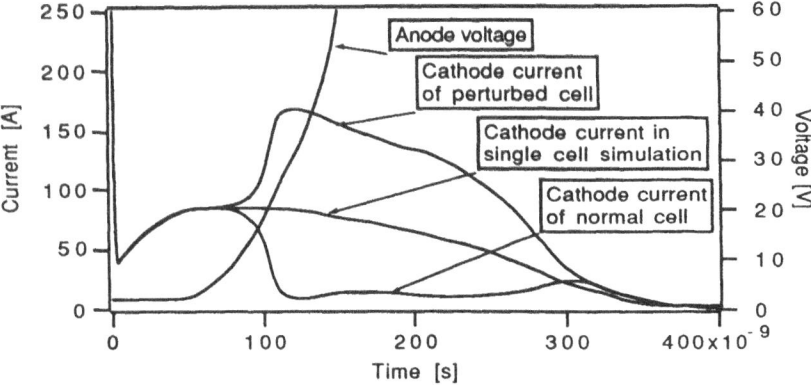

Fig. 25. Cathode currents and anode voltage in a turn-off simulation of two neighbouring cells (a slightly perturbed and a normal one) under the same conditions as used in Fig. 19. The cathode current from the single cell simulation in Fig. 19 is included for reference.

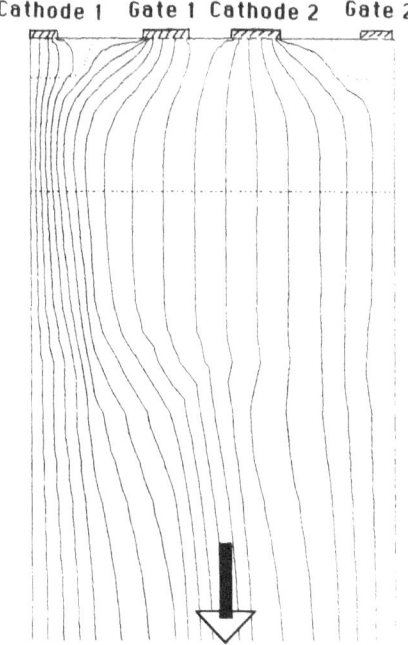

Fig. 26. Current flow lines at 120 nsec for the simulation in Fig. 25. The perturbed cell carries the whole cathode current of the 2-cell device.

We have now seen that an inhomogeneous current distribution will occur whenever the gate current is limited such that part of the device current continues to flow over the cathode during turn-off. The limit for the gate current is given by the total gate resistance, consisting of a device internal part, i.e. the resistance between the gate contact and the center of the emitter, and a part outside the device,

$$R_{gate} = R_{internal} + R_{external}, \tag{11}$$

In the examples above, the internal resistance was small (very small emitters). If the width of the emitters is increased, however, it will become more and more important. In normal GTO's, the gate current is limited solely by the internal resistance, and it is known that the design of the p-base is very important for the GTO.

Note that, at the beginning of the storage period, the internal resistance is modulated by the plasma of free carriers. During the storage period, however, the free carriers are removed from the p-base region and the internal resistance will increase. This means that, even if the maximum gate current exceeds the anode current at the beginning of the turn-off sequence, the gate current can be limited to a value lower than I_{anode} at the end of the storage period, which, as we have seen, leads to an instability.

A more detailed discussion of the influence of the device geometry is given in [18], where the effects of other parameters, such as delay times and type of perturbations, are discussed also.

4.3 Development of a Filament in a Thyristor

So far, we have studied only the onset of the current redistribution. The question remains of what happens after an initial current redistribution has occurred. In this section, we will address this issue and try to answer the following questions:

1.   Does the current redistribution continue beyond an initial phase?
2.   Is there any internal damping mechanism of the current redistribution?
3.   How does a GTO with a standard gate drive ($\beta_{eff} = 5$) finally turn off despite the inhomogeneous current distribution?

Now, we will consider normal GTO conditions in the simulations. The consequences of the results for an MCT will be briefly discussed in the conclusion below. Also, we use a different structure than above with 80 μm wide emitter, representing a standard GTO. We shall include an inductive load with a free wheeling (ideal) diode and a realistic gate drive (see Fig. 27c). Note that we have used an external clamping of the gate voltage (the diode in the gate circuit). In this way, we can model the avalanche breakdown of the gate-cathode junction without including the impact ionization mechanism, which in these types of simulations requires a lot of extra calculation time. The snubber protection in Fig. 27c is only included in the simulations when explicitly stated.

In order to study the whole turn-off sequence of a large GTO for the inhomogeneous case, it is not enough to simulate one thyristor cell. When an inhomogeneity has developed, the interaction between the cells becomes important, both the internal interaction between neighbours and the effect the inhomogeneous cell has on the voltage of the whole device. Therefore, we need to simulate a number of cells simultaneously.

Consider the device in Fig. 27a. It consists of a high number of cells and, as in the previous sections, we assume that there is one perturbed cell, surrounded by normal GTO cells. Actually, this device is simulated by replacing the (fully) homogeneous part of the device by a single cell (cell 3) multiplied by the number of cells it replaces, as shown in Fig. 27b. Internal interaction of neighbours is considered between the nearest neighbour (cell 2) and the perturbed cell (cell 1). Figure 27c shows how this equivalent device is implemented into our simulation program.

We perturb the cells (slight differences in emitter doping) such that cell 1 has the highest on-state current and that its neighbour has a slightly higher on-state current than the normal cell as well. Thereby, we expect to achieve a situation where, after an initial redistribution, only a few cells carry the current (cell 1 and 2). Then we shall investigate if the current redistribution continues, i.e., if the current will redistribute between cell 1 and 2 as well.

Figures 28 and 29 show the current and voltage characteristics of anode and gate for a turn-off simulation of 12 (10+2) GTO cells. Despite a β_{eff} of around 2.5 where we expect filamentation, the device turns off perfectly.

Figure 30 shows the cathode currents of the different cells. There is significant current redistribution! First, we have an initial current redistribution that transfers the current to the two "worst" cells as intended. However, after this first phase, the redistribution continues between the remaining current carrying cells. The instant at which this starts is when the gate current in the normal cells begins to fall. Since the total gate current increases still, there will be a strong rise in gate current in the cells that are still on. Figure 31 shows in detail how this gate current increase occurs for the simulation of 12 GTO cells.

When the second last cell has turned off, all the current is carried by the single cell still being on. Since the gate current is redistributed to this last cell also, it is, thus, enabled to turn off when the clamping voltage is reached. The current flow inside the device at two timepoints (marked a and b in Figs. 28, 30 and 31) is shown in Figs. 32 and 33.

However, there is a limit to how much current a single cell may turn-off. The voltage at the gate cannot exceed the breakdown voltage of the cathode-gate junction. When this voltage is reached, the further increase in gate current is delivered by an avalanche current which does not contribute to the turn-off of the cell. Therefore, when the current of the last cell becomes too high, it will not be able to turn off, as we shall see below.

In the present simulations we investigate only the mechanisms which are not related to heating or dynamic avalanche (impact ionization created by a dynamical increase in the electric field). We must therefore keep in mind that, when the final current redistribution takes place at

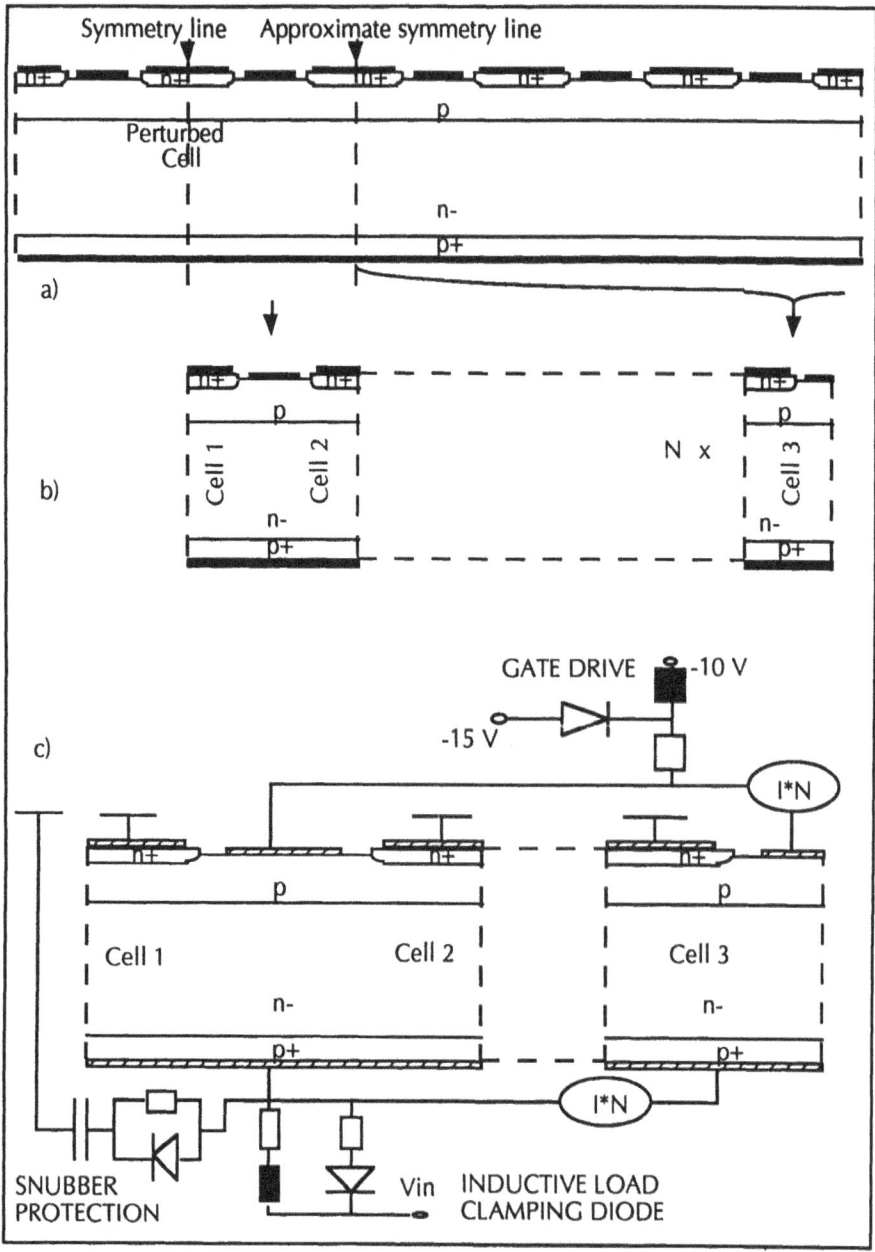

Fig. 27. (a) Part of the multi-cell GTO under investigation. (b) Equivalent device, taking advantage of the symmetry lines. Interaction between cell 2 and its normal neighbour is neglected. (c) Implementation of the equivalent device in the simulation program and the external circuit. The component (I * N) is a current multiplier which multiplies the current from the normal cell by a factor N before it flows into the external circuit. Note that snubber protection is included only in some of the simulations.

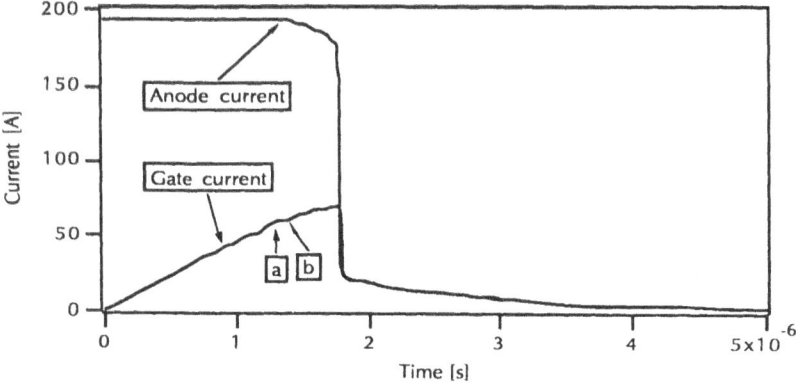

Fig. 28. Anode and gate currents of a turn-off of a GTO consisting of 12 cells. The points a) and b) mark the time-points for Figs. 32 and 33.

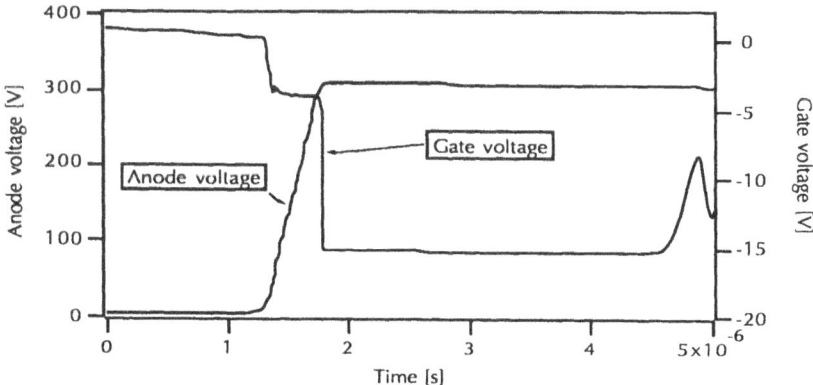

Fig. 29. Anode and gate voltages of the turn-off of the GTO consisting of 12 cells.

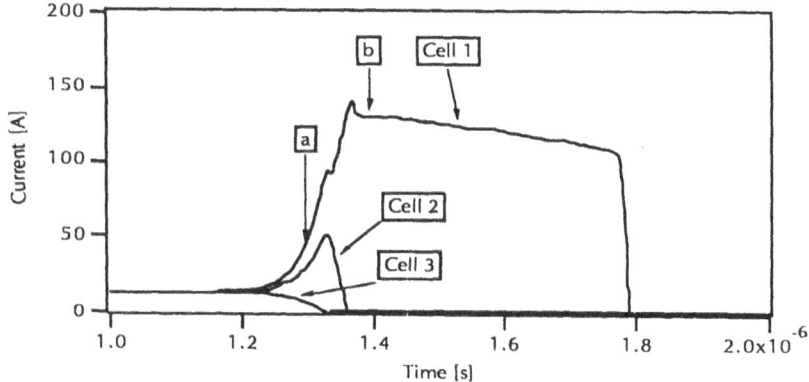

Fig. 30. Cathode current distribution between the cells during the turn-off simulation from Figs. 28 and 29. The points a) and b) mark the time-points for Figs. 32 and 33.

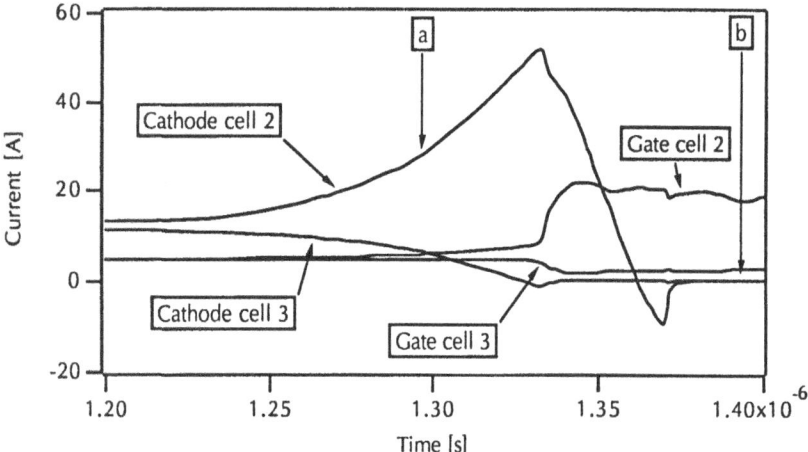

Fig. 31. Detail of the gate and cathode current characteristics for the turn-off from Figs. 28-30. Cell 1 and 2 have their gate contact in common and we can therefore not separate the gate currents of cell 1 and 2. The current denoted "Gate cell 2" in the figure is the sum of the gate currents from cell 1 and 2 divided by two.

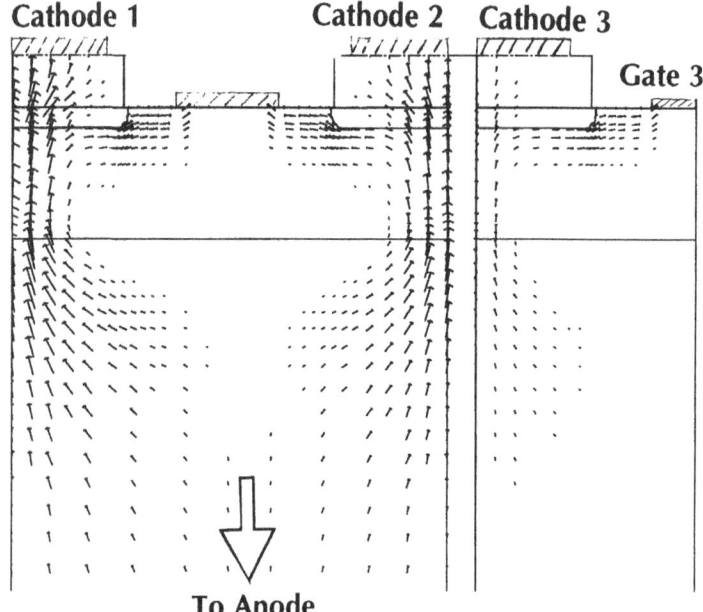

Fig. 32. Total current distribution at the time 1.30 μs (marked a. in Figs. 28 and 30) in the turn-off sequence of a 12 cell GTO. The arrow gives the direction of the current, the length of the arrow is proportional to the magnitude of the current. At this instant, the current is fairly equally shared between cell 1 and cell 2, whereas the normal cell 3 has a much lower current.

very high voltages, our simulations decribe only a generic behaviour and that a real device will experience local heating and dynamic avalanche under some of the conditions we simulate below.

Figures 34-36 show the results of a simulation where the total number of cells has been increased to 102 (100+2). Here, a similar current redistribution as in the previous case occurs, but the current level is too high for the last cell to turn off.

The simulations do not show any mechanisms which damp the current redistribution. However, the redistribution occurs at higher and higher voltages and the increase in anode voltage, dV/dt, becomes steeper as the current level of the remaining current carrying cells increases. Figures 37-39 show that the final current redistribution in the 102 cell device occurs at a higher anode voltage and a higher dV/dt than for the 12 cell device.

The higher voltage is necessary since a higher current must pass through the cell. Furthermore, because of the negative gate voltage, the resistance of a cell will be higher than in the device on-state, and an anode voltage much higher than the corresponding on-state voltage will be needed to force a high current through the cell.

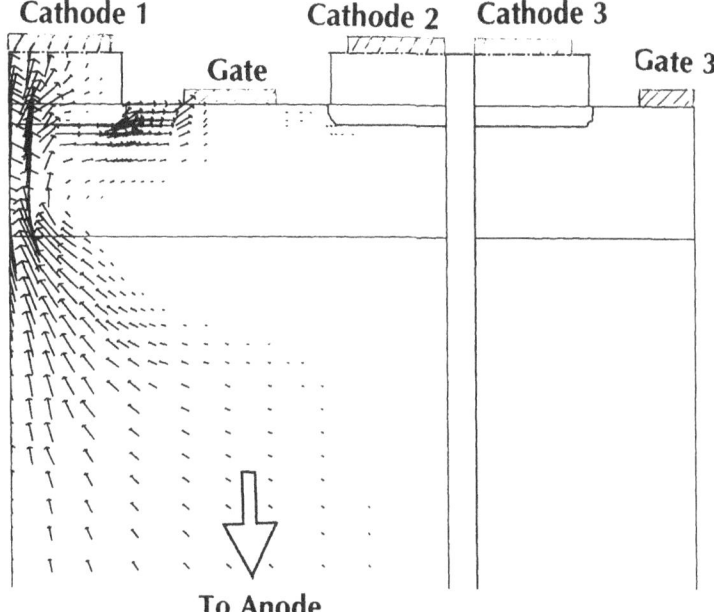

Fig. 33. Total current distribution at the time 1.38 μs (marked b. in Figs. 28 and 30) in the turn-off sequence of a 12 cell GTO. At this instant, the current is completely carried by cell 1. Cell 3 is off whereas in cell 2 some small current still is flowing through the gate.

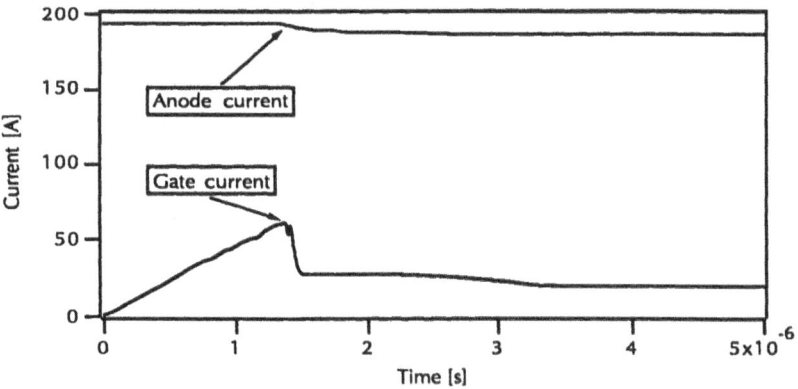

Fig. 34. Currents characteristics of a turn-off of a GTO consisting of 102 cells.

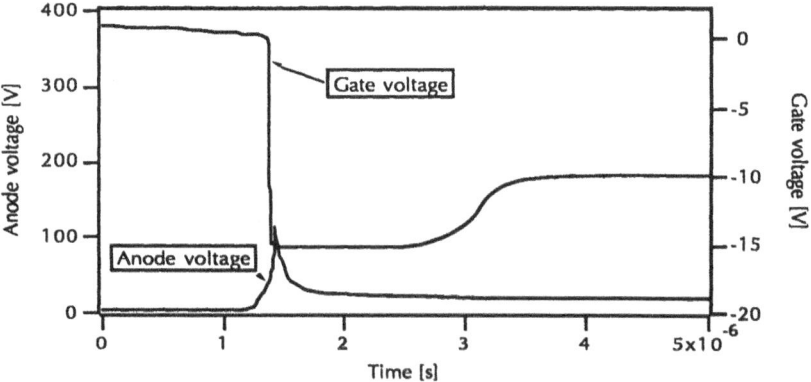

Fig. 35. Voltage characteristics of a turn-off of a GTO consisting of 102 cells (corresponding to Fig. 34).

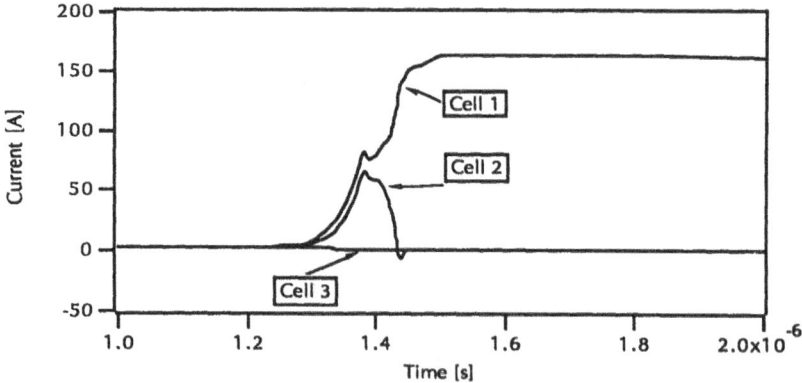

Fig. 36. Cathode current of the different cells during the turn-off simulation (Fig. 34, 102 GTO cells).

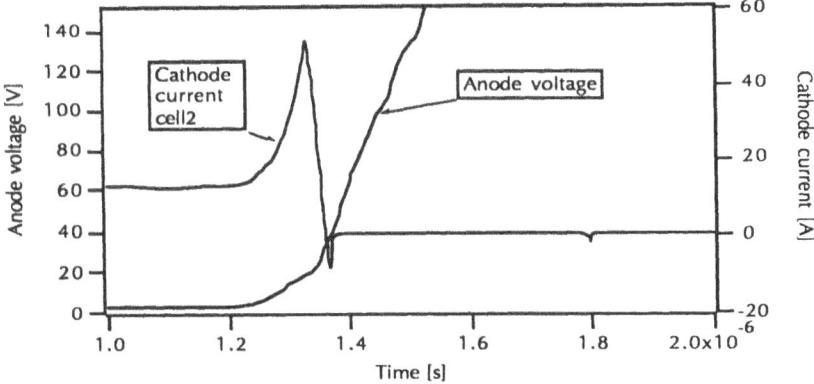

Fig. 37. Detail of the characteristics of the 12 cell turn-off (Fig. 30). The current redistribution between cell 1 and 2, i.e., the period when the cathode current of cell 2 falls, occurs at an anode voltage between 20 and 40 V.

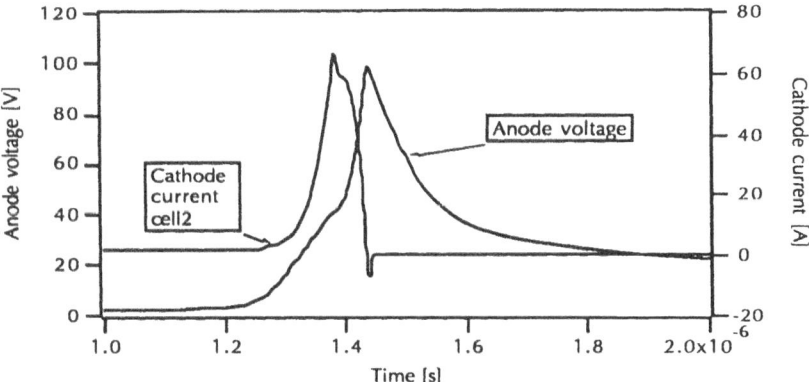

Figure 38. Detail of the characteristics of the 102 cell turn-off (Fig. 36). The current redistribution between cell 1 and cell 2, i.e., the period when the cathode current of cell 2 falls, occurs at an anode voltage between 50 and 100 V, which is significantely higher than for the 12 cell device (Fig. 37).

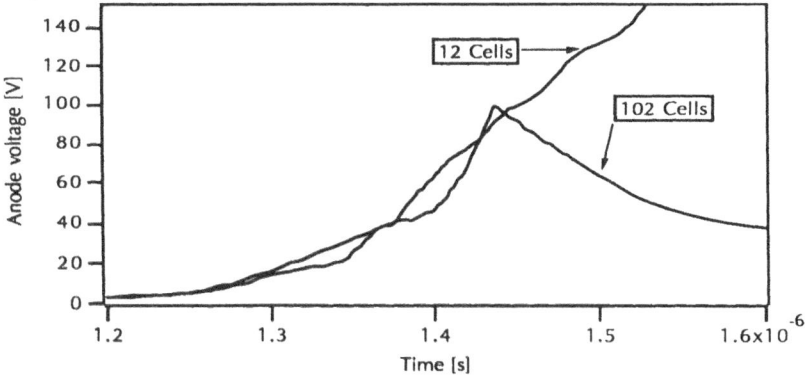

Fig. 39. Comparison of anode voltages between the 12 and the 102 cell GTO turn-off. A much higher dV/dt occurs in the 102 cell GTO in the final redistribution phase.

The increase in dV/dt is a consequence of an acceleration of the whole turn-off process at higher currents. This depends on a number of factors, but the main reason is that a higher current removes the plasma of free carriers much faster. A comparision of Figs. 37 and 38 demonstrates this effect very clearly: when the current in the cell is higher, the dI/dt during the fall time is increased.

A consequence of the last observation is that if the voltage-rise, dV/dt, is limited, the current redistribution will be interrupted. The familiar capacitive snubber protection circuitry, used in all GTO applications has exactly this effect. The parallel capacitor will limit the voltage increase over the switching device.

If we apply a small protection snubber circuit to the 102 cell GTO in our simulations, the current redistribution is interrupted by the snubber and the device can turn-off safely (Figs. 40 and 42).

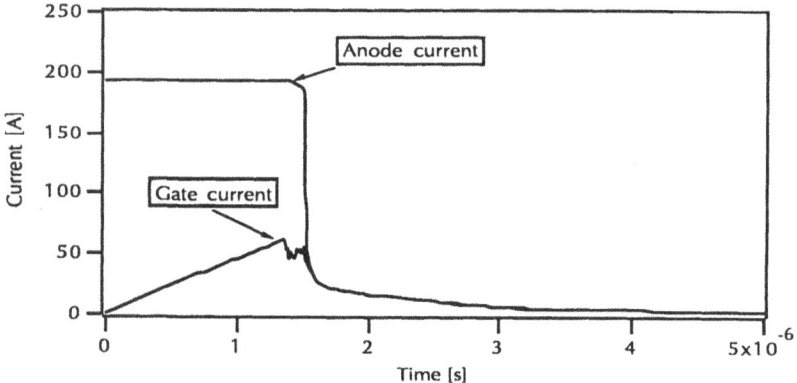

Fig. 40. Currents characteristics of a turn-off of the GTO consisting of 102 cells with a small capacitive snubber circuit (70 nF).

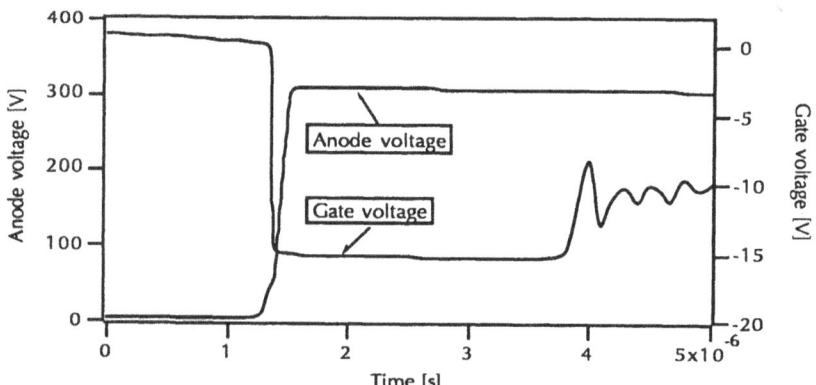

Fig. 41. Voltage characteristics of a turn-off of the GTO consisting of 102 cells with a small capacitive snubber circuit (70 nF).

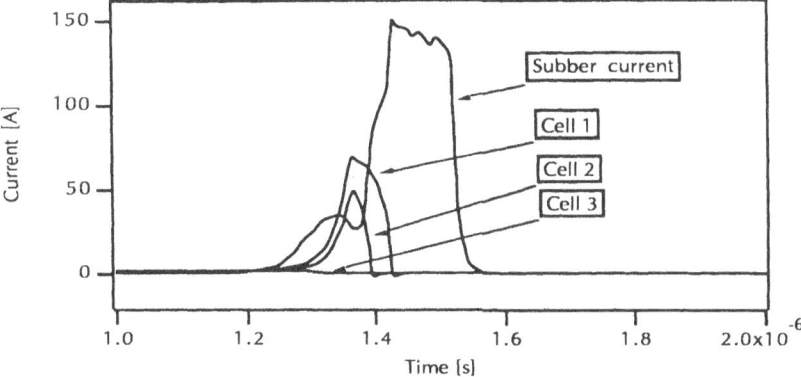

Fig. 42. Cathode current distribution between the cells during the turn-off simulation from Figs. 40 and 41 (102 GTO cells, snubber). The snubber interrupts the current redistribution and turn-off becomes possible.

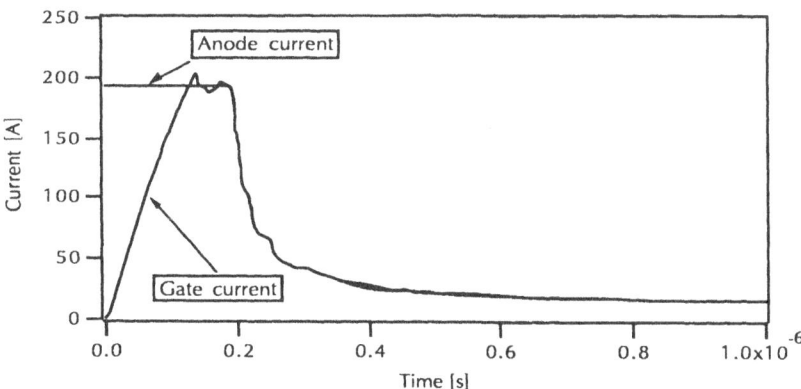

Fig. 43. Currents characteristics of a turn-off of the GTO consisting of 102 cells when $\beta_{eff} < \beta_{crit}$.

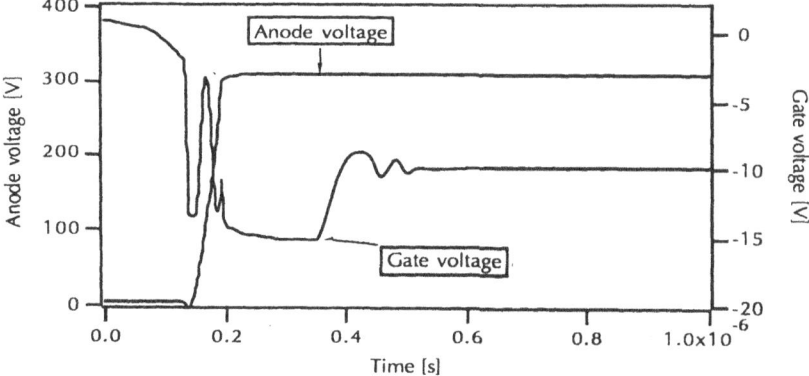

Fig. 44. Voltages characteristics of a turn-off of the GTO consisting of 102 cells when $\beta_{eff} < \beta_{crit}$.

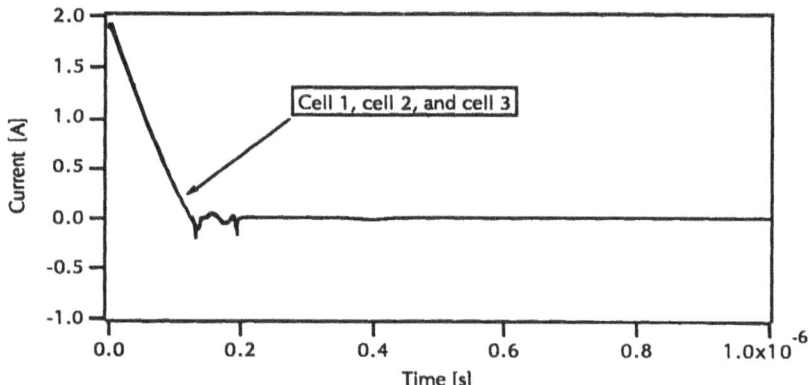

Fig. 45. Cathode current distribution between the cells during the turn-off simulation from Figs. 43 and 44 (102 GTO cells, $\beta_{eff} < \beta_{crit.}$). The turn-off is completely homogeneous.

Finally, in this section we check that the stability criterion from Section 4.2, $\beta_{eff} < \beta_{crit}$, is valid for our 102 cell GTO. The gate inductance is decreased, allowing for a fast current rise in the gate which insures that the whole device current is carried by the gate at the end of the storage period. The results are shown in Figs. 43-45; the turn-off is now stable against filamentation and the turn-off process is homogeneous.

5. CONCLUSION

We have shown two examples of how numerical device simulation can be used to understand and model the current filamentation instability in power semiconductor devices.

In Section 3, we described the thermal second breakdown (TSB) in bipolar transistors. We gave a brief explanation of its causes and showed a complete numerical simulation of the development of a filament in a simple case.

In Section 4, we gave a detailed analysis of the current filamentation in gate turn-off thyristors (GTO, MCT). These devices have a fine-structured cathode, consisting of a large number of cells (or stripes). During the turn-off process, the current is redistributed among these cells and, eventually, the whole device current will be carried by only a few cells.

We first simulated the onset of the current redistribution and explained how and why it occurs. We also gave a criterion for stability by showing that, if a very hard gate drive is used, giving a turn-off gain less than 1.0 (i.e. the gate current is equal to or exceeds the anode current during the turn-off process), the current redistribution does not occur.

We also simulated the turn-off process beyond its initial phase and showed how a current filament develops in a multiple cell GTO structure. We found no internal mechanisms damping the process. The current redistribution does not stop after an initial phase; it continues between the cells that are still on. Such a continuing redistribution, however, requires an increased rise in anode voltage. Therefore, if the anode voltage-increase is limited by external protection circuitry (capacitive snubber circuit), the current redistribution will be interrupted.

When an inhomogeneous current distribution has developed, some cells carry a considerable amount of excess current. However, since the gate current is also distributed to these cells, they will still be able to turn-off (unless the excess current is to high). In Section 4.3, we showed simulations illustrating and verifying this.

The redistribution of the gate current is possible because all the cells in a GTO have a common gate contact. In the MCT device, however, the gate of each cell is separately connected by an integrated MOSFET and, therefore, the *gate current* can only be redistributed between nearest neighbouring cells. Thus, the MCT will be more sensitive to current redistribution than the GTO, and protective snubber circuits will not be as effective. On the other hand, the stability criterion ($\beta_{eff} < 1.0$) is easier fulfilled in the MCT since the (thyristor-) gate current flows through the integrated MOSFET and will not be handled by an external circuit (a very fast rise of the MOS-gate voltage is required though[5]).

REFERENCES

1   H. Haken ed., *Cooperative Effects, Progress in Synergetics*, North Holland, Amsterdam, 1974

2   E. Schöll, *Nonequilibrium Phase Transitions in Semiconductors*, Springer-Verlag, 1987

3   J.B. Gunn, "Microwave oscillation of current in III-V semiconductors", *Solid State Commun.*, vol. 2, p. 88, 1963

4   S. Selberherr, *Analysis and Simulation of Semiconductor Devices*, Springer Verlag, Wien, 1986

5   K. Lilja, "Analysis and numerical simulation of current filamentation in power semiconductor devices," Ph.D. Thesis, Swiss Federal Institute of Technology, Zürich, to appear

6   M. Pinto, C. Rafferty, R. Dutton, *PISCES-II: Poisson and Continuity Equation Solver*, Stanford University, 1984

7   S. M. Sze, *Physics of Semiconductor Devices*, John Wiley & Sons, 1981

8   H. A. Schaft, "Second breakdown - a comprehensive review",
    Proceedings of the IEEE, vol. 55, pp. 1272-1288, 1967

9   L. Dunn, K. I. Nuttall, "An investigation of the voltage sustained
    by epitaxial bipolar transistors in current mode second
    breakdown," *Int. J. Electron.*, vol. 45, p. 353, 1978

10 H. Melchior, M. J. Strutt, "Secondary breakdown in transistors",
    Proceedings of the IEEE, vol. 52, p. 439, 1964

11 P. Hower, G. Reddi, "Avalanche injection and second breakdown
    in transistors," *IEEE Trans. Electr. Devices*, vol. ED-17,pp. 320-
    335, 1970

12 M. Jovanovic´, "A transistor model for numerical computation of
    forward-bias second-breakdown boundary," *PESC´90 Record,
    Proceedings of the Power Electronics Specialist Conference*,
    pp. 154-161, 1990

13 see e.g. A. Blicher, *Thyristor Physics*, Springer Verlag, 1976

14 V. A. K. Temple, "MOS-controlled thyristors - a new class of
    power devices," *IEEE Trans. El. Dev.*, ED-33, 10, pp. 1609-1618,
    1986

15 M. Stoisiek, K. Oppermann, G. Wachutka, "Turn-off behaviour of
    GTOs with small emitter elements," *AEÜ*, Band 43, pp. 320-327,
    1989

16 P. Palmer, C. Johnson, "Measurement of the redistribution of
    current in GTO thyristors during turn-off," *European Power
    Electronics Conference*, Aachen, 1989

17 H. Bleichner, M. Rosling, J. Vobecky, M. Lundqvist, E.
    Nordlander, "A comparative study of the carrier distributions in
    dynamically operating GTO:s by means of two optically probed
    measurement methods", *Proceedings ISPSD´90*, pp. 246-251,
    1990

18 K. Lilja, H. Grüning, "Onset of current filamentation in GTO
    devices," *PESC´90 Record, Proceedings of the Power Electronics
    Specialist Conf.*, pp. 398-406, 1990

DISCUSSION

A. Silard (TU Bucharest, Bucharest R)

   What is the role of temperature during the redistribution of
current at turn-off for a GTO ?

K. Lilja
Temperature was not included in the thyristor simulations. It is expected that the lowering of mobility at increasing temperature will damp the speed of current redistribution. Since the current redistribution occurs in a very short time period, the device cannot be very hot locally.

A. Silard
This argument may be valid for the storage time when there is little power dissipation and temperature rise. Considering the fall time with large power dissipation and fast temperature rise, the stability criteria for gated turn-off are linked to the intrinsic temperature of the lightly doped n-base. When this temperature is reached, microplasmas will form, leading to electrothermal failure. I think if you can stop the redistribution during the storage time, the critical phase of turn-off will take place during the fall time when you have a fast temperature rise. Ohashi and Nakagawa have shown that their devices were destroyed somewhere between 200 and 300 ^{o}C, which is slightly above the intrinsic temperature of the lightly doped n-base.

K. Lilja
Certainly, the temperature affects an instability which has already developed, but in my opinion, it is not the cause of it. Similarly, the temperature rise is not the cause of a current redistribution.

H. Grüning (ABB Corp. Research, Baden)
Field Controlled Thyristors (FCTh) behave like GTO's that the electron injection from the cathode can be cut off at once. The 1.6 cm^2 FCTh devices tested were limited by dynamic avalanche and no thermal problems during the tail phase were observed. The conclusion is that if a GTO exhibits a homogeneous current distribution at the end of the storage time, it will be safe unless it is driven into dynamic avalanche.

D. Silber (TU Bremen, Bremen D)
Were your transistor simulations performed under open base or shunted base conditions ?

K. Lilja
We have restricted ourselves to open base simulations. Work on thermal effects is being started but the physical models for higher temperatures are still very uncertain.

B.J. Baliga (North Carolina State University, Raleigh NC)
Have you also looked at emitter switching which gives very high turn-off gain in contrast to your simulations ?

K. Lilja
The emitter switched thyristor is a very interesting device, but I have not performed such simulations.

COMPARATIVE STUDY OF HIGH POWER DEVICES IN CONVERTER CIRCUITS

H.-Ch. Skudelny, A. Mertens, and J.-G. Langer

Rheinisch Westfälische Techn. Hochschule
Aachen, Germany

ABSTRACT

Power converters have reached a high standard in technical applications, ranging from switched mode power supplies and highly dynamic servo drives in the kilowatt range up to high power drives operating at several megawatts. All these applications are realized on the base of conventional power devices like thyristors, Gate Turn-Off thyristors (GTO's) and transistors. New devices like the Insulated Gate Bipolar Transistor (IGBT), MOS Controlled Thyristor (MCT), Static Induction Thyristor

---

NOMENCLATURE

A	area
I_{TAV}	maximum average current
I_{TGQM}	repetitive controllable turn-off current (GTO)
I_{TRMS}	maximum root mean square value of thyristor current
J	current density
P_c	conduction power loss
P_S	switching power, switching power loss
P_t	total power loss
Q	charge
R_{th}	thermal resistance
T	temperature period
V_{BD}	breakdown voltage
V_d	on-state voltage drop
w_d	drift region
$W_{on,off}$	energy loss
$w_{on,off}$	energy loss per Ampere
τ	carrier lifetime

Power Semiconductor Devices and Circuits, Edited by A.A. Jaecklin
Plenum Press, New York, 1992

(SITh) or Field Controlled Thyristor (FCTh) have recently become available or are in a state of development. The major goals of ongoing innovation in this field are reduction of cost of a semiconductor switch and increase of switching frequency. While the former is achieved by simplification of gate drive, auxiliary circuitry and cooling equipment, the latter necessitates lower commutation losses and leads to a reduction of noise and of the size of passive components. It is the goal of this paper to explore the potentials of improvement that are opened by new semiconductor devices. The discussion is confined to the power range above 10 kVA. In this power range, the voltage fed three phase inverter covers a very broad range of applications and is therefore regarded as a standard topology

Following a brief review of some principles of power semiconductors, the different devices are discussed in detail with special emphasis on the circuit utilization of the device. Among the characteristics that influence switching frequency and cost are switching speed, static and dynamic losses, structure of gate drive circuits and expenditure for snubbers. The possibility of protection against destructive operating conditions is another point of discussion when a choice between different devices has to be made. So-called soft switching converter topologies have been introduced recently with the benefits of higher switching frequency and lower Electro Magnetic Interference (EMI) at the same time. Since this field has found much interest in academic and industrial research, the behavior of semiconductor devices in these new converter circuits is also considered.

In conclusion, the characteristic power and frequency ranges are identified where each of the devices can be used best.

1. INTRODUCTION

Power Converters have reached a high standard in technical applications. Industrial drives in the power range up to several megawatts with d.c. motors as well as with induction motors and synchronous machines are in use. Electrolysis, induction heating and high voltage d.c. transmission (HVDC) are other examples of technical applications. In the lower power range, highly dynamic servo drives are capable of reverting their speed in less than one revolution. Switched mode power supplies can be reduced to almost credit card size by applying switching frequencies in the megahertz range.

All these applications are realized on the basis of conventional power devices like thyristors, GTOs and transistors. Research reports in technical conferences and journals focus on a variety of new devices with exotic names like Insulated Gate Biplar Transistor (IGBT), Static Induction Thyristor (SITh) or MOS-Controlled Thyristor (MCT) and new circuit configurations for using them. However, the question is: Is there really a need for new devices and new circuits?

To the design engineer some limitations of the present converter concepts are obvious:

- cost: In many applications, the advantage of power electronics cannot be utilized since the costs are too high. As the cost of a semiconductor switch is about three times as much as the cost of the semiconductor itself, a reduction of the auxiliary components like drive circuits, snubbers and cooling equipment is highly desirable.

- noise: In most power converters switching frequencies of about 1 kHz are used which is close to the highest sensitivity of the human ear. This is why the audible noise of power converters is a topic of growing importance. Increasing the switching frequency would help to reduce this disadvantage.

- EMI: In addition, electromagnetic interference (EMI) occurs due to high values of dV/dt and dI/dt in power converters. These problems increase with the power range of the converters and with the switching speed of the devices.

Overcoming these problems is certainly an objective of future development. In addition, some further improvements are desirable.

- Although the power efficiency of present converters is quite high, a further increase of the efficiency would be appreciated.

- Increasing the frequency reduces the size of magnetic and dielectric components.

- Increasing the feeding frequency for a.c. motors results in a higher speed which is advantageous for some applications, e.g. wood processing, grinding, centrifuging. As the output power increases with the motor speed, drives with high power density may be realized. For conventional motors, the power density is about 0.2 kW/kg. It has been demonstrated that 1 kW/kg is quite realistic for medium size motors. Even higher values are claimed in literature.

In the following we will investigate the interactions between device characteristics and circuit behaviour in order to find out which device improvements are most desirable. The discussion will be confined to the power range above 10 kW.

2. OPERATING PRINCIPLES AND PHYSICAL CONSTRAINTS OF POWER SEMICONDUCTOR DEVICES

A switching power device operates in two steady states: blocking high voltages at low current or conducting large currents at low voltage drops. The transients between the on-state and off-state in real devices are as important as the steady-state behaviour.

2.1 Blocking State

Blocking high voltages requires a reverse biased p-n junction. The breakdown voltage V_{BD} of an asymmetric p-n junction as shown in Fig. 1 depends on the doping density of the lightly doped drift region, N_D [1]:

$$V_{BD} \propto N_D^{\frac{3}{4}} \qquad (2.1)$$

The minimum width of the drift region, w_d, is given by

$$w_d \propto V_{BD}^{\frac{7}{6}} . \qquad (2.2)$$

If the depletion region is allowed to strech to the n+-buffer layer (punchthrough, see Fig. 1b) and the doping density N_D is further reduced, then the width of the drift region can be reduced to roughly 60% for the same blocking capability.

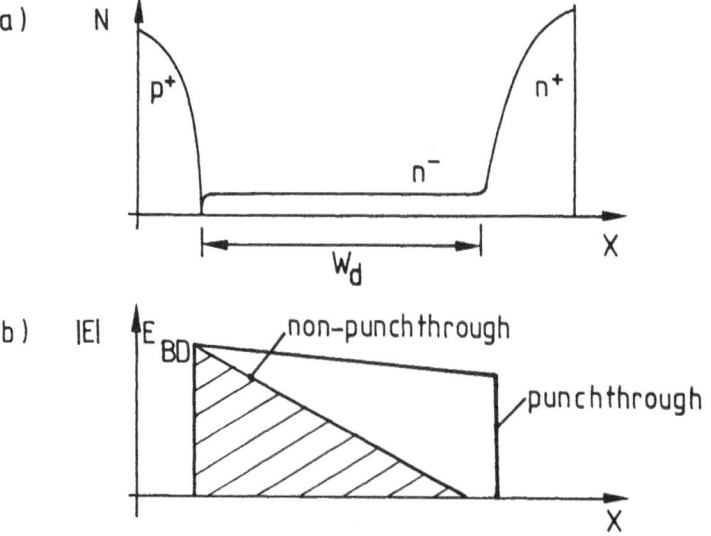

Fig. 1. Doping profile (a) and electric field in the off-state (b) of a generic high voltage p-n junction.

2.2 Conducting State

In devices with unipolar conduction mechanism (field effect transistors, FET), only majority carriers contribute to the current, and the device behaves like an ohmic resistor with the conductivity being determined by w_d and N_D. An estimate of the voltage drop of the drift region V_d, based on the above relations, gives

$$V_d \propto j \bullet V_{BD}^{\frac{5}{2}} \qquad (2.3)$$

with the current density j. In practice, there is an additional constraint given by thermal limitations. Assuming that a constant amount of heat per device area flow to by the cooling equipment, the relation between forward voltage drop V_d and breakdown voltage V_{BD} is almost linear:

$$V_d \propto V_{BD}^{\frac{5}{4}} \qquad (2.4)$$

At the same time, the maximum current density decreases hyperbolically:

$$j \propto V_{BD}^{-\frac{5}{4}} \qquad (2.5)$$

and thus the chip area necessary for a given current increases nearly linearly with the blocking voltage.

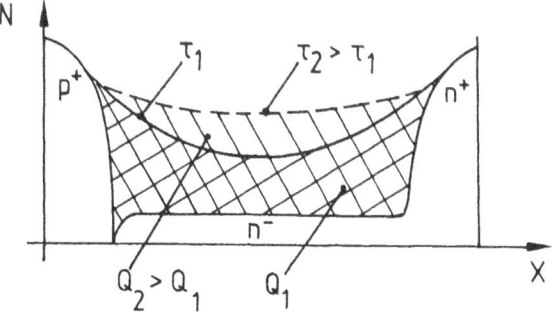

Fig. 2. Carrier concentration in a high voltage p-n junction in the conducting state, showing high-level carrier injection into the drift region.

In bipolar devices, the conducting state is characterized by high-level carrier injection from the p+ and n+ regions into the lightly doped drift region, causing a conductivity modulation (Fig. 2). The drift region is flooded with excess carriers representing a stored charge Q. As a result, the on-state voltage at the same current density is much smaller in bipolar devices. The on-state voltage can be influenced by adjusting the carrier lifetime τ in the drift region. Figure 3 shows the basic behaviour of on-state voltage versus wafer thickness d (which determines the blocking capability) at different values of carrier lifetime and constant current density. It is obvious that high blocking voltages and low voltage drops require large carrier lifetimes. An analysis based on scaling laws revealed the relationship[1]:

$$\tau \propto V_{BD}^{\frac{7}{3}}$$
(2.6)

under the constraint of constant on-state voltage. Fig. 3 was calculated for a non-punchthrough device.

In devices with an n+- buffer layer, the breakdown voltage can be made roughly 80 % higher for the same silicon thickness while maintaining on-state voltage and carrier lifetime[2]. The same increase in blocking capability, using a non-punchthrough design, would result in an increase in carrier lifetime by almost a factor of 4 (forward voltage drop kept constant).

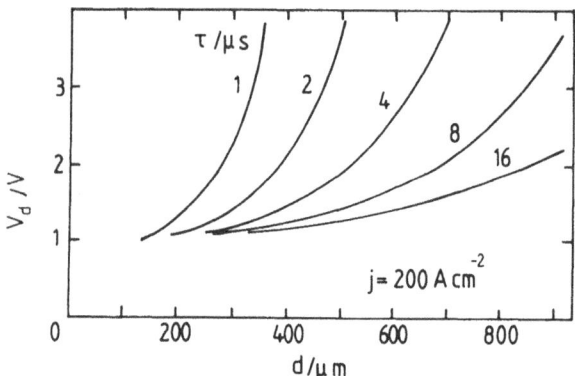

Fig. 3. On-state voltage drop V_d at a current density of 200 A/cm² as a function of silicon thickness d for various carrier lifetimes τ [2].

A simplified relation between the carrier lifetime τ and the stored charge Q is given by the fundamental equation of the charge control description

$$j = \frac{Q}{A \cdot \tau}$$
(2.7)

where A is the active device area. At a given current, the amount of stored charge is approximately proportional to the carrier lifetime.

2.3 Turn-on and Turn-off Transients

In a real device, the transients between on-state and off-state can not be abrupt. Rise and fall times of current and voltage are finite and depend on the device and on the circuit environment.

2.3.1 Majority Carrier Devices. When turning on a unipolar device, a conducting channel of, say, n-type behaviour has to be opened in the p$^+$-region of the structure in Fig. 1a). After that, the depletion region of the blocking junction is reduced until the on-state voltage is reached. The junction capacitance of the blocking junction is discharged by dumping the stored energy into the transient on-state resistance.

At turn-off, the conducting channel has to be removed. The depletion region is built up while charging up the junction capacitance correspondingly. As a result, current and voltage overlap both at turn-on and turn-off, causing switching losses.

2.3.2 Minority Carrier Devices. Since there is no stored charge associated with a unipolar conduction mechanism, the switching transients are relatively fast. When turning on a bipolar device, it takes significant time to build up stored charge after the blocking voltage has disappeared. Therefore, the on-state voltage is significantly higher in the first microseconds of current flow.

At turn-off, the stored charge has to be removed from the forward blocking junction before the device can withstand blocking voltage. This is usually accomplished by a negative gate current. After the storage time t_s (Fig. 4), when the p-n junction is free of excess carriers and ready to block, the current falls rapidly during t_f. The charge stored inside the device cannot be removed completely and causes a tail current to flow during t_{tail}. Both storage time and tail time can last several microseconds.

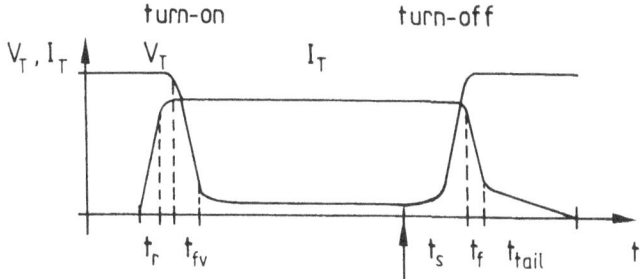

Fig. 4. Idealized waveforms of voltage and current during turn-on and turn-off of a bipolar device, indicating current rise time t_r, voltage fall time t_{fV}, storage time t_s, current fall time t_f and current tail time t_{tail}.

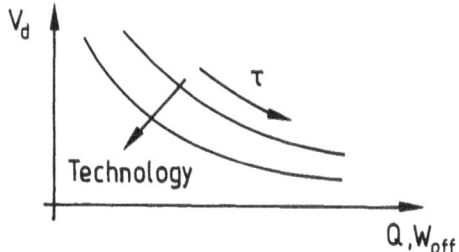

Fig. 5. Typical trade-off curves between forward voltage drop and switching performance.

For a given device, the switching times and switching losses are strongly correlated with the stored charge Q. It is therefore desirable to have short carrier lifetimes in order to achieve fast switching and low commutation losses, which is contradictory to the requirement of low on-state voltages. As result, a trade-off between switching performance and on-state voltage exists as shown in Fig. 5. The position of the trade-off curve however depends on the device type and on technology.

As we discussed previously, a punchthrough design is favourable in order to achieve low on-state losses. At the same time, the stored charge for given blocking voltage and on-state voltage is significantly lower (almost factor of 4).

2.3.3 Influence of Circuit Environment. Switching waveforms and losses are extremely dependent on the circuit environment. This is illustrated in Fig. 6, showing typical circuits, waveforms and switching loci for clamped inductive load, switching with snubber networks and resonant switching.

If snubber circuits are used, the overlap of voltage and current is widely reduced and so are the switching losses. However, the energy stored in the snubber components has to be dissipated in every cycle, increasing the total losses and deteriorating the efficiency. Therefore, snubber circuits will be avoided if possible. Shaping the switching waveforms by means of resonant components results in zero voltage (ZVS) or zero current switching (ZCS). Compared with conventional snubbers, the device losses can be further reduced because either turn-on losses (ZVS) or turn-off losses (ZCS) are zero, and the snubbing component (C in zero voltage switching, L in zero current switching) can often be made larger than in conventional snubbers. In addition, the energy of the resonant components is recovered and thus efficiency can be increased.

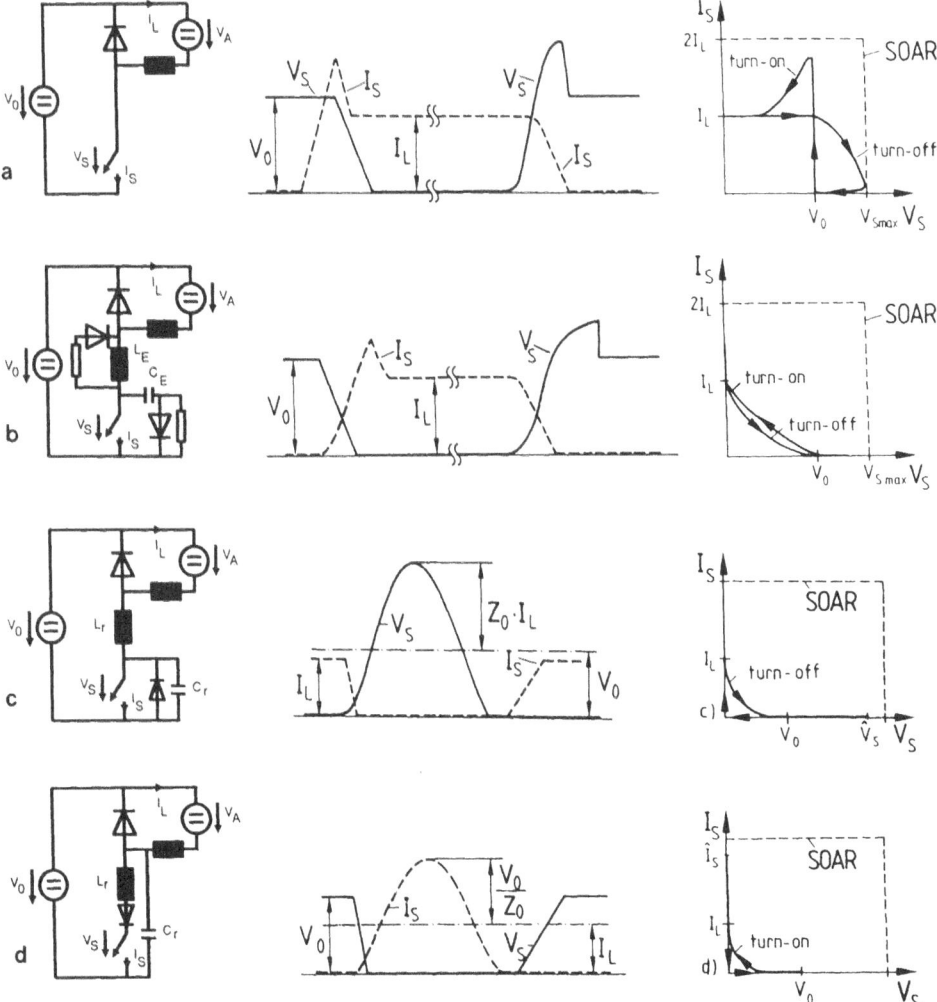

Fig. 6. Circuits, switching waveforms and switching loci for clamped inductive load (a), switching with turn-on and turn-off snubber (b), resonant switching using zero-voltage-switching (c) or zero-current-switching (d)[3].

2.4 Failure Mechanism

Failure of power semiconductor devices can always be attributed to exceeding a maximum temperature.

Static failure mechanisms include classical overvoltage, overcurrent and exceeding the maximum device temperature due to switching losses.

Dynamic failure mechanisms are manifold and include dynamic over-voltages, dynamic avalanche during turn-off and lateral effects like current filaments or emitter current crowding that lead to localized overtemperatures. Effects like these limit the turn-off current rating or safe operating areas of many devices.

To avoid destructive operating conditions, the circuit designer has to protect the device from transient and static overvoltages and from overcurrents and short circuits.

2.5 Gate Structures

In majority carrier devices, there are two different gating principles, the MOSFET and the Junction-FET (JFET) or Static Induction Transistor (SIT) (Fig. 7a and 7b). In a MOSFET, an inversion layer is generated in the p region by applying a positive gate-to-source voltage. Thus, an n-type channel is opened through the p-region. In a Junction FET, the width of a depletion layer between the p-type gate and the n-type bulk can be controlled by the gate-source voltage. If the depletion region fills the space between two adjacent gate regions, the device is turned off, and the depletion region stretching inside the bulk can take over the blocking voltage. The MOSFET gate is voltage controlled and requires current only for charging and discharging the gate capacitance. In order to reduce on-state losses, the channel is made as short as technology allows.

In a Bipolar Junction Transistor (BJT, Figure 7c), a small base current causes a large collector current. The current gain β depends on the base width. For large blocking voltages (up to 1000 V), the base has to be fairly wide in order to avoid reach-through, and therefore β is in the region of 5 to 10 only.

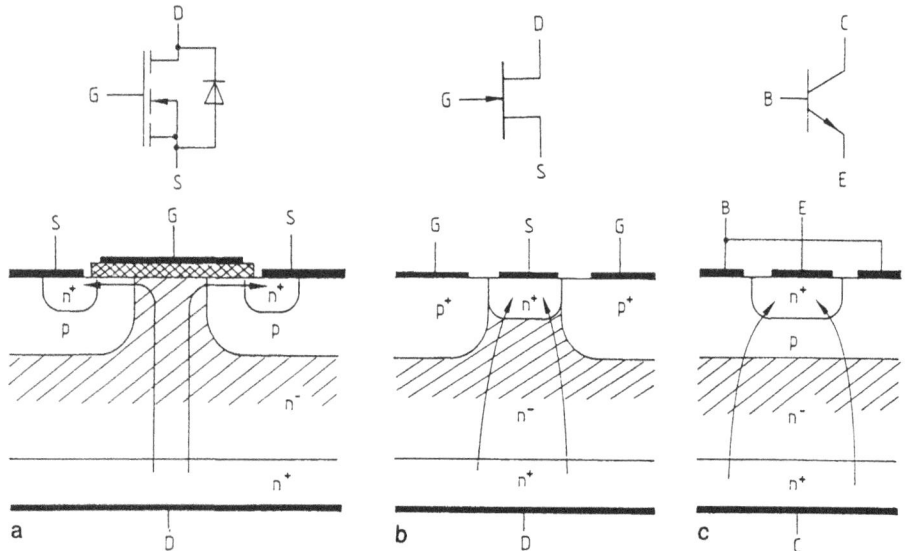

Fig. 7. Basic structures and circuit symbols of MOSFET (a), Junction-FET or SIT (b) and BJT (c), indicating the direction of current flow (arrows) and the location of the depletion region in the blocking state (hatched area).

3. COMPARISON OF POWER SEMICONDUCTOR DEVICES

In the following chapter, the different power devices are compared from a circuit designer's point of view. After a brief introduction of the power devices known today, their availability and prices are discussed. Device characteristics like on-state losses, switching times, switching losses and thermal resistance determine the maximum frequency of operation in a three phase PWM voltage source inverter. Extra expense for snubber circuits, gate drive and protection are other features that may influence the choice of semiconductor components. A lookout to the improvements we can expect from new resonant inverter topologies closes this chapter.

3.1 Devices, Power Range and Prices

The gate structures, discussed in the previous section, are the basis of a systematics of power semiconductor devices (Fig. 8). We have seen some of the basic limitations of these simple devices that led to the development of more complex structures. On the bipolar side, the monolithic Darlington was introduced to reduce the problem of low current gains (Fig. 9a). The back-to-back connection of two bipolar transistors is a model for the thyristor whose structure is shown in Fig. 9b). This model demonstrates the typical latch up behaviour. No continuous gate current is needed but the device can not be turned off via the gate. Improvement of thyristors led to the GTO which can be turned off with a negative gate current. It has basically the same structure but requires some technological changes, i.e. a finer gate-cathode interdigitation and a reduction of the sheet resistance of the p-base.

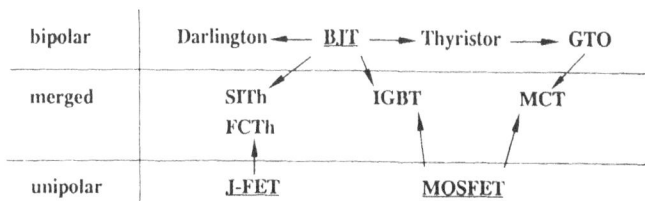

Fig. 8. Systematics of power semiconductor devices.

The fact that bipolar devices require relatively high gate power makes it desirable to combine unipolar gate structures with the bipolar conduction mechanism. In this sense, the Insulated Gate Bipolar Transistor (IGBT) is a MOSFET with an additional p-layer at the drain which operates as an emitter when current flows, thus introducing a conductivity modulation. The JFET or SIT is related to the Field Controlled Thyristor (FCTh) or Static Induction Thyristor (SITh) in the same manner. However, the gate is not completely isolated from the

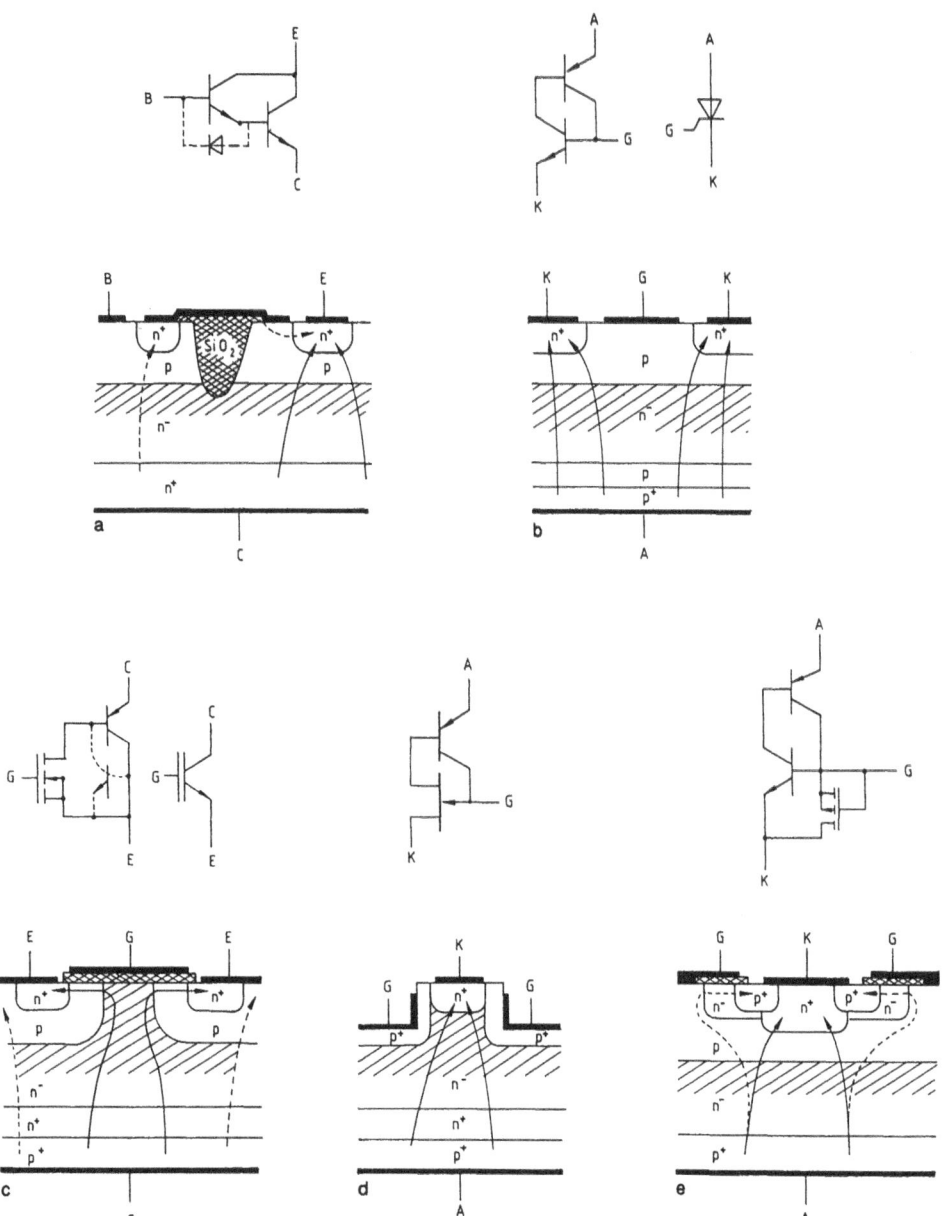

Fig. 9. Basic structure and circuit symbols of monolithic Darlington (a), thyristor (b), IGBT (c), FCTh (d) and MCT (e). Dashed arrows show path of auxiliary current flow.

bulk and thus charge extraction via the gate is possible. The MOS Controlled Thyristor (MCT) is a GTO with MOS switched emitters. The negative gate current necessary for turn off is flowing through a MOSFET that shorts the n-emitter, driven by the gate-cathode junction voltage. Both FCTh and MCT are candidates for replacing conventional GTOs in the future.

Figure 10 provides an overview of the voltage and current ratings of the devices available in the marketplace. The current ratings shown are nominal values in case of transistors and maximum average currents in case of fast thyristors. For GTOs, both the maximum average current I_{TAV} and the maximum controllable current I_{TGQM} are shown.

The limitation of MOSFETs towards high switching powers is a result of their maximum chip size, in combination with a drop of current density with increasing blocking voltages as discussed in section 2.2. Furthermore, the internal diode causes problems at voltages above 500 V because it can not be made fast enough.

The limits for the IGBT are expected to be shifted towards higher voltages (short dashed line). A comparison between GTOs and frequency thyristors reveals that both devices are available at similar current ratings but much higher voltage ratings are realized with GTOs. In the future, the blocking voltage capability of GTOs will be extended further up to 8 or 9 kV [24].

The price per unit of switching power is shown in Fig. 11 as a function of switching power for different devices (summer 1991). We considered power transistor modules including antiparallel diodes and thyristors in hockey puck packages. Switching power P_S is calculated as the product of blocking voltage and the nominal current or the maximum average currents in case of transistors or thyristors respectively. For fair comparison with turn-off devices, the maximum controllable current of GTOs was also considered (dashed line). The regions indicated in the figure are meant to give an orientation only. Prices are subject to negotiation and depend strongly on the volume of an order. We tried to consider prices for about 1000 pieces in case of transistors and about 100 pieces for thyristors. The MOSFET seems to be the most expensive device. Prices for IGBTs are not yet stable, but are expected to settle at about 1.2 to 1.3 times the price for BJTs. In terms of controllable current, GTOs are seen to be cheaper than BJTs (at higher power levels). However, there is still a factor of two when comparing GTO prices with frequency thyristors and a factor of four to the price of phase controlled thyristors.

3.2 Conduction Loss

Figure 12 shows the typical voltage drop of the devices when conducting the nominal current. The current density of turn-off devices is usually in the range of 25 to 100 A/cm^2.

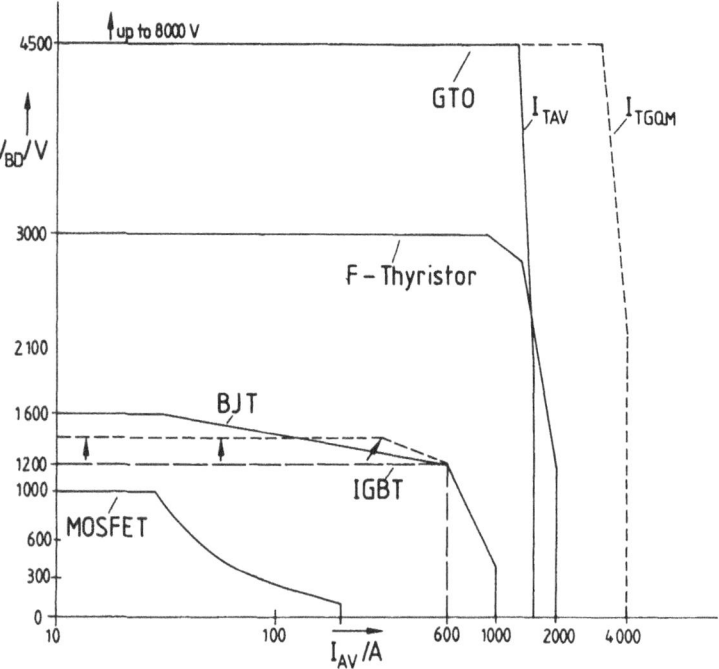

Fig. 10. Voltage and current ratings of power semiconductor devices available today.

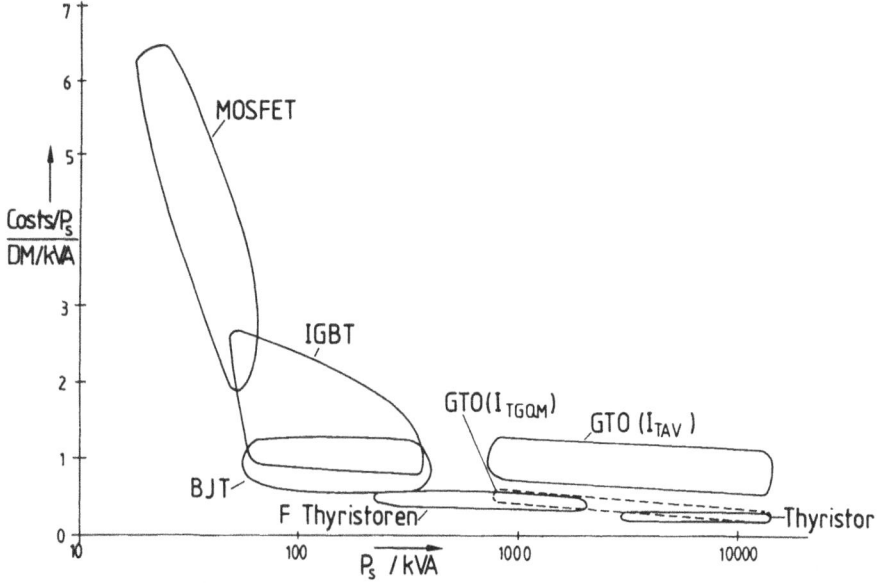

Fig. 11. Price per unit of switching power as a function of switching power Ps for devices available on the market (summer 1991).

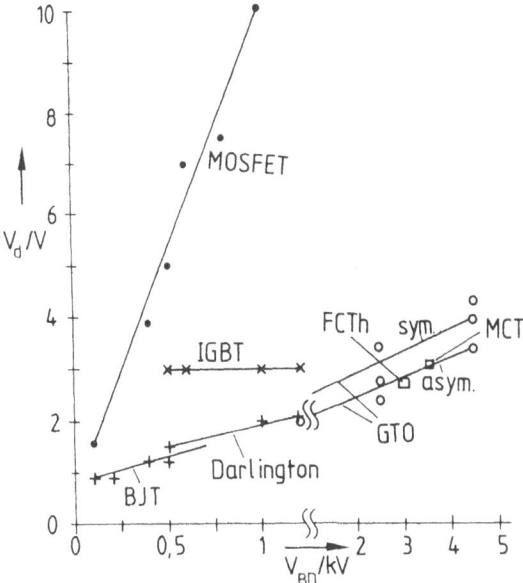

Fig. 12. Typical voltage drop at nominal device current ($T_J = 25$ °C) - Measurements at ISEA[1] and values from data books [3-11,15].

Remembering eq. (2.5), Fig. 12. confirms for MOSFETs that the current density of unipolar devices decreases rapidly with increasing blocking voltage. MOSFETs with blocking voltages greater than 500 V cannot be used at the nominal current, because of their high voltage drop.

Below a blocking voltage of 600 V, bipolar junction transistors are of interest, because they have a very low voltage drop.

At higher blocking voltages, thyristors are preferred. They have a voltage drop in the order of 2 to 4 V which is strongly dependent on the carrier lifetime (see Fig. 3).

The on-state voltage drop is also dependent on the junction temperature. The R_{DSon} of MOSFETs is about 50 % higher at a junction temperature of 120 °C. The voltage drop of bipolar devices increases or decreases with increasing temperature. The behaviour of devices from different manufactures may be quite different.

3.3 Switching Times and Energy Losses

The turn-on and turn-off time of the devices are summarized in Table 1. The turn-on time increases with increased blocking capability of the devices.

---

[1] Institute for Power Electronics and Electrical Drives, Aachen Technical University.

Table 1. Switching times of semiconductor devices.

Device	$t_{on}/\mu s$	$t_{off}/\mu s$	$t_f/\mu s$	$t_{tail}/\mu s$
MOSFET	0.1 - 0.5	0.1 - 1	0.03 - 0.2	-
IGBT	1 - 2	1 - 4	0.1 - 0.2	1 - 2
DBJT	1 - 4	10 - 25	0.2 - 1	-
GTO	5 - 10	10 - 30	0.5 - 2	5 - 80

The switching times, especially the turn-off time of the current-controlled devices are longer than those of the voltage-controlled devices. The biggest part of the turn-off time of DBJTs (Darlington-Transistors) and GTOs is the storage time. Due to the short switching times, voltage controlled devices are suitable for higher switching frequencies.

The comparison of the switching energy combines loss measurements at ISEA [3,14,16,17,27], from the literature[18,15] and data book specifications. Measurements of turn-on and turn-off energy losses show that the energy losses are almost proportional to the current [3,14,15,16,17]. As an example, the measured turn-off losses of GTOs are shown in Fig. 13a [17]. Fig. 13b shows the waveforms of voltage and current at turn-off of a 2000 A/2500 V GTO.

It is therefore convenient to compare the energy loss per Ampere:

$$w_{on} = \frac{W_{on}}{I} \tag{3.1}$$

$$w_{off} = \frac{W_{off}}{I} \tag{3.2}$$

of the different devices (Fig. 14). The values in Fig. 14 are valid for switching against a dc voltage of approximately half the breakdown voltage. Only the FCTh was measured at one third of the breakdown voltage.

No snubbers were used for the Darlington-BJT, the IGBTs and the FCTh. The GTOs were switched with minimum snubbers without exceeding the dI/dt and dV/dt limits. Typical values are given in Table 2. Figure 14 illustrates the advantage of the IGBT when compared to the Darlington-Transistor.

For the GTO, it has to be mentioned that the turn-off energy loss of new 3.5 kV devices are reduced to 1000 μJ/A [19].

The switching energy loss of the FCTh is comparable to that of a good GTO. The advantage of the FCTh is that it does not need any snubber circuit which yields less total losses if semiconductor and snubber losses are added.

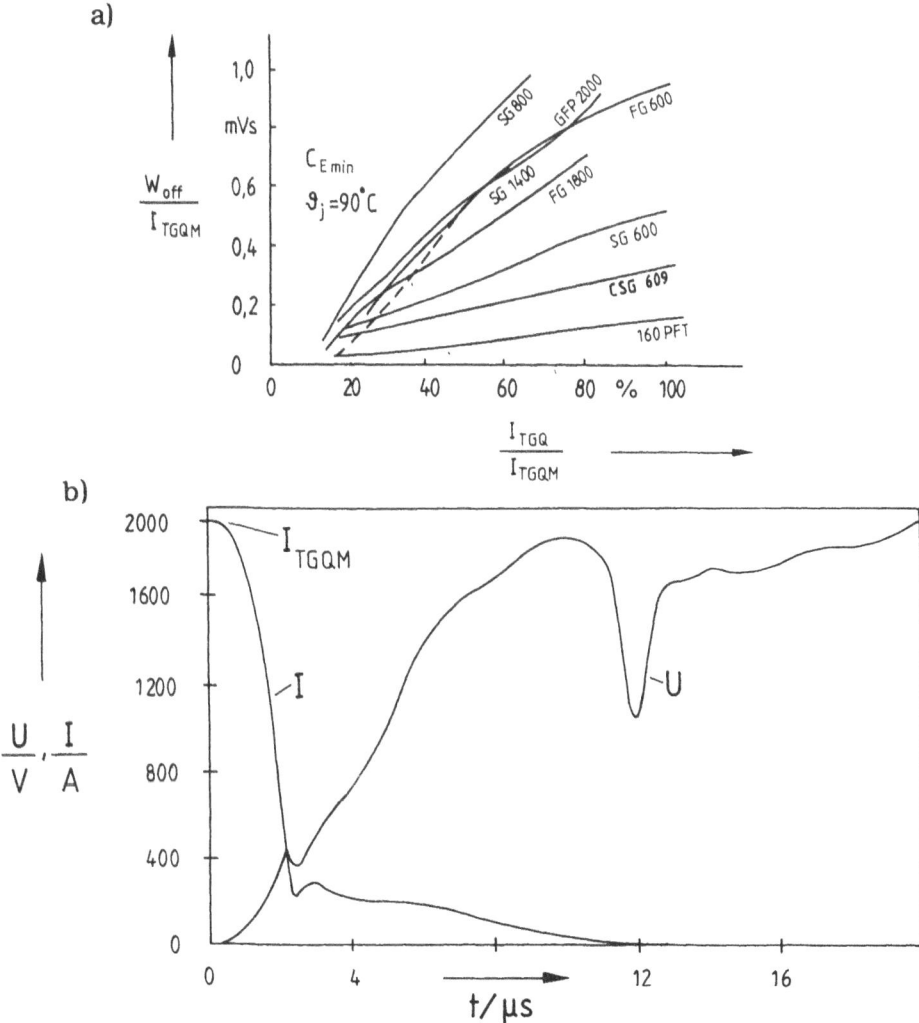

Fig. 13. a) Turn-off energy loss of different GTOs [17](CSG 609 measured at ABB)
b) Turn-off waveforms of a 2000 A/2500 V GTO.

Table 2. Values of dI/dt and dV/dt for GTOs.

	conventional	fast
dI/dt	300 A/µs	600 A/µs
dV/dt	1000 V/µs	2000 V/µs

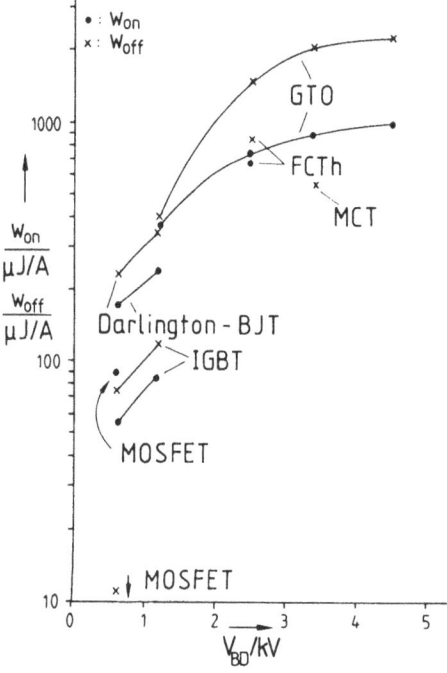

Fig. 14. Turn-on and turn-off energy loss per Ampere of various semiconductor devices.

3.4 Thermal Resistance

The power which can be removed from a semiconductor chip is determined by the temperature difference between junction and case

$$\Delta T = T_j - T_c, \tag{3.3}$$

and the thermal resistance R_{th} of the package.

From the data sheet specifications, a thermal constant K_T was found which is the temperature difference ΔT occurring when conducting the nominal current I_N without switching.

$$K_T = R_{th} \cdot V_d \cdot I_N \tag{3.4}$$

Typical values of K_T are given in Table 3.

Table 3. Thermal constant K_T for semiconductor devices (GTO: $I_N = I_{TGQM}$)[4-11].

	K_T/K
MOSFET	45 - 65
IGBT	60 - 70
DBJT	45 - 65
GTO	75 - 190

Table 3 demonstrates that GTOs cannot be used continuously at their repetitive controllable turn-off current I_{TGQM} because the thermal constant is too high. Normally their average current I_{TAV} is about 30% and their RMS-current I_{TRMS} is about 45% of the turn-off current I_{TGQM}.

The thermal resistance, being only a second order function of breakdown voltage, decreases hyperbolically with increasing current I_{TGQM}. This is illustrated in Fig. 15. However, the forward voltage drop V_d in-

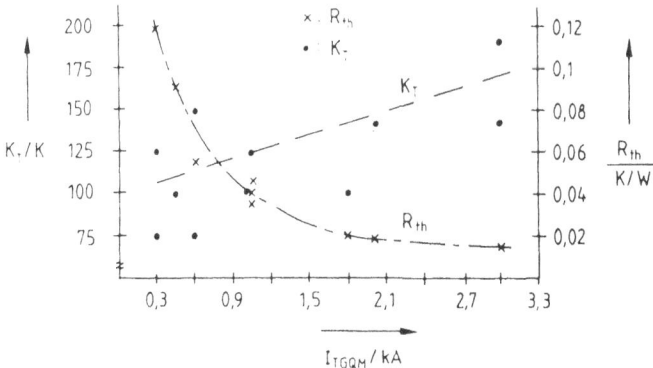

Fig. 15. Thermal resistance R_{th} and thermal constant K_T as a function of turn-off current I_{TGQM}. Data of asymmetrical GTOs from [11]. (For K_T only the minimum and maximum values are shown).

creases with the blocking voltage V_{BD} as shown in Fig. 12. As a result, the thermal constant K_T increases with the blocking voltage (Fig. 15).

3.5 Switching Frequency

With the results of sections 3.2, 3.3 and 3.4, we evaluate the maximum switching frequencies, f_s, of the devices when used in a PWM inverter with sinewave output current with period T

$$I(t) = \hat{I} \bullet \sin (2 \pi \frac{t}{T}).$$

The peak current is limited by the device current rating ($\hat{I} \leq I_N$).
First we consider the conduction power loss P_C:

$$P_C = V_0 \bullet \bar{I} + R_0 \bullet \bar{I}^2 \leq V_{dN} \bullet \bar{I} ,$$

with $V_{dN} = V_0 + R_0 \bullet I_N$.   (3.5)
The average current in an inverter switch equals

$$\bar{I} = \hat{I} (\frac{1}{2\pi} + \frac{m \bullet \cos\varphi}{8})$$ (3.6)

for $m \leq 1$ and $f_s/f \geq 5$ [20] where f is the output frequency. The worst
case occurs for a modulation depth m = 1 and a power factor $\cos\varphi$ = 1:

$$\bar{I} = 0{,}284 \; \hat{I} .$$ (3.7)

Assuming the switching energy loss is proportional to the current,
the switching power loss can be written as:

$$P_S = \frac{1}{T} \sum_{i=1}^{\frac{T}{2\,T_s}} \left[\left(w_{on} + w_{off} \right) \bullet \hat{I} \bullet \sin \left(2\; p\; \frac{i\; T_s}{T} \right) \right],$$ (3.8)

for $m \leq 1$ and f_s/f being an integer larger than 5. For $T/T_S \rightarrow 0$
eq. (3.8) reduces to

$$P_S = f_S \bullet (w_{on} + w_{off}) \bullet \hat{I}/\pi.$$

The total power loss of the devices turns out to be

$$P_t = P_c + P_s = \left(V_d \bullet 0.284 + \left(w_{on} + w_{off} \right) \bullet \frac{f_s}{p} \right) \bullet \hat{I}$$ (3.9)

With the temperature difference

$$\Delta T = R_{th} \bullet P_t$$ (3.10)

and equations 3.4 and 3.9 the maximum switching frequency is found

$$f_{Smax} = \frac{p \cdot V_d}{w_{on} + w_{off}} \cdot \left[\frac{DT}{K_T} \cdot \frac{I_N}{\hat{I}} - 0.284\right]. \qquad (3.11)$$

It is often proposed that switching losses and conduction losses should be approximately equal. In this case, the switching frequency f_{Seq} has to be:

$$f_{Seq} = \frac{0.892 \cdot V_d}{w_{on} + w_{off}} \qquad (3.12)$$

The switching frequencies attainable with different semiconductor devices in hard switching PWM operation are given in Table 4.

The maximum switching frequency f_{Smax} was evaluated assuming

$\Delta T = 40$ K, $I_N/\hat{I} = 1$ for IGBT, DBJT ; $I_N/\hat{I} = 2$ for GTO.

The temperature difference has to be decreased to 20 K for high power GTOs because the thermal resistance between junction and case ($R_{thj-c} = 0.016$ K/W) is lower than between case and ambient.

With increasing power, the sensitivity of equation 3.11 increases too. The result is also dependent on the temperature dependence of the on-state voltage drop V_d and the switching losses.

Table 4. Switching frequencies of devices; [1] K_T is unknown.

Devices	f_{Seq}/kHz	f_{Smax}/kHz
IGBT	12 - 21	15 - 23
DBJT	3.0 - 5	5
GTO	1 - 3	0.2 - 5
FCTh [1]	1.75	2

3.6 Snubber Circuits

Many research activities have been carried out concerning snubber circuits in order to reduce the switching energy loss in the device so that the switching frequency can be increased.

Using the conventional RCD-turn-off snubber and RLD-turn-on snubber (see chapter 2), the total losses increase with increased snubber values L and C. Therefore, regenerative snubber circuits were developed.

In these circuits, the energy stored in the snubber components L and C is fed back into the circuit and is not dissipated[17]. The disadvantage of regenerative snubbers is the high number of components and the complexity of the circuits.

Using snubber circuits, restrictions on the switching times have to be accepted. A turn-off snubber needs a minimum on-time of the switch to reset the snubber-capacitor, a turn-on snubber needs a minimum off-time to demagnetize the series inductor.

Because of the disadvantages of snubbers described before,

- increased total losses or increased complexity,
- restrictions on the switching times,

snubber circuits are avoided in practice if possible. In case of MOSFET, IGBT and DBJT, snubber circuits are not used. These devices are used at switching frequencies and current levels which can be handled without snubber circuits.

An overvoltage protection is realized by connecting a fast capacitor close to each inverter leg to minimize the parasitic inductance (Fig. 16a) or by the overvoltage protection circuit shown in Fig. 16b.

GTOs always need a turn-on and turn-off snubber circuit. To minimize the disadvantages described before, normally the minimum snubber values possible not to exceed the dI/dt and dV/dt ratings are chosen.

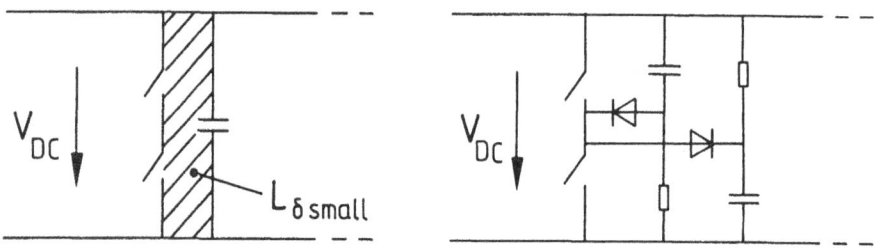

Fig. 16.   a) Overvoltage protection with a blocking capacitor
b) Overvoltage protection circuit.

The most important snubber circuit for inverter legs is shown in Fig. 17 [21]. The circuit consists of a minimum number of elements and is partially regenerative and partially dissipative. The circuit can be extended to a fully regenerative circuit by replacing the resistor with a

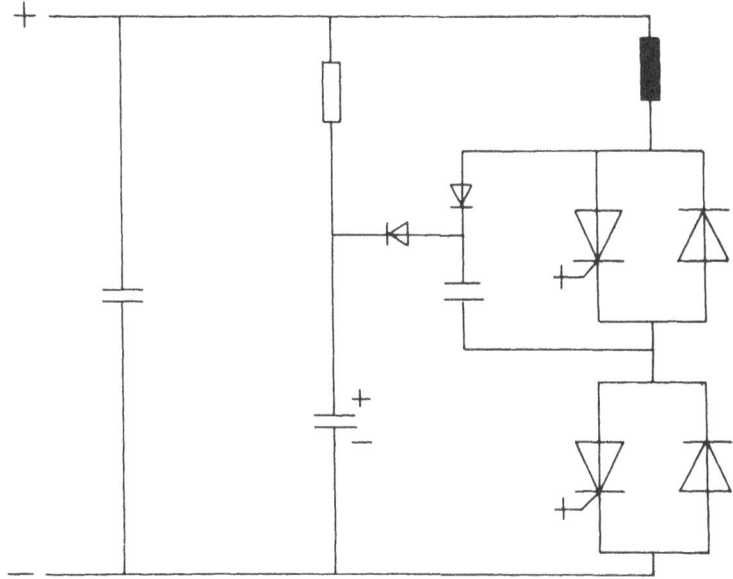

Fig. 17. Optimal snubber circuit for inverter legs.

dc to dc converter, feeding the energy back to the dc voltage source. Today this circuit is used in most high power GTO-inverters.

In summary, it can be pointed out that snubbers are only used if necessary and are not used to increase the switching frequency. Therefore, the switching frequencies shown in Table 4 can be regarded as the limit of today's power electronic devices in hard switching applications.

3.7 Gate-Drive-Circuit

The expense for the driver circuits of the different devices discussed differs extremly. For the voltage controlled devices, a simple circuit as shown in Fig. 18 is used to turn the device on and off.

The circuit consists of a galvanic isolation, an amplification circuit and the driver stage. The driver stage consists of two switches to connect each supply voltage terminal to the control terminal. The positive supply voltage V_D is in the range of 10 to 20 V and the negative voltage V_S can be 0 V In case of MOSFETs or down to - 15 V for IGBTs, especially if the blocking voltage exceeds 600 V. The power needed is less than 3 W.

The complexity and the expense of the driver of current controlled devices (BJT and GTO) increase with increased voltage and current ratings. Figure 19 shows a typical base- or gate current-waveform.

A current pulse I_{ton} of 1 to 5 % of the load current is needed to turn on the devices properly. Especially for high power devices, the dI/dt must be high so that the device turns on homogeneously. Therefore, a driver voltage V_D of minimum 15 V and a low parasitic inductance is needed.

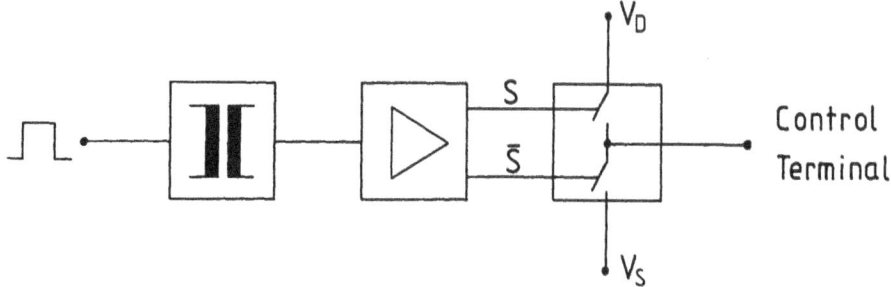

Fig. 18. Driver circuit for voltage controlled devices.

The on-state gate current I_{on} depends on the current gain of the BJT or the threshold current of the GTO. The supply voltage V_{Don} for the current supplied during on-state can be small, i.e. 5 V, because no high dI/dt is needed.

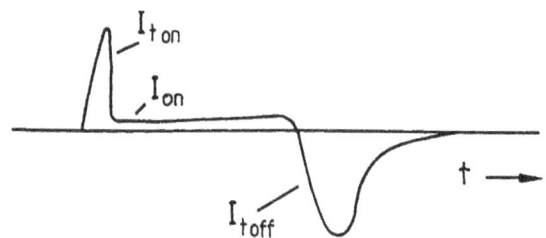

Fig. 19. Typical waveform of driver current for current-controlled devices.

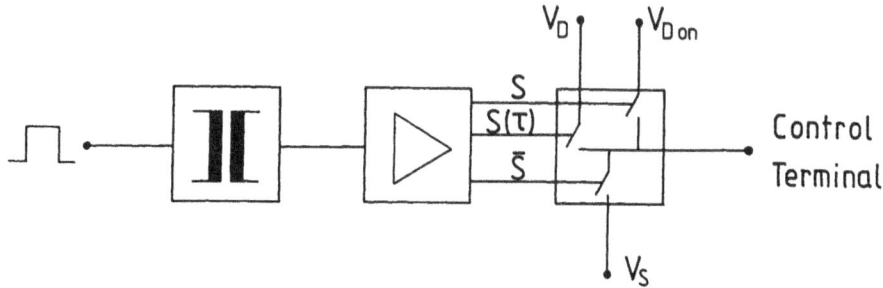

Fig. 20. Driver circuit for current controlled devices.

The negative base current pulse reaches 20 % to 35 % of the current to be turned off for GTOs. In the case of BJTs and DBJT the negative current is in the range of 10 % to 20 % but can also be 100 %

of the turn-off current, depending on how fast the device should be turned off. A high negative dI/dt and a high negative base current lead to a short storage time. Therefore, the voltage V_S has to be similar to V_D. The principle driver circuit is shown in Fig. 20. The output driver stage needs one voltage source and one switch more than the one shown in Fig. 19 but its required power is less.

In bipolar transistor and Darlington devices, the on-state driver current I_{on} is usually reduced by using an anti-saturation diode (Fig. 21). Thus also the storage time is reduced.

The two positive supply voltages are only used if V_D has to be above 10 V and I_{on} is above a few Ampere.

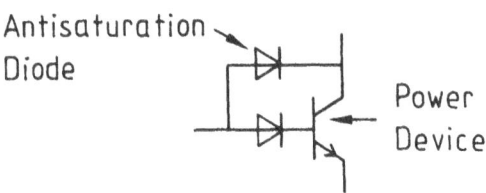

Fig. 21. Connection of anti-saturation diode.

The base drive power can be written as

$$P = V_D \bullet Q_{ton} \bullet f_s + V_{Don} \bullet I_{on} \bullet a + V_S \bullet Q_{toff} \bullet f_s$$

with $Q_{ton} = \int I_{ton}\, dt$ for one pulse and duty cycle a.

The driver losses increase with increased device ratings and are always higher than those of a voltage controlled device.

Special demands are posed by the FCTh. While its counterpart, the SITh, is turned off with a gate current similar to that of a GTO, the FCTh needs a negative gate current that exceeds the turn-off anode current. The gate current pulse has to be very steep in order to turn off the device homogeneously. This can be achieved by a very low inductive gate connection and by choosing V_S at about 30 V, not exceeding the negative gate-cathode breakdown voltage.

The MCT as a voltage controlled device does not need a continuous gate current. However, since the gate voltage slope has to be rather steep both at turn-on and turn-off, and the gate capacitance of a high current device is large, it also requires relatively high peak gate currents.

3.8 Protection Strategies

In this section, we want to give a brief overview of the possibilities of protecting a device against dynamic overcurrents and dynamic overvoltages. Only dynamic cases are taken into account, because slow events are easily protected by measuring the voltage and the current of the link.

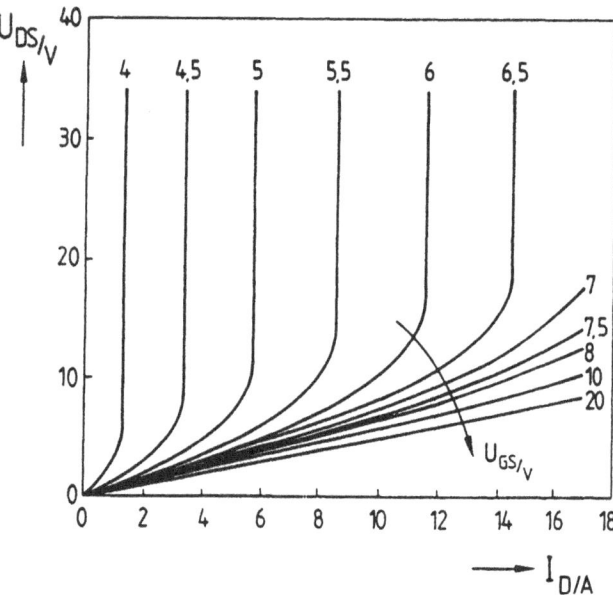

Fig. 22. Output characteristic of a MOSFET.

Dynamic overshoots have to be sensed quickly and the driver circuit has to react immediately. A commonly used method to sense the current in a switch uses the fact that the on-state voltage drop increases with current. Moreover the saturation voltage is dependent on the gate voltage, in case of the MOSFET (Fig. 22) and the IGBT, or on the base current in case of BJTs. By limiting the gate voltage in case of unipolar gates or the gate current in case of bipolar gates, the short circuit current is limited. A fast reacting protection circuit turns the device off before the device temperature reaches dangerous values.

Figure 23 shows the realisation of the sensing circuit. During off state of the device, the high blocking voltage is blocked by a diode. During conduction of the switch the voltage is compared with a reference voltage.

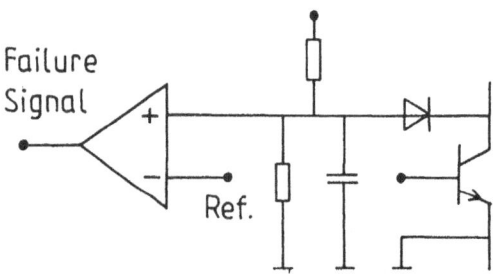

Fig. 23. Realisation of the on-state voltage measurement.

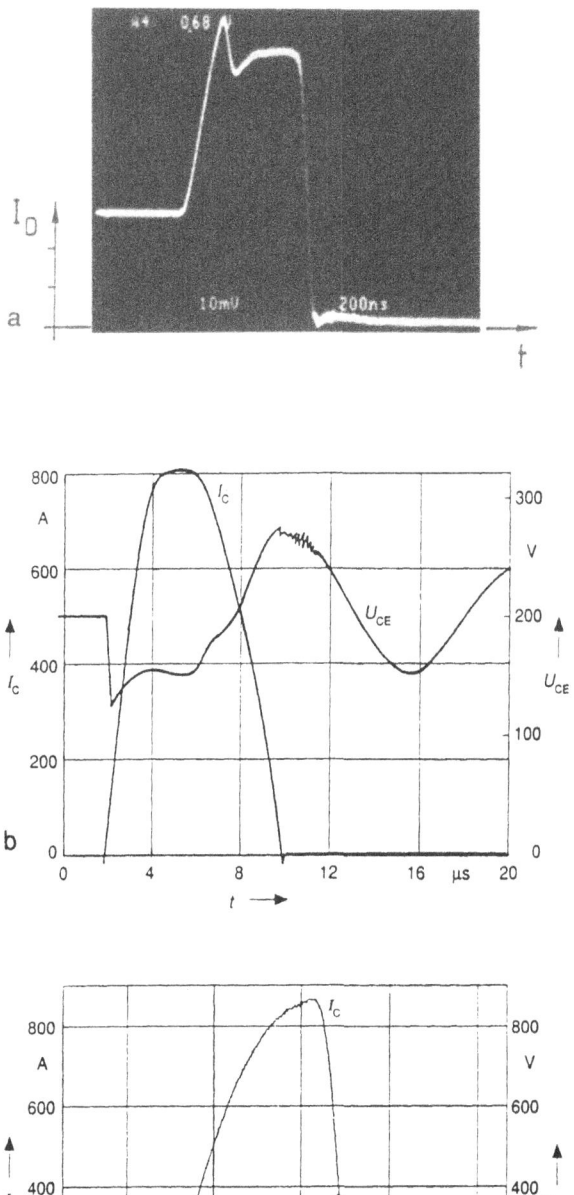

Fig. 24. Current and voltage waveform under short circuit condition
   a) MOSFET (10A/500V); 2.5A/div, 200ns/div [22]
   b) Darlington-BJT (480A/500V) [14]
   c) IGBT (400A/500V) [14].

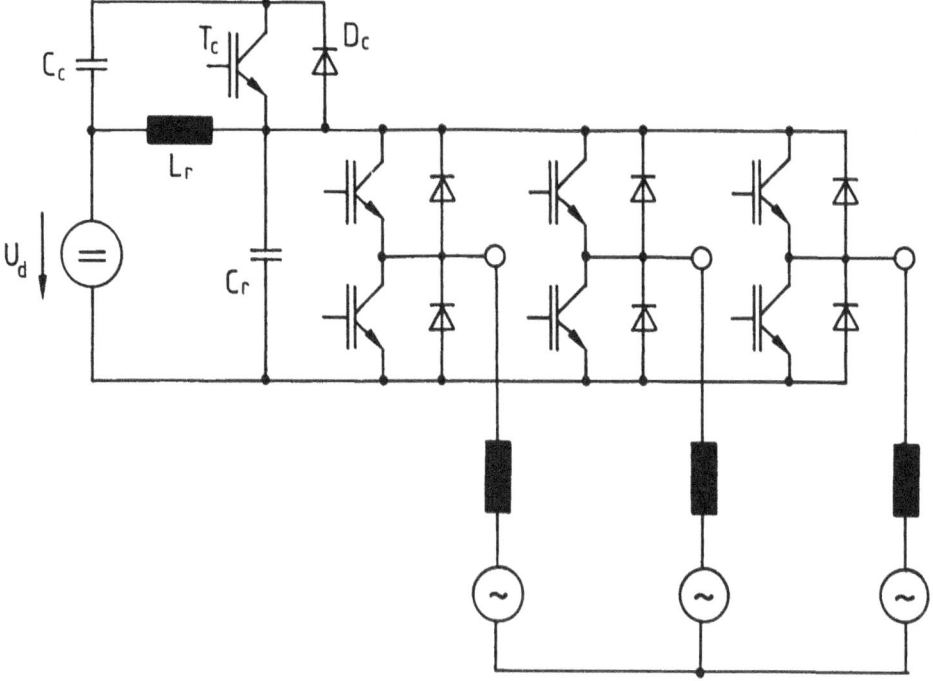

Fig. 25. Circuit diagram of an actively clamped resonant DC link inverter.

In [22] and [14], it is shown that the switch can be turned off properly under short circuit conditions. Figure 24 shows the current and voltage waveforms using a MOSFET (a) and a Darlington-BJT (b). The temperature dependence of the on-state voltage has to be taken into account when choosing the reference voltage V_{Ref}.

Using IGBTs, the maximum current under short circuit condition can be 4 to 10 times the nominal current (Fig. 24c). The IGBT cannot be handled as easy as MOSFET and BJT because no manufacturer guarantees this operation mode for more than a few switching cycles.

In [23], an overcurrent protection for GTOs is proposed. The turn-off current is sensed during turn-off of the device by measuring the storage time. If the storage time exceeds a certain limit, the device is turned on again, and a fuse connected in series with the device breaks the short circuit.

A retriggerable overcurrent protection, however, is only possible with fast direct current sensing.

Protection circuits against overvoltage were mentioned in chapter 3.5. Another method which is often used with MOSFETs and BJT is connecting a breakover diode from the drain/collector to the gate/base. If the blocking voltage exceeds the breakdown voltage, the device is turned on and works in its active operating area.

3.9 Power Semiconductor Devices in Resonant Converters

In this section, the application of power semiconductor devices in ZVS and ZCS resonant converters is discussed. Basic waveforms of such circuits have been depicted in Fig. 6c and 6d. The benefit of this technique is that voltage and current waveforms are shaped such that switching losses are largely reduced and, consequently, higher switching frequencies can be reached. Disadvantages of resonant converters are higher peak voltage or peak current stress, a relatively high amount of energy circulating between the resonant components and restrictions in the control of power flow. The most mature resonant inverter topology, presenting an alternative to voltage source PWM inverters, is the Resonant DC Link Inverter (RDCLI) depicted in Fig. 25 [25,26,27]. This circuit belongs to the class of zero voltage switching inverters.

3.9.1 Zero Voltage Switching Applications. In zero voltage switching circuits, a capacitor C_r is connected directly in parallel to the switch, serving as a snubber during turn-off. The energy stored in C_r is recovered by resonating with an inductor L_r before the device is turned on again. Therefore, the parasitic output capacitance of unipolar devices (MOSFET, JFET) can be discharged together with C_r, resulting in zero turn-on losses.

Since there is no stored charge involved in the conduction process, a MOSFET or JFET can turn off without a blocking voltage. As a result, current falls while voltage is still very low and turn-off losses can be reduced by at least an order of magnitude (Fig. 26).

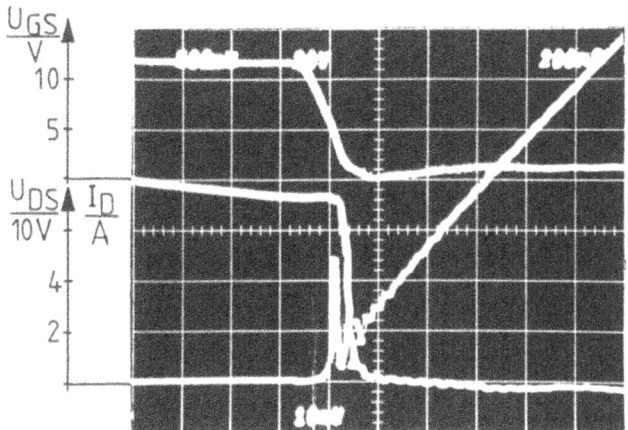

Fig. 26. Turn-off waveforms of a MOSFET under zero voltage switching conditions (200 ns/div).

Devices with conductivity modulation exhibit a different behaviour. Since sweep-out of excess carriers in the drift region is linked to a blocking voltage building up, the stored charge is removed from the device only when the voltage across C_r is rising. This is why the tail current of IGBTs is significantly larger under ZVS, compared with switching of clamped inductive load (Fig. 27). The result is that turn-off losses can be reduced by a factor of 3 to 5 only (Fig. 28). A similar behaviour is found for BJTs switching at zero voltage, the difference being that at least some of the charge can be extracted by negative gate current and does not contribute to the turn-off losses. A reduction of turn-off losses by about a factor of ten can be expected [28].

GTOs are used only with turn-off snubbers. In ZVS circuits, the snubber C_r can be placed directly across the device with minimum inductance. Thus the peak voltage during current fall time, caused by parasitic impedance of the snubber circuit, is eliminated [29]. Since turn-off losses depend on the snubber capacitor used, there is a potential for further reduction of turn-off losses by using a larger C_r. The problem of snubber losses is eliminated because the energy in C_r is recovered.

With the MCT exhibiting a similar turn-off behaviour as IGBTs, the reduction of turn-off losses is also in the same range (factor of 3 [30]). The FCTh has been operated under zero voltage switching in previous laboratory tests. Switching losses could hardly be measured because the tail current was significantly reduced, resulting in very high relative errors. However, it can be stated that turn-off losses are reduced by at least one order of magnitude [16].

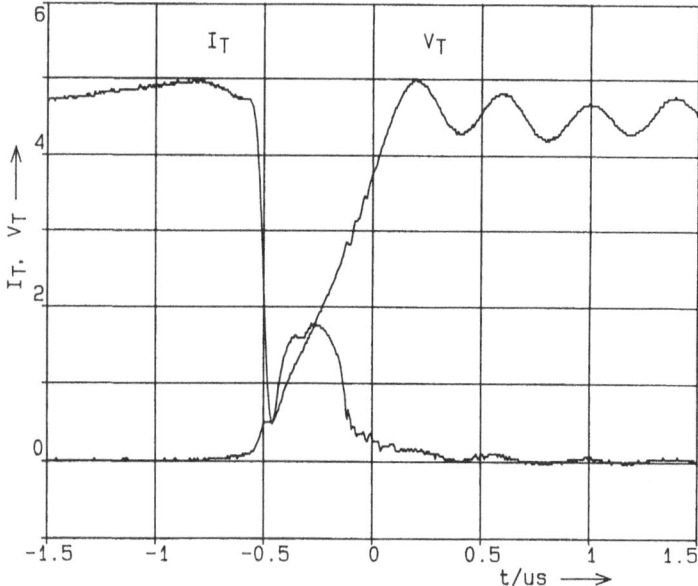

Fig. 27. Turn-off waveforms of a 50 A 1200 V IGBT under ZVS; I_T: 10 A/div, V_T: 200 V/div.

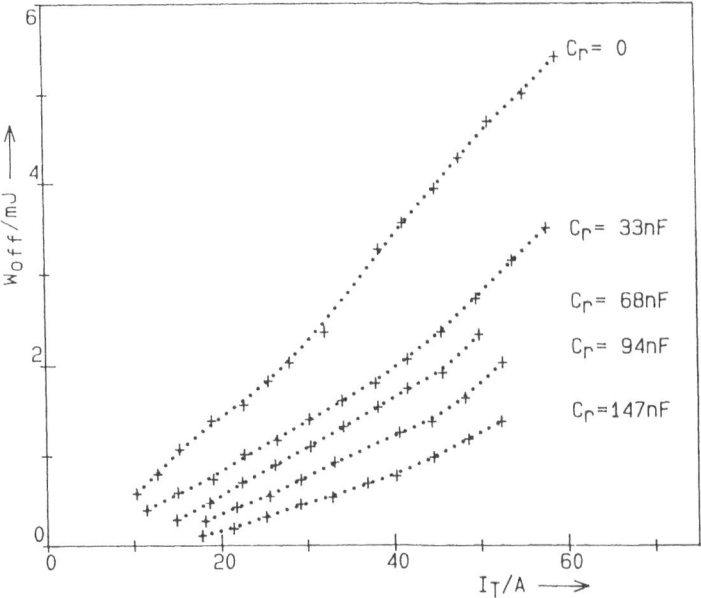

Fig. 28. Measured turn-off losses of a 50 A 1200 V IGBT as a function of turn-off current for various snubber capacitors C_r.

When turning on bipolar devices under zero voltage conditions, the conductivity modulation does not start immediately when the device takes over current with high dI/dt from the antiparallel diode. A dynamic saturation voltage appears, similar to the forward recovery phenomenon of a diode. Figure 29 shows the dynamic peak voltage during ZVS turn-on of a 10 A 2000 V FCTh as a function of dI/dt. Similar figures have been found for first generation IGBTs as well as for MCTs [30].

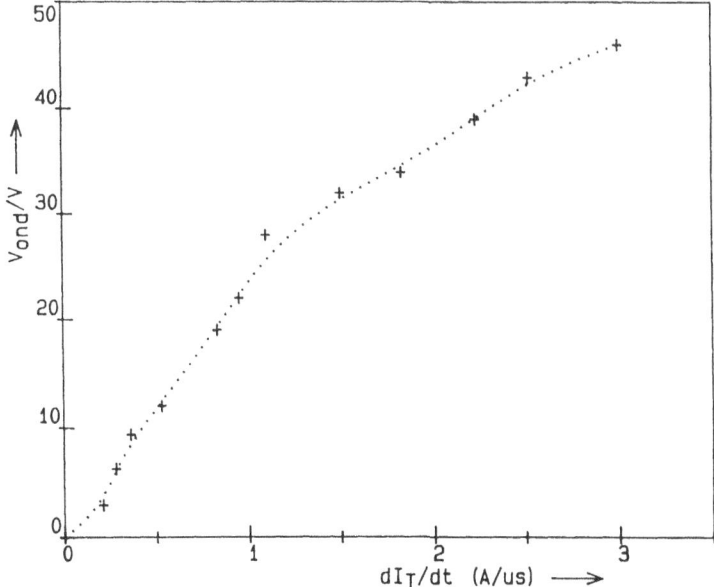

Fig. 29. Dynamic saturation voltage (forward recovery) of a FCTh (2000 V, 10 A) as a function of dI/dt.

An additional stress is caused by dissipative discharge of C_r into the device when the peak voltage is over.

Storage times of purely bipolar devices (BJT, GTO) cause a limitation of the switching frequency achievable. High peak gate currents can be applied to reduce the storage times. This however poses high demands on the drive circuits.

3.9.2 Zero Current Switching Applications.

In most zero current switching applications, a series diode is necessary in order to obtain a reverse blocking switch. The turn-off behaviour of the switch is then governed by the reverse recovery of the series diode. Also, the series diode causes additional conduction losses.

Another drawback of bipolar devices when used in ZCS circuits is illustrated in Fig. 30. Due to the MOS-gate, the stored charge can not be extracted while the switch is reverse biased. If the hold-off time between the zero crossings of current and voltage is reasonably small, there is still some charge present that causes a tail current when forward voltage is reapplied. This results in turn-off losses that can not be neglected. However, we have shown in laboratory experiments that it is possible to build a 10 kW chopper operating at 120 kHz with one 50 A 1200 V IGBT used in ZCS mode (Fig. 30).

Investigations on the behaviour of GTOs in ZCS [31] have shown that even a reverse blocking GTO has to be used with a series and an antiparallel diode (Fig. 31). The antiparallel diode provides a connection between cathode and anode such that the gate current can extract carriers from both the n regions adjacent to the gate.

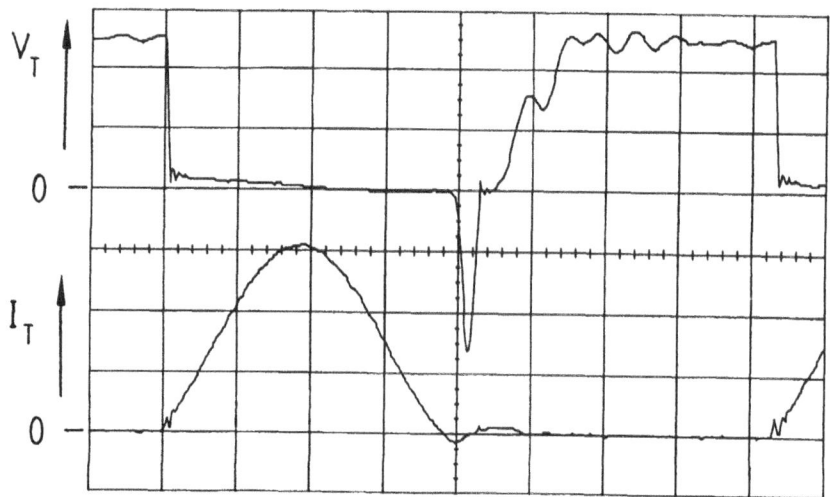

Fig. 30. Switching waveforms of an IGBT under ZCS:
I_T: 25 A/div, V_T: 200 V/div.

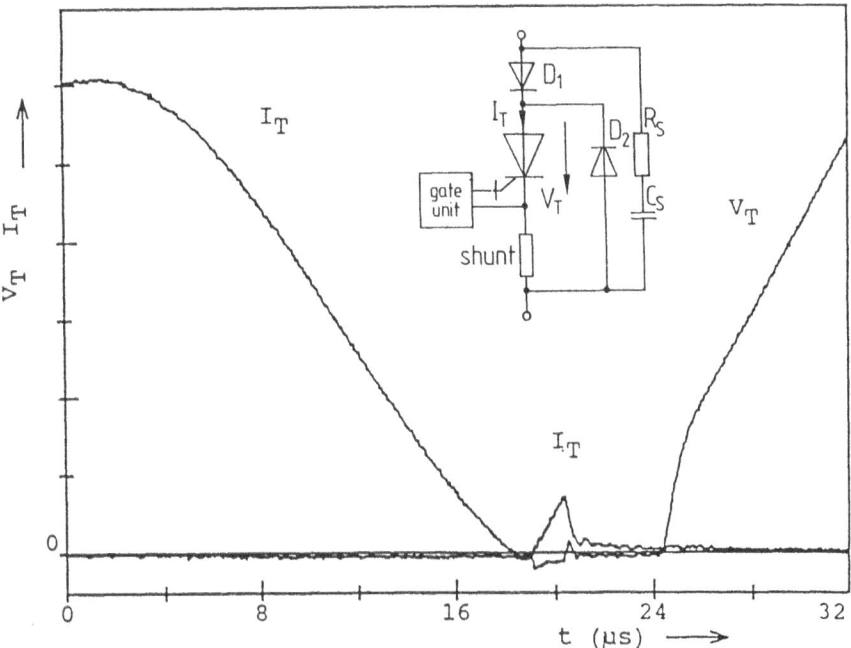

Fig. 31. Turn-off waveforms of a GTO under ZCS;
I_T: 50 A/div, V_T: 100 V/div.

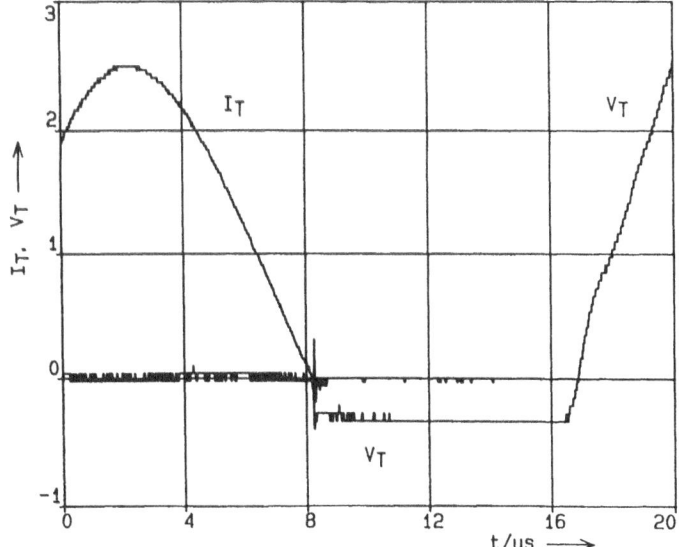

Fig. 32. Turn-off waveforms of an asymmetrical FCTh under ZCS;
I_T: 5 A/div, V_T: 100 V/div.

Thus, the GTO is returned to the forward blocking state while the switch is reverse biased. Figure 31 shows ZCS turn-off waveforms of a GTO, the peak in I_T being a part of the gate current flowing through the antiparallel diode and the anode. Turn-off losses could be reduced to less than one tenth of conventional turn-off losses. Also, the gate drive requirements are much easier to meet.

The structure of the FCTh provides a connection between the n-regions around the gate inside the device. Also, a symmetrical design can make the FCTh reverse blocking. In consequence, this device can be used without additional diodes in ZCS circuits, making it an ideal candidate for such applications. Figure 32 is showing a ZCS turn-off of an asymmetrical FCTh (a series diode had to be used). In contrast to the GTO, there is no tail current visible during voltage rise time, resulting in zero turn-off losses.

3.10 Summary

An estimation of maximum switching frequencies is summarized in Fig. 33. The region of application of the devices is shown in the plane of switching frequency vs. output power of a three phase PWM inverter. This figure has been published by a leading manufacturer of power devices and shows good agreement with our own estimations.

The MOSFET is physically limited to low voltages because of its forward voltage drop. It is the device best suited for low power applications, combining low losses, high speed switching, ease of control and protection. Further improvements of R_{DSon} for low voltage devices are expected from a reduction of the cell pitch.

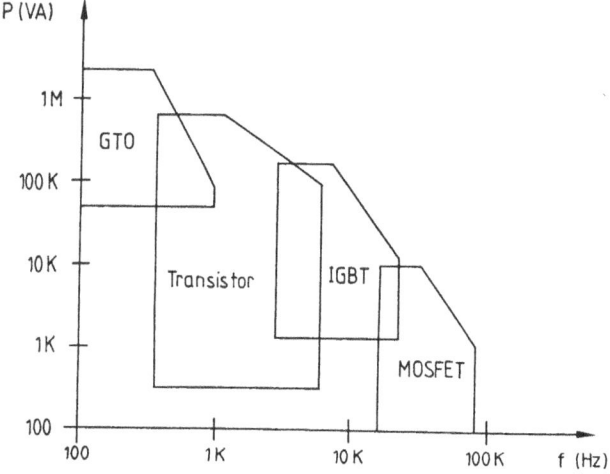

Fig. 33. Regions of application for different power semiconductor devices [11].

In the medium power range, the IGBT is in a process of replacing the Darlington-BJT. The benefits are low gate drive requirements, fast and low-loss switching and only moderately higher prices. The drawbacks of higher conduction losses and so far unsolved problems of protection keep open a field for future applications of BJTs, especially in low voltage, high current applications. IGBTs for voltages between 1500 V and 2000 V are under development.

For high power devices, GTOs are still the only choice. They are presently being improved, striving for higher voltages and lower turn-off losses with smaller snubber capacitors. To our opinion, this device will not be replaced in the near future. However, the following improvements are highly desirable:

- snubberless switching,
- higher switching frequencies (~ 2 kHz),
- possibility of paralleling devices,
- possibility of series connection,
- lower gate drive power consumption,
- short-circuit ruggedness.

Candidates for future high power applications are FCTh and MCT. Both devices have the potential to realize most of the above improvements to a certain extent. The high degree of technical advantage, resulting in higher efficiency, smaller volume, modular converter design, cheaper gate drives and higher reliability, would certainly justify a higher price for these new devices.

4. IDEAL SWITCHES

In the previous chapter, the present status of power semiconductor devices has been discussed. However, there is a continuous development in this field. New devices with improved characteristics will appear and find application.

The question may arise of which kind these improvements will be and to what extend they may pertain. Simply, which are the characteristics of an ideal switch and which are the physical limits for it.

The desired static characteristics are high blocking voltage, low on-state voltage, high forward current and high control gain. As discussed in chapter 2, there are physical limits to the voltage and current rating. The present status of the different device types and the expected margin for improvements have been discussed in chapter 3.

It is interesting to discuss the switching characteristics of an ideal device.

With reference to Fig. 4 the ideal turn-on implies switching on as quickly as possible, in the ideal case in zero time.

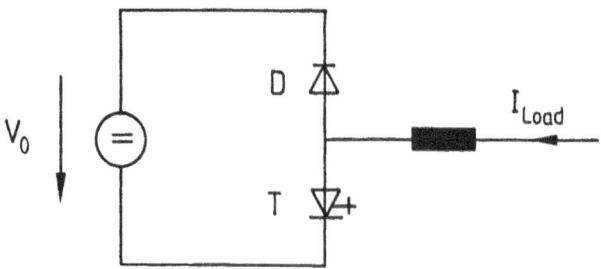

Fig. 34. Turning on an ideal switch T.

However, in most practical applications the active switch is connected to a diode to form a commutation group, including an inductive load, as shown in Fig. 34. If the switch T switches on in zero time, the voltage V_0 has to be blocked by diode D instantaneously. This is not possible with presently available diodes. What would happen is a high dI/dt in the commutation circuit, limited only by the stray inductance of the circuit. There would be a reverse recovery current in the diode which might destroy the diode by overvoltage at current snap-off. A snubber capacitor is necessary to protect the diode, and the ideal switch is stressed by the reverse recovery current in addition to the load current. The benefit of avoiding the turn-on losses of the switch T is offset by the necessary additional current stress. Hence, it is better to limit the dI/dt by non-ideal turn-on of the switch T.

From this consideration one has to conclude that the margin for improvement for the turn-on of the switch is determined by the turn-off characteristic of the diode. In other words, the improvement of the switch has to start with the diode.

Soft switching arrangements offer the possibility of reducing turn-on losses at the expense of additional passive components. Therefore, the benefit of lower losses due faster turn-on times would not be very significant. In case of ZCS circuits, an improvement of the diodes would be helpful because the reverse recovery causes oscillations and overvoltages.

A similar consideration holds for the turn-off characteristics: If an ideal switch is assumed to switch off in zero time, it would have an infinite value of dI/dt. In connection with the stray inductance of the circuit, an infinite voltage would destroy the device. This is why a well defined small turn-off time is necessary. Assuming linear decay of the current at turn-off, one obtains:

$$V_{Stress} = L_{Stray} \frac{\Delta I}{\Delta t}$$

L_{Stray} is the stray inductance in the loop that contains the device and the most adjacent voltage source, normally the blocking capacitor.

From this equation, one concludes that for high current devices the stray inductance has to be as small as possible. This is increasingly difficult in the kA range. One also sees that for increasing the switching power, it is better to raise the voltage than the current.

Apparently, there are limits to the reduction of L_{Stray}. If one does not accept a further increase of the voltage stress one has to increase the switching time Δt.

Table 5 gives some figures which illustrate the above discussion.

Table 5. Minimum switching times at a given peak voltage stress; example for high power GTO.

Type of the loop	L_{Stray}	ΔI	V_{Stress}	Δt
wired connection	1 μH	3 kA	600 V	5 μs
low inductance design	0.2 μH	3 kA	600 V	1 μs
estimated limit for flat-pack device	0.05 μH	3 kA	600 V	0.25 μs

As discussed previously, an increase of switching time increases the switching losses. Consequently, the maximum switching frequency is to be reduced with increasing current rating.

In ZVS circuits, the stray inductance can be minimized by placing snubbers right across the devices. Stray inductance is then governed by the intrinsic impedance of capacitor and device package. Improved packages with lower stray inductance would help to overcome some of the limitations found in soft switching converters.

Steep turn-off current slopes do not occur in ZCS applications.

As there is an interest in increasing the switching frequency beyond the present limits the parallel connection of smaller devices would help. Probably, modules of parallel connected devices with integrated snubbers and gate units will be useful in many applications.

When discussing the turn-off behaviour, the problem of tail current has to be addressed. Most semiconductor power devices exhibit a tail current from the anode to the gate terminal after turn-off. It decays in several μs depending on the carrier life-time and the dV/dt. As the voltage is high, the tail current contributes considerably to the losses. Particularly for GTOs, these losses are the greatest part of the overall switching losses. If the tail current can be reduced in size and time, an immediate benefit would result without any disadvantage in the performance. In section 3.9 it was shown that stored charge and tail current are the device characteristics that inhibit a very large reduction of turn-off losses in soft switching circuits.

Summarizing this chapter, the following statements are made:

- Improving the static parameters of semiconductor power devices would be appreciated by the users.

- As for the dynamic behaviour an ideal device is not desirable:
  • The turn-on time should be matched to the recovery-characteristic of the diode.
  • The turn-off time should be matched to the overvoltage caused by the stray inductance of the commutation circuit.
- Reducing the tail current, however, would be a good objective for further development.

REFERENCES

1   P. Roggwiller, R. Sittig, "Physical limitations and exploratory devices", *Semiconductor Devices for Power Conditioning*, Plenum Press, New York 1982, pp. 215-238.

2   P. De Bruyne, J. Vitins, R. Sittig, "Reverse conducting thyristors", *Semiconductor Devices for Power Conditioning*, Plenum Press, New York 1982, pp. 151-173.

3   G. Langer, "Vergleich von Gleichstromumrichtern mit hart und weich schaltenden Ventilen", Dissertation RWTH Aachen 1991.

4   Siemens: *Sipmos Halbleiter*, Datenbuch 1991/92.

5   Toshiba: *GTR-Module*, 1989; *IGBT-Module*, 1990.

6   ABB: *IGBT-Module*, 1988.

7   AEG: *LTR-Powerblocks*, 1987; *Fast Thyristors*, 1988.

8   International Rectifier, *Gate Turn-Off Thyristors*, 1988.

9   Westcode, *Gate Turn-Off Thyristors*, 1990.

10   Marconi, *Gate Turn-Off Thyristors*, 1989.

11   Powerex, *GTO Thyristors*, 1991.

12   B.J. Baliga, "Evolution of MOS-bipolar power semiconductor technology", *Proc. of the IEEE*, vol. 76, no. 4, pp. 409-418, April 1988.

13   F. Bauer et al., "Design aspects of MOS-controlled thyristor elements: technology, simulation, and experimental results", *IEEE Trans. on Electron Devices*, vol. ED-38, no. 7, pp. 1605-1611, July 1991.

14   A. Ackva, Th. Reckhorn, M. Posluszny, "Kurzschlußfestigkeit und systemeigene Sicherheit durch optimale Ansteuerung: Eine Treiberstufe für Bipolartransistoren und IGBT im Vergleich", *etz-Archiv Bd.* 12 (1990) Nr. 9, pp. 291-295.

15   F. Bauer, "The MOS controlled thyristor and its limits", *Power Semiconductor Devices and Circuits*, ed. A. A. Jaecklin, Plenum Press, New York, 1992.

16   G. Venkataramanan, A. Mertens, H.-Ch. Skudelny, H. Grüning, "Switching characteristics of field controlled thyristors", *EPE-MADEP '91*, Florenz 1991, pp. 220-225.

17   G. Fregien, "Ein Beitrag zur Anwendung von Hochleistungs-Abschaltthyristoren", Dissertation RWTH Aachen, 1988.

18   R. Bayerer, J. Teigelkötter, S. Wollenberg, "Advanced module IGBT technology for high dynamic motor drives and SMPS", *PCI & MOTORCON 88, Electronic Motion Control System International Conference*, München, June 1988, pp. 1-12.

19   H. Matsuda, T. Fujiwara, K. Nishitani, "Design optimization for improving high power GTO switching characteristics with alloy free technology", *Proc. ISPSD'90, Tokyo* 1990, pp. 240-245.

20   H.W. van Broeck, "Vergleich von spannungseinprägenden Wechselrichtern mit zwei und drei Zweigpaaren zur Speisung einer Drehstromasynchronmaschine unter Verwendung der Pulsweitenmodulation hoher Taktzahl", Dissertation RWTH Aachen 1985.

21   R. Marquardt, "Beschaltung für elektronische Zweigpaare in Antiparallelschaltung", *Offenlegungsschrift DE 3244623 A1*, 1984.

22   Th. Kalker, "Overcurrent protection for power-MOSFETs", *EPE '89*, Aachen 1989, pp. 181-184.

23   G. Fregien, "Overcurrent protection for GTO-thyristors", *EPE '87*, Grenoble, France 1987, pp. 431-436.

24   M. Kekura, H. Akiyama, M. Tani, S.-I. Yamada, "8000 V 1000 A gate turn-off thyristor with low on-state voltage and low switching loss", *IEEE-PESC'89, Milwaukee WI 1989*, pp. 330-336.

25   D.M. Divan, G. Skibinski, "Zero switching loss inverters for high power applications", *Proc. IEEE-IAS Ann. Meeting 1987*, pp. 627-634.

26   A. Mertens, D.M. Divan, "A high frequency resonant DC link inverter using IGBTs", *Proc. IPEC, Tokyo 1990*, pp. 152-160.

27   A. Mertens, "Design of a 20 kVA resonant DC link inverter on the base of experimental device evaluation", *Proc. EPE 1991*, vol. 4, Florence, Italy, pp. 172-177.

28   G.L. Skibinski, D.M. Divan, "Characterization of power transistors under zero voltage switching", *Proc. IEEE-IAS Ann. Meeting 1987*, Atlanta GA, pp. 493-503.

29   G.L. Skibinski, D.M. Divan, "GTO characterization for zero voltage switching applications", *Proc. IEEE-PESC '91*, Cambridge MA, pp. 437-443.

30   H.R. Chang, A.V. Radun, "Performance of 500 V, 450 A parallel MOS-controlled thyristors (MCTs) in a resonant DC-link circuit", *Proc. of IEEE-IAS Ann. Meeting 1990*, Seattle WA, pp. 1613-1617.

31   A. Mertens, H.-Ch. Skudelny, P. Caldeira, T. Lipo, "Characterization of GTOs under different modes of zero current switching", *IEEE-PESC, 1991*, pp. 444-452.

DISCUSSION

B.J. Baliga (NCSU, Raleigh NC)
How important is the problem of voltage overshoot at turning-on of diodes and transistors ?

H.Ch. Skudelny
The dynamic overshoot is a problem only in some applications ; this is treated in more detail in the paper.

T. Ohmi (Tohoku University, Sendai J)
Is the cost of the gate drive circuit included in the cost comparison between different converter types ? Will the IGBT still be more expensive than the BJT when taking into account the gate drive ?

H.Ch. Skudelny

Since gate drive circuits are still under development, there was no good price basis available and, therefore, these costs were not included in our estimates. Of course, the gate drive circuit is basically cheaper for unipolar than for bipolar devices.

J.M. Peter (SGS Thomson, Rousset F)

One important cause for losses are the on-state losses. If you consider small powers like 1 or 2 kW, the bipolar transistor has a very low voltage drop and low total conduction losses - if you are able to develop a good drive. What do you consider to be the best trade-off ?

H.Ch. Skudelny

If the operating voltage is low and bipolars with a sophisticated base drive, like a "super-Darlington", are used, very low conduction losses can be obtained.

N. Zommer (ABB IXYS, San Jose CA)

The frequency ranges for IGBT and MOSFET you have shown are quite limited. If you run these switches up to 4 MHz or more in a zero-voltage-switching mode, the capacitances become important with respect to losses. What are your comments ?

H.Ch. Skudelny / H. G. Langer (RWTH Aachen, Aachen D)

The figures presented apply for hard switching only (inductive load). To be clear, the evaluation of the frequency performance was based on rated currents. At lower currents, the frequency may be increased. The importance of the capacitance for zero-voltage-switching and its influence of the frequency is mentioned in the paper. However, the behaviour in resonant circuits has not been dealt with in detail.

H. Grüning (ABB Corp. Research, Baden)

Since the switching frequency in discussion depends on the voltage switched, how do you account for voltage derating ?

H.Ch. Skudelny

The figures measured have been obtained at half break down voltage of the devices. The derating is nearly in inverse proportion to the voltage.

T.A. Lipo (U. Wisconsin, Madison WI)

Can you comment on using conventional thyristors in your zero current switching application ?

H.Ch. Skudelny / A. Mertens (RWTH Aachen, Aachen D)

Conventional thyristors are not expected to have a similar behaviour in this case and were not investigated. We have done some experiments with GTO's in zero-current-switching circuits and achieved hold-off times much shorter than with regular thyristors in similar applications. We expect that we would not be able to extend the frequency limit using conventional thyristors.

D. Silber (U. Bremen, Bremen D)

Did you use the GTO in this case as a gate-assisted turn-off device ?

A. Mertens

Yes, the current was forced to zero by an external circuit while the gate drive was used to return the device to its blocking state.

HVDC VALVES WITH LIGHT-TRIGGERED THYRISTORS

Bo E. Danielsson

Asea Brown Boveri Power Systems
Ludvika, Sweden

ABSTRACT

Conventional series-connected High Voltage Direct Current (HVDC) thyristors are triggered electrically (referred to as ETTs). Using Light-Triggered Thyristors (LTTs) instead, valve electronics can be considerably simplified. Extensive comparative tests have shown that for a cost-effective solution the LTT should be self-protected, and full-sized rather than of the smaller, auxiliary type. The obvious light sources, Light-Emitting Diodes (LEDs) or Laser Diodes (LDs), are relatively weak and require a very sensitive LTT gate. A principal concern was that the LTT presented unexpected turn-on difficulties when triggering from AC voltage at low firing angle. Both LED and LD triggering of 45 cm^2, 7 kV self-protected LTTs are currently being tested in a 125 kV, 1050 A test valve in the Konti-Skan HVDC transmission link (Danish side). The light sources are in both cases exceedingly powerful and specially developed for this project. So far, the LED solution has evinced unacceptable degradation, most probably due to the high LED current. The LD system exhibits tolerable degradation margins but appears to be too expensive at present to pose a competitive threat to ETTs. A new, promising, cost-efficient solution, based on commercial 50 mW continuous-wave (cw) LDs, is under test on a laboratory scale.

1. INTRODUCTION

In many electrical power transmission applications, the High Voltage Direct Current (HVDC) technique offers substantial advantages compared to the exclusive use of Alternating Current (AC). In the case of very long overhead transmission lines (longer than 500 kilometres),

HVDC will be the most economical solution, and in other applications like high power submarine cables and back-to-back connection of electrical networks with different frequency and/or phase angle the HVDC technique appears to be the only reasonable one. In HVDC links, 50 or 60 Hz alternating current is converted into direct current and vice versa. The AC/DC converter is subjected to very high voltages and each valve in the converter bridge must be able to withstand several hundred kV, see Fig. 1. The line current is normally in the range of 1000 - 4000 A.

As indicated in Fig. 1, modern HVDC valves are based on thyristors. The voltage capability of a single thyristor is limited to less than 10 kV

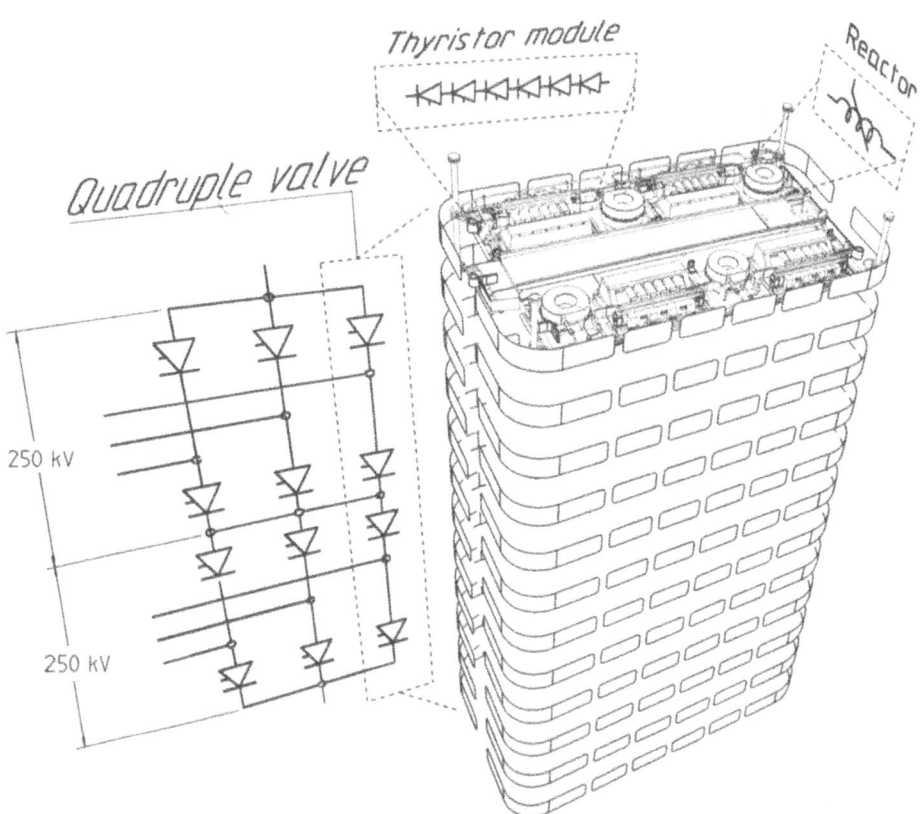

Fig. 1. Thyristor valve (quadruple valve) for HVDC transmission systems. The thyristors are arranged in modules with current-limiting reactors in between.

and a large number of thyristors must therefore be connected in series to support the voltage. The thyristor string is often organized in thyristor modules with current-limiting inductors in between. Most thyristors are located at a high electrical potential, thus making certain demands upon the valve design:

a) the necessary means for firing and protection of the thyristors must be energized by valve current at each particular thyristor level

b) triggering signals between the firing and protection means and the control system at the ground potential must be transmitted optically.

Even though a rather sophisticated but cost-efficient and reliable solution based on a compact Thyristor Control Unit (TCU) has been evolved for conventional, Electrically Triggered Thyristors (ETTs)[1], firing the thyristors directly by an optical signal without use of a TCU has been considered as an advantageous approach for simplifying valve electronics in order to possibly reduce converter cost and further improve valve reliability.

Use of Light-Triggered Thyristors (LTTs) has been discussed since the first development of HVDC thyristor valves in the 1960s, but this concept has up to now been economically inferior to the well-established conventional solution using ETTs. So far, HVDC valves utilizing LTTs have only been used in test installations[2,3,4] and in an HVDC transmission in USA/Canada[5], but they appear to be gaining increasing acceptance in Japan.

A main concern when using LTTs is that sufficiently inexpensive and reliable light sources can only produce comparatively weak triggering signals. The LTT gate must therefore be exceedingly sensitive, still allowing safe triggering. The gate area for initial turn-on must then be very small in order to reduce the risk of spurious capacitive triggering at fast voltage transients which, in turn, may reduce the inrush current capability of the LTT. These conditions were discussed in detail by Silber et al.[6] and several different solutions were presented for the delicate design and optimization of an LTT device. For the LTTs in our investigations, the concept of a multistage amplifying gate for obtaining good du_D/dt and di_T/dt properties despite the weak optical triggering signals of the latter, was adopted by the supplier[7].

In order to take full advantage of the LTT concept, the sophisticated TCU protection feature should be omitted. Consequently, the LTT should be self-protected against voltage triggering, i.e., the thyristor must not be damaged by spurious triggering due to an excessively high off-state voltage (leakage current), an excessively fast du_D/dt (displacement current) or an unintended forward blocking voltage during the reverse recovery period (stored charge). Self-protected LTTs have now been announced by several major thyristor manufacturers[7-13]. Most of them, however, are still at the prototype stage.

A summary of our investigations on different circuit configurations, based on LTTs, and of our laboratory and field test experiences of the solution chosen was presented very recently[4]. To present a more complete background, these investigations and experiences will be briefly dealt with here and the solution selected will then be treated in more detail. An unexpected triggering reluctance at AC voltage, discovered in realistic application tests, will be explained as being the result of fundamental physical effects. Comments will also be made on a new design utilizing recent improvements in optical components.

2. VALVE REQUIREMENTS

The HVDC valve design is based on requirements emanating from installation specifications and circuit calculations. The main aspects are:

a) The converter must be able to operate at steady state under specified conditions within rather wide current and voltage ranges. As an example, valve current and voltage waveforms at normal rectifier operation ($\alpha = 15°$), resulting in thermal equilibrium, are shown in Fig. 2a and 2b.

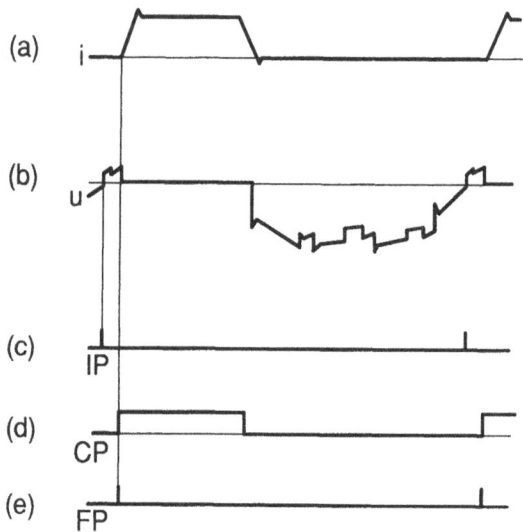

Fig. 2. (a) Thyristor current and (b) voltage waveforms with corresponding logic pulses (c)-(e) for an HVDC rectifier. IP = indicating pulse, CP = control pulse and FP = firing pulse.

b) Uninterrupted operation must continue during transitory AC faults and disturbances. Other sudden events like lightning strokes or earth faults close to the converter may interrupt operation momentarily but they must not damage the converter valves.

c) The valve must withstand factory tests specified at higher AC and DC levels than can be obtained in service conditions in order to ensure sufficient design margins.

The valve requirements, in turn, form the basis of the thyristor design. Because of the large number of thyristors in a single HVDC installation, the thyristors are generally optimized for that particular project. Development and manufacture of these HVDC thyristors are accomplished by extensive measurements and tests in synthetic test circuits in accordance with the specific thyristor requirements. The complete set of established data is finally confirmed in the HVDC Thyristor Specification. In the case of an advanced LTT, basic properties like turn-on behaviour, inrush current and voltage rate-of-rise capabilities are of particular importance and must be carefully investigated and determined.

Fulfillment of the valve requirements is ensured by testing the valve structure in accordance with standard procedures[14,15] and additional internally specified tests. Any valve design based on LTTs must therefore be subjected to a complete type test program including the tests referred to above in order to be approved for HVDC use.

Finally, in replacing the well-established and proven ETT solution in HVDC applications, the LTT valve must be examined in the course of commercial testing for a sufficiently long time in order to check long-term properties and gain field experience.

The type test procedures can be classified as operational tests or dielectric tests. Important test items are briefly commented on below. Approximate test values, valid for a 7 kV thyristor, are indicated in the text.

2.1 Operational Tests

Operational tests are usually performed using a 6-pulse back-to-back test circuit for best representation of realistic operational stresses of components.

a) Periodic firing at different delay angles. In addition to a heat run test at normal converter firing angle (typically 15°) and a regular inverter operation test (typically 18° commutation margin), two important extremes have to be tested. At $\alpha = \alpha_{min}$ (typically 5°), the applied anode to cathode voltage of some thyristors might be too low for triggering because of uneven voltage distribution along the thyristor string (see Fig. 3). At $\alpha = 90°$ the anode voltage is at maximum, which will create the maximum repetitive inrush current stresses (several hundred A in amplitude, but lasting only about 10 µs).

b) Operation at low AC voltage. Triggering difficulties may occur for the same reason as for $\alpha = \alpha_{min}$.

c) Operation at intermittent direct current. The current may become intermittent during the 120° conduction period at very low current levels because of a certain AC ripple. The firing system must then refire the thyristors several times during the conduction interval.

d) Surge current tests. If a short circuit should occur across a converter valve, a surge current of fundamental frequency, but about

10 times higher in amplitude than the line current, will pass through the thyristors in another valve. The thyristors must then be able to block immediately after the surge pulse. They must also be able to withstand a number of subsequent current surges (normally three or four) without damage if the blocking command should fail.

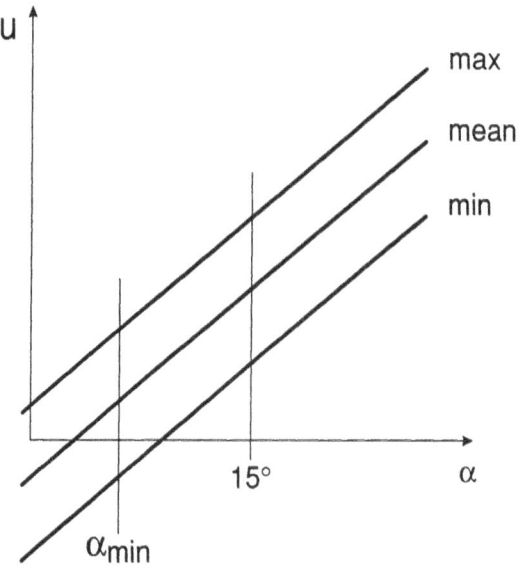

Fig. 3. Uneven thyristor voltage distribution resulting in low or negative thyristor voltage at $\alpha_{\min}$.

e) Impulse voltage during valve reverse recovery. The thyristors could be self-triggered by the stored charge if a forward blocking voltage occurs accidentally across the valve before the thyristors recover after the conducting period. The thyristors must sustain such self-triggering under the specified subsequent current conditions, otherwise adequate protective firing must be installed. Note: A test of this capability is not specified in the standard procedures. Typical voltage and current waveforms upon recovery protection triggering are shown in Fig. 4.

2.2 Dielectric Tests

Full (non-scaled) voltage is required in the dielectric tests as they are performed on a complete (single) valve. No repetitive current is involved in these tests.

a) AC and DC voltage test. The thyristor string must withstand voltages that exceed the rated values with specified safety margins.

b) Switching impulse test including non-periodic firing. Transitory impulse voltages may reach the converter when, for example, AC network switches are operating. The voltage amplitude is limited by the valve surge arrester, but, in rare cases, the amplitude can be high enough to cause protective firing of the valve. The thyristors must then withstand the subsequent inrush current (several kA in amplitude). The voltage stresses are normally represented in the test by a 250/2500 µs standard impulse waveshape voltage.

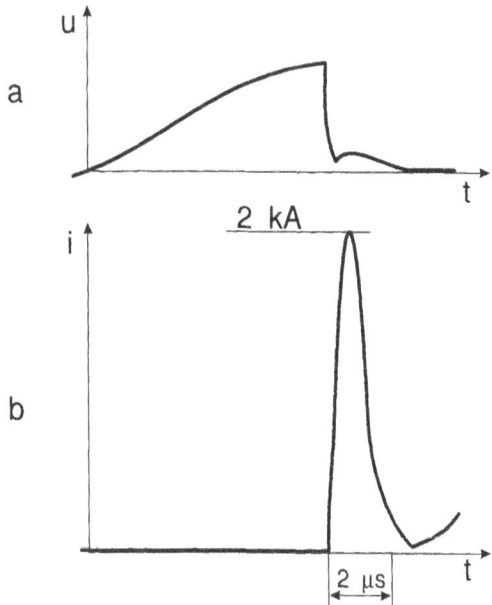

Fig. 4. (a) Voltage and (b) inrush current waveforms upon recovery protection triggering.

c) Lightning impulse test. Lightning strokes near the converter can create very fast impulse voltages across the valve (directly or by induction). The resulting du_D/dt stresses must not damage the thyristors. In cases of very high du_D/dt, protective firing may be activated. Testing is normally performed with a 1.2/50 µs standard impulse waveshape voltage. The corresponding du_D/dt is reduced by the reactor and the RC snubber to a few kV/µs.

d) Steep front test. Earth faults close to the converter will result in very steep impulse voltages across the valves. The voltage rate-of-rise is normally specified to be 1200 kV/µs resulting in higher du_D/dt stresses on the thyristors (about 10 kV/µs, but of a limited amplitude) than in the lightning impulse case.

3. LTT CIRCUIT ALTERNATIVES

LTTs can be used in the HVDC valve in two ways, different in principle, see Fig. 5, i.e., either as an auxiliary thyristor that supplies gate pulses to a full-size ETT (main thyristor) or as a full-size thyristor replacing the ETT. The conventional ETT solution is also shown in the figure for comparison (the actual circuitry is in all cases greatly reduced for the sake of simplicity).

In all three cases, an optical indicating pulse (IP) is generated at each thyristor level as soon as the anode to cathode voltage across the thyristor has reached an appropriate level for safe turn-on, as indicated in Fig. 2c. When the converter control system subsequently orders the valve to conduct, i.e., when the system generates the control pulse (CP) in Fig. 2d, the valve control system will send optical firing pulses (FP), Fig. 2e, to all thyristor levels of the string. In the ETT case the FP triggers the TCU to release a gate pulse from a precharged capacitor. This optical/electrical conversion of the regular triggering signals does not occur in the auxiliary LTT case, but the protection and monitoring functions of the TCU will still remain. In the full-size LTT case, finally, the protection function is integrated in the thyristor itself so that the TCU can be replaced by a simple thyristor monitoring unit (TMU).

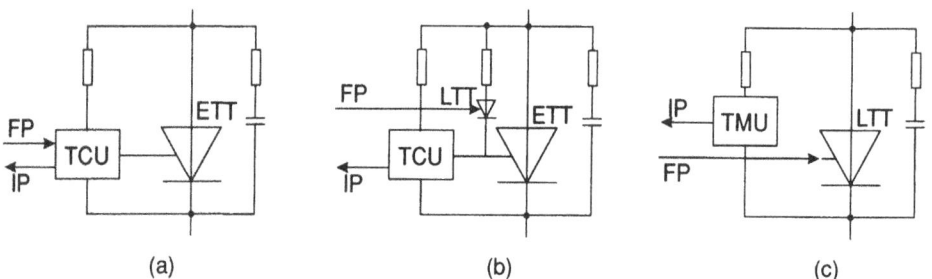

(a)   (b)   (c)

Fig. 5. (a) Indirect optical triggering of an ETT; (b) and (c) direct optical triggering of an auxiliary LTT and a full-size LTT, respectively. In the latter case, the Thyristor Control Unit (TCU) is replaced by a simpler Thyristor Monitoring Unit (TMU).

3.1 Evaluation of LTT Circuit Solutions

Obvious advantages of the auxiliary LTT concept are[6]:

a) auxiliary LTTs can be standardized for easy combination with different ETT types.

b) these LTTs will not be heated up by load current and they will therefore operate at a lower and more constant temperature than the full-size LTT. This condition should be advantageous for the auxiliary solutions as it will facilitate proper design of the gate structure.

The full-size LTT concept, on the other hand, will offer a simpler and probably more cost-efficient solution, particularly if the LTT is self-protected.

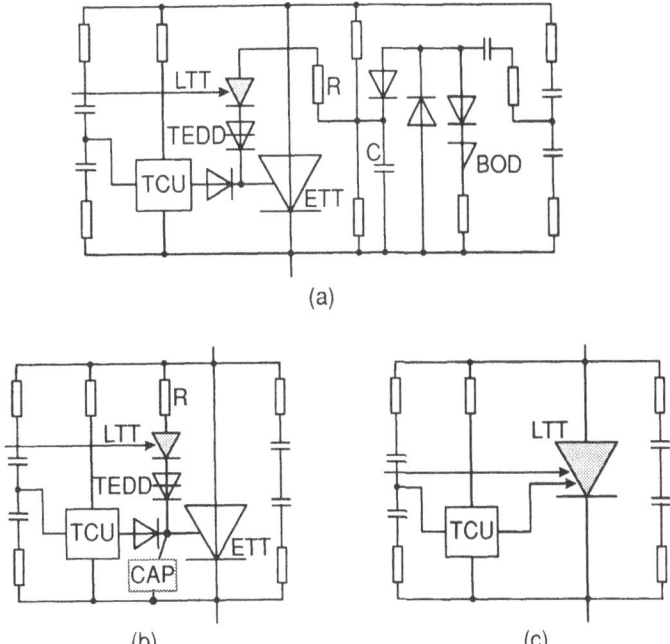

(a)

(b)   (c)

Fig. 6. Different LTT alternatives for comparative evaluation. (a) Low voltage auxiliary LTT (solution A) with charging circuits, voltage grading elements and voltage-limiting Break-Over Diode (BOD): sufficient gate current to the ETT is ensured by means of the transient surge suppressor (TEDD). (b) Full voltage auxiliary LTT (solution B): spurious ETT firing by displacement currents from the LTT is prevented by a capacitive current suppressor (CAP). (c) Full-size LTT (solution C): protective triggering is effected optically by the TCU.

In order to evaluate possible use of LTTs in HVDC valves, three different LTT solutions, as set out in Fig. 6, were examined using laboratory measurements as well as extensive operational and dielectric tests. Two variants of the auxiliary LTT concept (solutions A and B) were investigated. The third solution (C) utilized a full-size LTT (without self-protection at that early point in time). All LTTs were triggered by dual infrared Light-Emitting Diodes (LEDs) connected to the LTT by optical fibre bundles. Hard driving of selected samples of HVDC standard type LEDs for TCU triggering resulted in satisfactory turn-on (30 mW/8 μs optical pulses, 5 times overdrive). The three LTT solutions are described in more detail below:

a) Solution A is characterized by (1) voltage grading circuits for reducing the LTT voltage and (2) charging circuits for storing gate pulse energy in the capacitor C. The LTT voltage is limited by a Break-Over Diode (BOD), thus protecting the LTT against dangerous overvoltages. By adjusting the voltage grading elements, the same type of LTT can be used for different thyristor level ratings. On triggering the LTT, the stored charge in capacitor C will generate a gate pulse to the ETT. The ETT is prevented from firing at a too low capacitor charge by the transient surge suppressor (TEDD). The gate current amplitude is limited to about 50 A by the resistor R. Protective triggering is effected by the standard TCU.

b) The circuitry in solution B is considerably simplified as compared to solution A. When triggering the LTT by an optical signal at a higher anode to cathode voltage than the TEDD switching voltage (35 V), the ETT anode to gate voltage will provide a gate pulse for firing the ETT. The amplitude of the gate pulse will be greatly dependent on the voltage level, and in particular on the difference in voltage turn-on transients for the two thyristors. In practice, the amplitude is limited in the worst case to about 200 A by the resistor R. Protective triggering is effected by a standard TCU in this case too.

c) Obviously, solution C is the simplest one as the main thyristor is triggered directly by the optical signal. As mentioned above, the LTT in this comparative investigation is not of the self-protected type. Protective triggering is therefore brought about by a separate light pulse from the modified TCU in which the electrical gate pulse feeds an output LED.

3.2 Brief Summary of Results

Solution A would probably work satisfactorily in an HVDC converter. The ETT always receives reliable gate pulses, even at low delay angle or low AC voltage operation (possibly with the exception of intermittent direct current operation in which the refiring pulses can be rather weak). The turn-on delay time is nearly independent of the ETT voltage (but of course longer than for the conventional ETT solution as the delay time for the auxiliary LTT is included). Some negative aspects:

a) A possible drawback is that, in the case of distorted voltage waveforms, a gate pulse could be released from the capacitor C without triggering the ETT, i.e., if the anode to cathode voltage is too low for turn-on.

b) The main objection to this solution is that the circuit is rather complex and costly and thus can hardly be said to simplify the valve electronics.

Solution B turned out to be questionable for HVDC valve applications:

a) The main drawback is that the displacement current emanating from the LTT at high du_D/dt may cause spurious firing of the ETT unless this capacitive current is not suppressed by special means (the dotted box in Fig. 6). This risk has also been reported by others[16]. Besides increasing the complexity of the total circuitry, the suppression circuit will bypass part of the gate current upon regular firing which, in turn, will increase the overall turn-on delay time, especially at low voltages. (A possible way to avoid this spurious triggering without using means for current suppression could be to reduce the area of the LTT, but this would at the same time reduce its current capability).

b) Another negative experience of solution B is that spurious triggering of a single ETT at the AC voltage test may result in voltage breakdown and destruction of both the ETT and the LTT in the subsequent negative half-period, since the thyristor voltage will then be double as high as in the undisturbed case.

Further simplification of solution B might be obtained by using an auxiliary LTT with integral self-protection[12], but this will not obviate the drawbacks mentioned above.

The test experiences of solution C were promising, except that some of these early LTT samples suffered from insufficient voltage and inrush current capabilities due to manufacturing imperfections. The results indicated:

a) Further work was to be done to increase the di_T/dt and du_D/dt capability margins despite the weak triggering power.

b) Optical protection triggering initiated by TCU circuits instead of self-protection did not prove to be an economical and reliable method.

The conclusion of this evaluation was then that:

a) Solution A should be avoided because of its complexity and because it probably does not provide any cost reduction.

b) Solution B should be excluded for the same reasons (although not to an equal extent), but also because of the risk of spurious triggering.

c) Solution C proved to be a promising concept but, to obtain an economical solution, the LTT should be of the self-protected type in order to avoid the need for costly external protection.

4. HVDC VALVE WITH FULL-SIZE SELF-PROTECTED LTTS

As a result of the evaluation of the various LTT concepts, subsequent valve development was mainly focused on efficient use of a full-size LTT with self-protecting features, improved gate sensitivity and improved di_T/dt vs. du_D/dt trade-off.

The LTT valve was designed to be field tested in a 125 kV, 1050 A test valve (Fig. 7). This single valve, originally equipped with 7 kV ETTs, was designed for commercial operation in the Konti-Skan HVDC transmission link (Danish side), replacing a mercury-arc valve. The main tasks of the design work were then to:

a) design appropriate optical systems for supplying the required firing pulses

b) design suitable valve electronics for monitoring the status of the thyristors and for controlling the firing pulses

c) verify the valve performance in operational and dielectric tests.

Fig. 7. The LTT test valve installed in the Danish terminal of the Konti-Skan HVDC transmission link.

Table 1. Main data for 45 cm^2 LTT for HVDC application. The given di_T/dt values are valid for both light triggering and overvoltage triggering.

Off-state voltage	7000	V
Line current	1600	A
On-state voltage	1.7 V	at 1600 A
Min. optical triggering power	10	mW at λ = 940 nm*)
du_D/dt capability	4000	V/μs to 7000 V
di_T/dt capability, single pulse	3000	A/μs
di_T/dt capability, 50 Hz	800	A/μs
Maximum operating temperature	80	°C

*) LED triggering; Laser triggering (840 nm) reduces the minimum value to about 5 mW because of the concentrated light spot.

4.1 The 7 kV Self-Protected LTT

A new 45 cm^2, 7 kV self-protected LTT was available in 1987 and it was primarily developed to fulfill the requirements for use in the Konti-Skan test valve. Important thyristor ratings quoted by the manufacturer (full utilization)[7] are shown in Table 1.

In order to obtain a reasonably short delay time, t_d, at turn-on, the thyristor should be triggered by a higher optical power than the specified minimum value. An overdrive factor of 3 resulting in $t_d \leq 4$ μs upon triggering from 1000 V proved to be sufficient in that respect, thus requiring an optical light power of at least 30 mW (λ = 940 nm) at the thyristor inlet. However, it was found in a separate study that the delay time decreases continuously with increasing optical power down to less than 0.2 μs at 100 times overdrive. At still higher overdrive (up to 5000 was tested) the light absorption will create a conduction channel directly in the gate region, thereby causing the turn-on delay to practically vanish.

The sensitivity of the inrush current capability to reduced gate overdrive was studied indirectly in a non-destructive test. The transient thermal response upon the inrush current pulses was then measured using an IR camera in the way described for ETTs[17]. No marked increase in the temperature stresses was observed even at low gate overdrive, as can be seen in Fig. 8. This satisfactory behaviour demonstrates the adequate action of the multistage amplifying gate. Thus the overdrive factor of 3 is quite sufficient to ensure good inrush current capability.

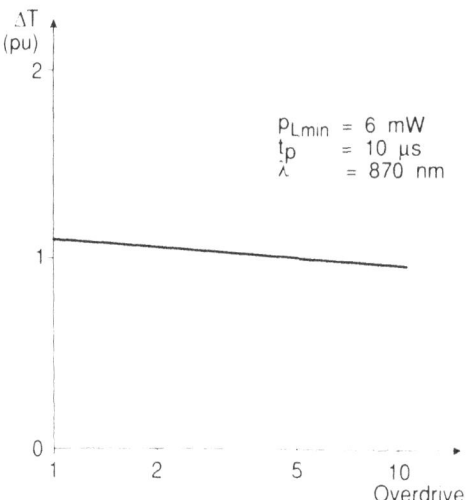

Fig. 8. Infrared camera measurement of the transient temperature increase ΔT in the gate region upon triggering for different gate overdrive factors; ΔT = 1 pu at overdrive factor of 5.

Self-protection triggering at overvoltage and high du_D/dt was intentionally designed to take place in the gate region in order to ensure good di_T/dt properties. The triggering voltage levels were guaranteed by the thyristor supplier being sufficiently higher than the corresponding TCU protection levels for ETTs, see Fig. 9. The even more delicate issue of self-protection capability upon triggering during the recovery period, however, was studied separately in a physical 12-pulse converter model, in which each branch of the bridge was represented by a single thyristor and all relevant circuit capacitances and inductances were properly scaled. After applying a DC voltage across the bridge, a load current pulse was conducted through the device under test. Earth faults at different times during reverse recovery were then simulated by triggering a spark gap, and the resulting voltage and current transients proved to be similar to those in Fig. 4. The LTT withstood self-triggering without damage at both room temperature and the maximum operating temperature.

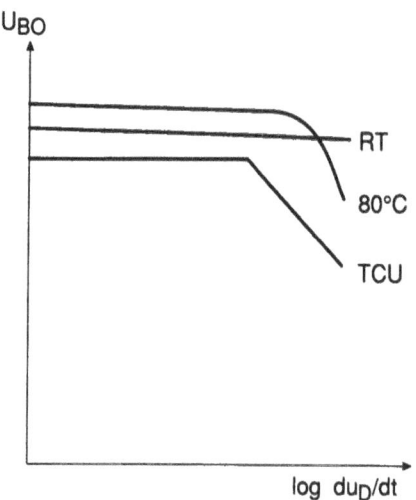

Fig. 9. Self-protection levels at room temperature (RT) and 80 °C, compared with the TCU protection characteristics.

Another encouraging experience gained from these converter model experiments was that the transients measured showed good agreement with regular computer calculations, see the example in Fig. 10, thus supporting the calculation technique. The calculations indicated that, in off-state conditions, earth faults could cause du_D/dt-triggering but probably no overvoltage triggering in this converter scheme.

4.2 Optical Triggering Systems

As indicated in the Introduction, any LTT concept has to compete with the well-established ETT solution. This implies, as a rough esti-mate, that the extra cost of the more powerful LTT optical system must not be higher than the cost reduction when replacing the TCU by a TMU. Our conclusion regarding LED triggering was then:

a) Not more than one LED can be afforded for each thyristor level.

b) The optical link must utilize a large-area fibre bundle of natural glass.

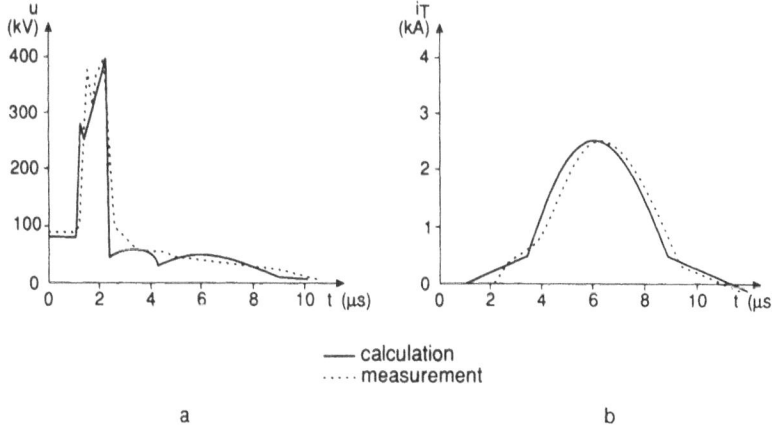

Fig. 10. Twelve-pulse converter modelling of (a) voltage and (b) current transients in protection triggering conditions.

The conclusion for laser diode triggering was:

a) Several thyristors must be triggered by a small number of laser diodes, using an optical star coupler.

b) A small-area monofibre of low-damping synthetic glass should be used for the optical link.

No LEDs or laser diodes with sufficient light power capability to ful-fill the above requirements were commercially available at that point in time (1987). New light sources with high optical output were therefore developed for our project:

a) A prototype LED based on the HVDC standard LED concept: the increased light power was mainly obtained by introducing aluminium into the chip structure and by improving the cooling design.

b) A prototype 1 W quasi-cw (continuous-wave) laser diode (from an external manufacturer): acceptable properties were obtained by improving the chip design, the laser cooling and the burn-in procedure.

LEDs of the Burrus type (i.e., an LED structure for which the optical coupling efficiency is improved by etching a dip into the substrate at the emitting region) were also considered, but this approach was rejected because the light power of commercially available samples was far too weak and because preliminary investigations indicated that the development of strong Burrus diodes would require too much effort.

Two optical systems based on LEDs and Laser Diodes (LDs), respectively, were designed for comparative evaluation, see Fig. 11.

In the LED approach in Fig. 11a, the LTT can be triggered satisfactorily by a single light source (30 mW is needed for 3 times overdrive) provided that the fibre length is limited to about 20 m. There is, however, only a small margin for light power degradation. In fact, separate degradation tests at increased frequency, corresponding to 30 years of operation (3.5 A/20 μs, 40° C, 2.5 kHz), indicated that the LED output will gradually degrade to about 35 % of its original value, see Fig. 12. The degradation of these prototypes, not observed for HVDC standard LEDs, is primarily due to the high drive current and it is probably further accelerated by the elevated ambient temperature. Tests at reduced current (2.9 A/10 μs) and room temperature indicated a considerably lower degradation.

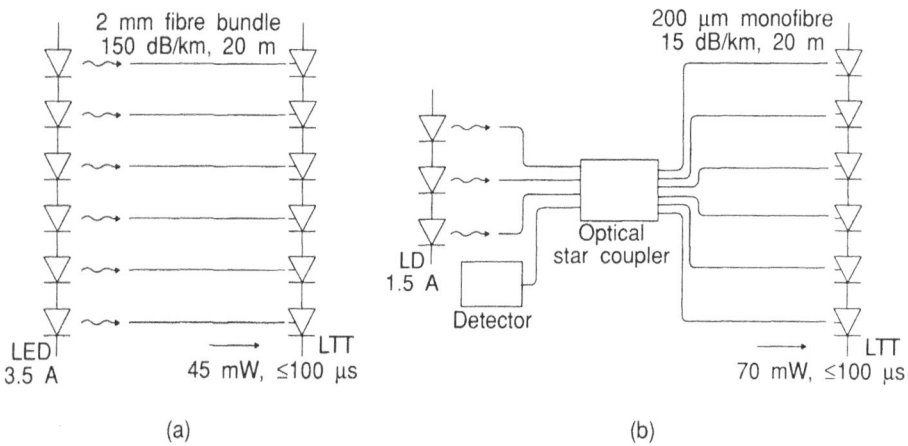

Fig. 11. Optical firing systems for full-sized self-protected LTTs, based on (a) Light-Emitting Diodes (LED) and (b) Laser Diodes (LD).

In the laser diode approach in Fig. 11b, on the other hand, three laser diodes are used for triggering six thyristors. The LDs can be driven at a fairly low current level to generate a more than satisfactory light pulse through each of the optical 6 x 6 star coupler outlets, despite the

substantial insertion light power loss. In fact, one laser diode is redun-
dant as only two of them are needed to generate sufficient light power
(70 mW will result in an overdrive of more than 10). Degradation tests
indicated that the optical power of the laser diodes in this application
will degrade to about 60 % after 30 years of operation, which is considerably lower than for the LEDs. Furthermore, the degradation can be
monitored continuously during the operation as one of the remaining
inlets of the star coupler is used for detecting the back-scattered light as
a measure for the incoming light power. The low-loss monofibre
includes polymer cladding to increase mechanical strength. Despite the
comparatively large diameter of the monofibre, this new light guide type
proved to meet all our mechanical and electrical HVDC requirements.
The length of the light guide is of minor significance because of its low
damping, thus, imposing practically no restrictions on the distance between the control equipment and the valve.

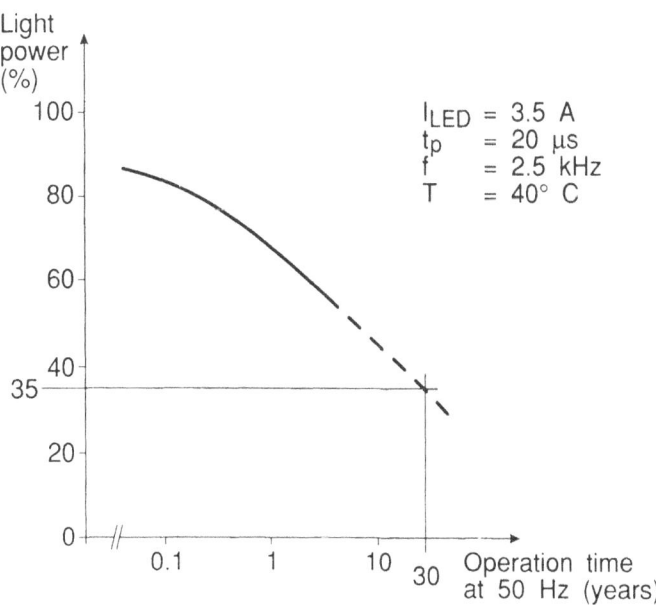

Fig. 12. LED degradation test at increased frequency. The x-axis is scaled to the time
that corresponds to continuous 50 Hz operation.

When using direct light triggering, a new principle for generating
firing pulses from the valve control electronics (Fig. 13) turned out to be
advantageous. Unlike the conventional solution in Fig. 2, high-frequency
Indicating Pulses (IP) are generated by the TMU as long as the LTT voltage exceeds about 110 V. The optical firing signal (FP) starts at the beginning of the Control Pulse (CP) command and will last as long as IP is
generated. The FP signal is then cut off as soon as all thyristors in the
valve have turned on, thus, reducing the stress on the light sources. The

firing pulse length will normally be about 20 µs, but operation at low delay angle or low AC voltage will require longer pulses. The maximum pulse length is limited to 100 µs.

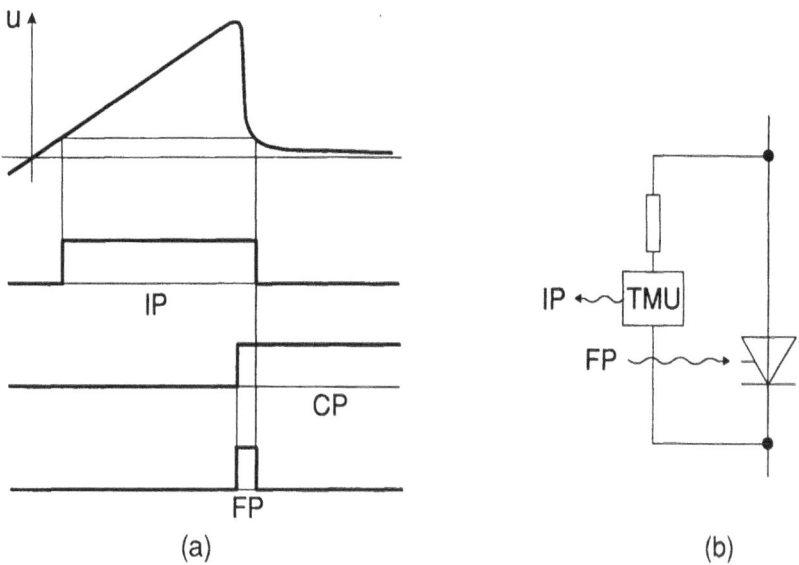

(a)   (b)

Fig. 13. Firing pulse principle (a) for the system based on full-size self-protected LTTs (b). High frequency Indicating Pulses (IP) are generated by the TMU as long as the thyristor voltage exceeds about 110 V. On Control Pulse command (CP), a Firing Pulse (FP) is supplied to the thyristor until it has turned on.

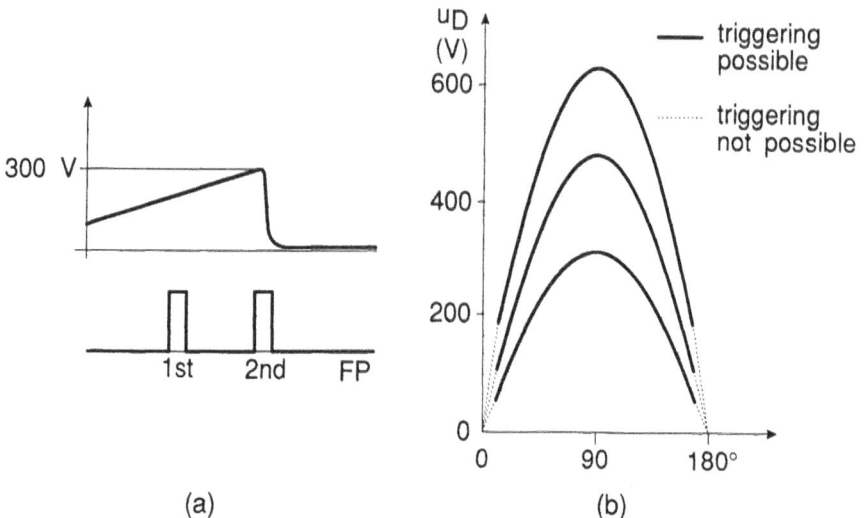

(a)   (b)

Fig. 14. Triggering from AC voltage in (a) preliminary operational type test and (b) laboratory test. A second firing pulse was required in case (a).

4.3 Firing Difficulties in AC Conditions

Surprisingly high voltage levels for enabling thyristor firing were observed in preliminary operational tests. In fact, the thyristors required typically 300 V to be fired with 10 μs pulses at 2 times overdrive, see Fig. 14a. As no firing difficulties had been observed in regular triggering tests at low DC voltage (50 V), this unexpected behaviour must be due to the actual operational conditions. Additional laboratory 50 Hz AC investigations on single devices showed that, in spite of the fact that the thyristor could be fired from less than 10 V if it was triggered from the top of the sine curve, it could not be fired from the AC ramp at a lower voltage than about 50 V if the amplitude was increased to 320 V (the same overdrive in both cases), see Fig. 14b. Still higher values were required if the amplitude was further increased as indicated in the diagram. Furthermore, these conditions proved to be symmetrical around the AC maximum, thus indicating that a possible influence of external capacitive currents was not a probable cause of this triggering reluctance.

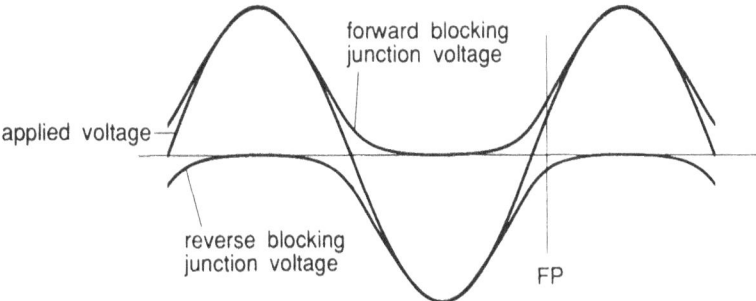

Fig. 15. Pn-junction voltages in AC conditions. The reverse blocking junction is reverse-biased when firing is requested (FP).

The origin of this phenomenon, perhaps not obvious at first glance, is instead that the forward and reverse blocking pn-junctions are not charged and discharged at the same rate as the applied AC voltage (Fig. 15). The reverse blocking junction is then still strongly reverse-biased if the firing signal is applied at an early stage of the rising thyristor voltage. This junction must first be sufficiently discharged by a positive anode current until complete firing of the thyristor can be accomplished. With this explanation in mind, the AC effect was physically demonstrated using a simple model of a pnp structure (see Fig. 16), in which discrete small diodes, capacitors and resistors represented the basic pn-junction properties (rectifying behaviour, charge and leakage, respectively). The applied voltage is supported solely by the appropriate pn-junction at the voltage peak, but, because of the rectifying properties of the diode, the corresponding capacitor cannot be discharged during the subsequent AC voltage decrease without charging the opposite one at the same time unless the leakage current through the resistor is high enough. The junction voltage decrease will then slow down correspondingly. This precharging effect is approximately symmetrical on both sides of the voltage peaks, thus explaining the symmetrical behaviour in Fig. 14b.

The AC effect has been studied theoretically under the simplified assumption that the leakage current can be disregarded. Using the step junction approximation, the width w_n of the depletion layer in the n-base depends on the absolute value of the junction voltage u_j (including the diffusion voltage) as

$$w_n = \sqrt{\frac{2\varepsilon}{q}\frac{1}{N_D}u_j} \quad , \tag{1}$$

where ε is the dielectric constant of silicon, q is the electron charge and N_D is the base doping concentration.

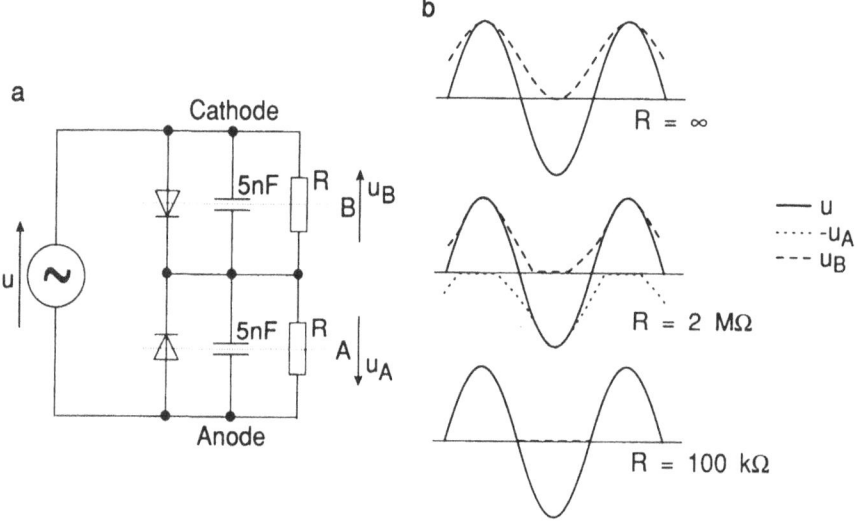

Fig. 16. (a) Simple physical model of a pnp structure subjected to AC voltage; B = forward blocking junction and A = reverse blocking junction (the thyristor cathode junction is not represented here because it is shunted by the emitter shorts). (b) Measured voltages for different values of the resistance R.

On altering the voltage u applied to the thyristor, two depletion layers, emerging from opposite pn-junctions, appear (Fig. 17). Since the neutral part between the depletion layers will move in pace with the AC voltage, the sum of the two depletion layer widths, w_{nA} and w_{nB} in Fig. 17, will remain unchanged. However, this sum equals w_{nB} at $u = U_0$ as w_{nA} will practically vanish when the voltage is supported by the forward blocking junction only. Let

$$u = U_o \sin \omega t \quad . \tag{2}$$

Then, disregarding the junction diffusion voltage and bearing in mind that w_n is proportional to $\sqrt{u_j}$ and u is the sum of the applied voltages u_A, u_B across the pn-junctions, we obtain:

$$\sqrt{u_A} + \sqrt{u_B} = \sqrt{U_o} \quad , \tag{3}$$

$$u_B - u_A = u \quad , \tag{4}$$

resulting in

$$u_B = \frac{(u + U_o)^2}{4U_o} \quad , \tag{5}$$

$$u_A = \frac{(u + U_o)^2}{4U_o} - u \quad . \tag{6}$$

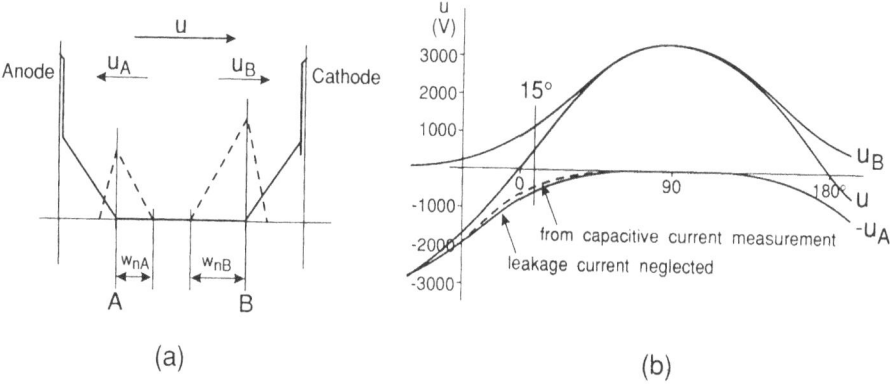

(a)   (b)

Fig. 17. Theoretical calculation of the pn-junction voltages of a thyristor subjected to AC voltage. (a) Thyristor structure and depletion layers. (b) Applied thyristor voltage u and applied voltages u_A and u_B across the junctions A and B, respectively. Two methods are used for calculating the reverse blocking junction voltage u_A.

The three voltages u_A, u_B and u versus time is shown in Fig 17b for a relevant case ($U_o = 3300$ V). It can be seen in this diagram that at the normal firing angle $\alpha = 15°$ the reverse blocking junction is still strongly reverse-biased (more than 500 V).

The assumption that the leakage current during the voltage alteration can be disregarded is only justified when the thermally generated leakage current from the depletion layer of a reverse-biased junction is much smaller than the associated displacement current. Otherwise, the generated electrons will be used for gradually building up the increasing neutral part of the n-region. Hence, as the leakage current will drastically increase with temperature, the AC effect will be greatly reduced at high temperatures.

The influence of any leakage currents at room temperature was then taken into consideration by studying the measured capacitive current from the LTT at AC voltage, see Fig. 18a. Since the current i_{cap} from the reverse blocking junction, A, is (junction area = A)

$$i_{cap} = - A\ C_d\ \frac{du_A}{dt}\ , \tag{7}$$

where the junction capacitance C_d per unit area is (step junction approximation)

$$C_d = \sqrt{\frac{q\varepsilon}{2}\ N_D}\ \frac{1}{u_A}\ , \tag{8}$$

the capacitive current can be used as a measure of the junction voltage u_A. The approximation in equation (8) was checked at different DC voltages using a capacitance meter. As indicated in Fig. 18b, the agreement between calculated and measured values was good for $u > 100$ V. Introducing equations (2)-(4), we finally obtain:

$$i_{cap} = \frac{A\ \sqrt{\frac{q\varepsilon}{2}\ N_D}}{\sqrt{u_A + \sqrt{U_o\ \sin\omega t + u_A}}}\ \omega\ U_o\ \cos\omega t\ . \tag{9}$$

u_A can now be solved for various phase angles $\alpha = \omega t$ using the corresponding measured i_{cap} and actual values of A and N_D . This semi-empirical relationship between u_A and α is in good agreement with the theoretical one, as shown in Fig. 17b, thus validating the assumption in equation (3) and thereby indicating that the leakage current has only a minor influence on the AC effect at room temperature.

In case reverse bias remains on the reverse blocking junction upon thyristor triggering, this junction must be discharged by anode current (holes injected from the anode and electrons passing through the forward blocking junction) before firing can be completed. The anode current, in turn, must be stimulated by the optical firing pulse. Therefore,

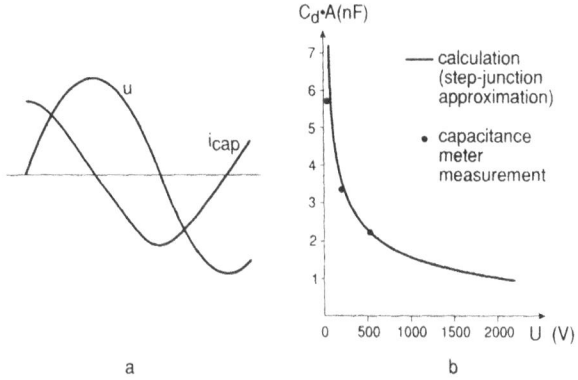

a   b

Fig. 18. (a) Measured capacitive current for an LTT subjected to AC voltage. (b) Calculated and measured pn-junction capacitance of the LTT.

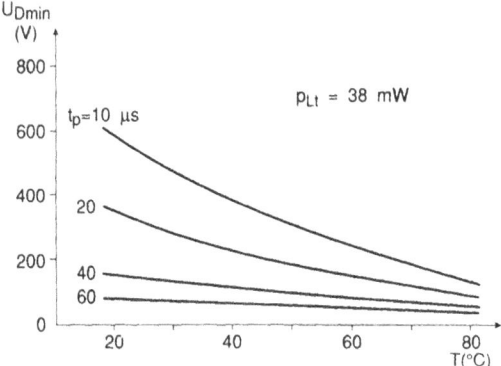

Fig. 19. The dependence of minimum off-state voltage U_{Dmin} (for triggering in AC conditions) on operating temperature at different pulse lengths.

the AC effect causes a substantial increase in the optical energy needed for turn-on. This can be achieved by increasing either the optical power or the firing pulse length. In case of LED triggering, the maximum optical power is limited to about 40 mW, and the pulse length must then be substantially increased to achieve turn-on at low U_{Dmin} values, see Fig. 19. However, adequate increase in pulse length is easily obtained by using the new firing pulse principle (Fig. 13). The diagram in Fig. 19 also shows that the AC effect is reduced at elevated thyristor temperature. This behaviour is mainly caused by the increase in leakage current at high temperature as discussed above.

The AC effect is perhaps not a major concern in HVDC applications as it can be neutralized by using sufficiently long firing pulses. However, different LTT samples showed different reluctance to triggering at AC voltages. One way used in our project to avoid negative effects of too high U_{Dmin} values is simply to reject improper samples in a 100% test. Another way, of course, is to further increase the optical energy, but this will result in even higher stresses on the light sources. Instead, a better approach would be to consider this effect already at the LTT design stage in order to reduce or, preferably, eliminate its influence on the turn-on behaviour.

4.4 Type Tests

Series connection of self-protected LTTs proved to work satisfactorily in the type tests. A main concern was if differences in turn-on delay time would cause dangerous voltage spikes on late-triggering samples, but no such effects were observed, not even in cases of forced self-protection triggering (maximum voltage overshoot was about 1 kV, resulting in a total voltage spike of about 9 kV). Sufficient inrush current capability at self-protection triggering was demonstrated, both upon overvoltage triggering and triggering in reverse recovery.

Comparing LED and laser diode triggering systems, no major differences were found. Both systems passed the final operational test, even if there turned out to be no margins for the LED triggering when operating at low delay angle or low AC voltage.

The fully equipped test valve (48 thyristor levels) passed all dielectric tests, including a forced self-firing test (the major part of the valve was optically fired on the top of a switching impulse while the rest was subjected to self-firing), and even in cases when self-firing occurred, no thyristor failure was observed. Note: In a special test with enhanced valve voltage (well above the type test requirements), forced self-triggering of the whole valve resulted in some cases in thyristor damage. Burn marks appeared in the gate region of these thyristors, thus indicating that the inrush current capability margin was rather small.

Consequently, on the basis of the type test results, the new HVDC valve design with full-size self-protected LTTs, monitored by simplified electronics and triggered by either LEDs or laser diodes, was technically approved for use in the test valve application.

4.5 Test Valve Experiences

The Konti-Skan LTT test valve is partly triggered by LEDs (36 thyristors) and partly by LDs (12 thyristors), see Fig. 20a. The valve entered into commercial operation in May 1988 and the total operation time until August 1991 is more than 12,000 hours. No forced outages or other severe operational disturbances due to the LTT technique have been observed up to now.

The main concern during the scheduled follow-up has been the continuous degradation of the light sources, see Fig. 20b, as expected from previous degradation test experience. The mean value of the light power from the LEDs is now quite close to its allowable bottom limit, and, in consequence of this trend, twelve LEDs generating light pulses that were too weak have now been replaced by new ones. Three non-working LEDs have been observed during the follow-up, one LED after 2500 hours and two LEDs after 10,500 hours operation time. These failures have most probably occurred in regular valve operation before the inspection. Two of the corresponding thyristors were in good condition, thus indicating that they did survive operation without regular light triggering due to the self-protecting action. The third one, however, was damaged because of a voltage flash-over at the wafer edge. Obviously, the voltage capability of that thyristor was too low to match continuous long term self-protecting operation.

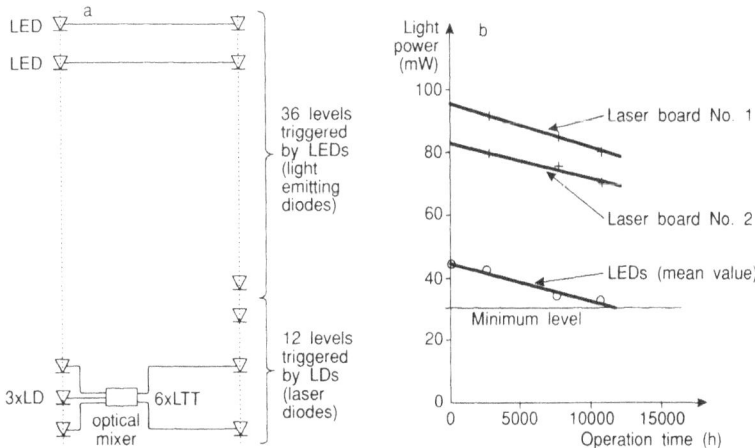

Fig. 20. (a) Optical triggering system layout for the Konti-Skan test valve. (b) Prototype light source output (thyristor end) versus test valve operation time. The minimum power level for safe triggering is indicated in the diagram.

The laser diodes have worked quite satisfactorily so far. Degradation appears to be faster than expected from the laboratory tests, but there are still wide margins to the minimum allowable light power level, and no thyristor problems have been observed.

5. CONCLUSIONS AND SOME FURTHER COMMENTS

An economically competitive LTT valve design must be based on a full-size LTT rather than on the auxiliary one. The LTT must be self-protected to avoid complicated electronic protection circuits. The concept reported here utilizing 45 cm² 7 kV LTTs, triggered by LEDs or laser diodes, is technically approved for the Konti-Skan test valve application. Full-scale commercial introduction, however, will call for further improvements, both concerning the thyristors (greater inrush current margins at self-protection and reduced triggering reluctance in AC conditions) and the light sources (lower degradation) as well. The economic aspects of the LTT solution in comparison with the cost-effective conventional ETT system can therefore not be fully clarified until these obstacles have been removed.

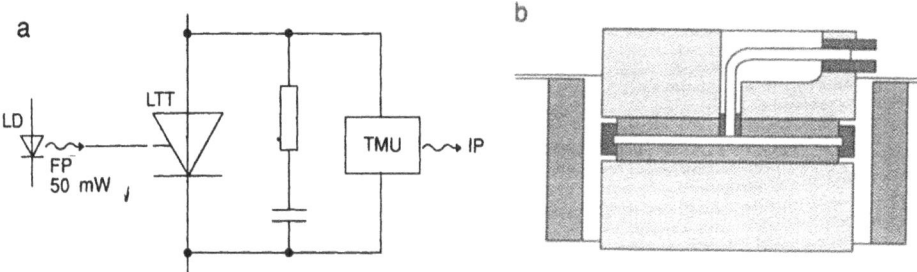

Fig. 21. Improved LTT system, based on commercial continuous-wave laser diodes and thyristors designed for laser triggering. (a) Outline of the basic arrangement. (b) Mechanical design of the LTT.

A main concern is still the cost and the reliability of the light sources. The degradation problem can, of course, be handled in a straightforward manner by using two or more LEDs per thyristor (or a more generous laser diode arrangement) to reduce their drive current, but this should be avoided for economic reasons. A better way, currently under development and schematically shown in Fig. 21, would be to design the thyristor for laser triggering only. The optical sensitivity can then be considerably improved due to the concentrated light beam, both regarding the wafer gate response and the optical feed-through efficiency. Furthermore, laser triggering has the additional important advantage that there will be practically no restrictions of the length of the light guides between the valves and the control equipment, as low-loss monofibres can be used. Commercially available 50 mW continuous-wave laser diodes (originally intended for office automation products) are used as light sources, one LD per thyristor. No light source redundancy is needed, as the thyristor will operate by self-triggering in case of missing regular firing pulses (in fact, a failing laser diode can be replaced by a new one without affecting valve operation). Even though the LDs are of the continuous-wave type, the light pulses are cut off by the TMU signals when the thyristors have turned on in order to save LD life time. Possible degradation is handled differently in this solution. Using a small photodiode included in the laser diode housing, the optical power is continuously monitored and regulated by a fast-response electronic circuit. The optical power is kept constant and the degradation, if any, will result in an increasing drive current. This regulation circuit will also perform the delicate task of providing a very steep light pulse front with an acceptable overshoot. On-going degradation tests are accelerated by an elevated ambient temperature and the results are promising so far.

A similar LTT technique has already been developed to be used in our thyristor switch for Controlled Series Compensation (CSC)[18]. This test switch, bypassing a series capacitor (see Fig. 22), will be installed in Kanawha River (USA) in the autumn of 1991. Although this switch operates somewhat differently compared to HVDC valves (no firing cut-off for technical reasons, and no repetitive triggering at AC voltage due to the continuous current operation), it is believed that this development and coming field test experience will form a valuable basis for HVDC applications as well.

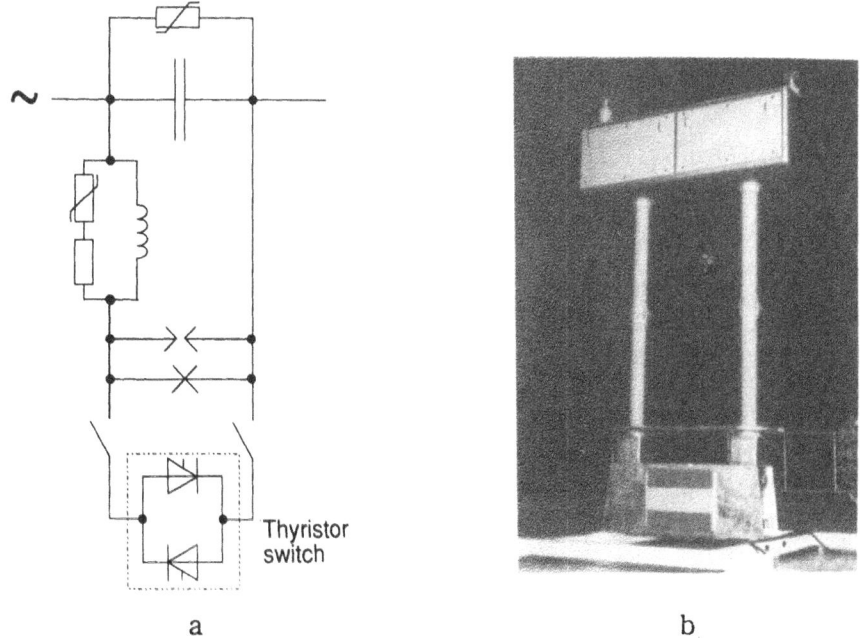

a   b

Fig. 22. Thyristor switch for Controlled Series Compensation (CSC). (a) Basic circuit diagram of the series capacitor by-pass arrangement. (b) Picture of the thyristor switch.

REFERENCES

1   Å. Ekstrœm and E. Albertsson, "Electrically triggered thyristors still have the edge", *Transmission and Distribution* (Oct 1986), pp. 65-69.

2   G. Addis, B.L. Dansky and R. Nakata, "Advanced HVDC valve", *Proc. IEE* no. 255 (1985), pp. 340-345.

3   K. Murabayashi, T. Takahashi, S. Tanabe, H. Ikegame, S. Motegi, S. Kobayashi and N. Itoh, "Development of a 500 kV light-triggered thyristor valve for HVDC transmission", *Proc. IEEE MONTECH '86*, pp. 161-167.

4   B.E. Danielsson, "HVDC valve with light-triggered thyristors", *Proc. IEE* no. 345 (1991), pp. 159-164.

5   Hydro-Quebec HVDC Project Administration; private communication.

6   D. Silber, H. Maeder and M. Fuellmann, "Light-activated thyristors", *Semiconductor Devices for Power Conditioning*, Plenum Press, New York (1982), pp. 49-78.

7   A. Åberg, "Some comments on the design of light fired power thyristors", *Proc. 13th Nordic Semicond. Meeting* (1988), pp. 77-80.

8   O.L. Harding, P.D. Taylor and P.J. Frith, "Recent advantages in high voltage thyristor design", *Proc. IEE* no. 255 (1985), pp. 315-319.

9   Y. Shimizu, R. Iyotani, N. Konishi and T. Yatsuo, "A high-voltage light-activated thyristor with a novel over-voltage self-protection structure", *IEEE Trans. El. Dev.*, ED-36 (1989), pp. 1001-1004.

10   F. Cibulka, L. Crane and J. Marks, "Field evaluation of industry's first self-protected, light-triggered thyristor", *IEEE Trans. Pow. Delivery*, PD-5 (1990), pp. 110-115.

11   H. Mitlehner, F. Pfirsch and H.J. Schulze, "A novel 8 kV light-triggered thyristor with overvoltage self protection", *Proc. 1990 Int. Symp. Pow. Semicon. Dev. & ICs* (1990), pp. 289-294.

12   L.O. Eriksson, V.A.K. Temple, L.B. Major and H. Mehta, "V_{BO} - protection of power thyristors using an external trigger thyristor", *1990 IEEE Ind. Appl. Soc. Annual Meeting Conf. Rec.* (1990), pp. 1648-1657.

13   K. Itoh, Y. Tsunoda, K. Akabane, H. Kashiwazaki and S. Horiuchi, "Development of large-capacity VBO-free light-triggered thyristors and their application to SVC valves", *Conf. Rec. 22nd Annual IEEE Pow. Electron. Spec. Conf.* (1991), pp. 453-459.

14   *IEC Standard Publication 700*, "Testing of semiconductor valves for high-voltage DC power transmission".

15   *IEEE Std. 857-1990*, "IEEE guide for test procedures for HVDC thyristor valves".

16   L.O. Eriksson and F.A. Selim, "A light triggered pilot thyristor for high voltage applications", *SATECH '86* (1986), Paper P2-1.5.

17   B.E. Danielsson, "Initial turn-on area of gate-controlled thyristors", *Solid State Electron.*, 22 (1979), pp. 659-662.

18   R.M. Maliszewski, B.M. Pasternack, H.N. Scherer Jr., M. Chamia, H. Frank and L. Paulsson, "Power flow control in a highly integrated transmission network", *CIGRE 1990 Session*, Paris, France (1990).

DISCUSSION

A. Jaecklin (ABB Corp. Research, Baden)

Can you compare your efforts to those of competitors, especially in Japan, working on similar concepts ?

B. Danielsson

To my knowledge, their thyristors are not self-protected. Typically, they use two LED's per thyristor and the optical path is relatively short. In contrast, we need around 50 m of light guides in a large HVDC installations, making it impossible to use LED's presently available.

A. Jaecklin

The concept of explaining your firing problems by two simultaneously blocking space charge layers is very interesting. Have you seen such an effect in conventional high voltage thyristors too ?

B. Danielsson

Yes, some conventional thyristors are not triggering at e.g. 150 V in our test circuit and need a second trigger pulse. Normally, you will not notice this effect because the gate current is too high.

P. Silber (TU Bremen, Bremen D)

What does your concept of self-protection include ?

B. Danielsson

Basically, three protection functions are included:
• overvoltage protection, that means triggering in case of a high leakage current,
• excessive dv/dt, indicated by large capacitive currents,
• too high amount of stored charges, leading to triggering during reverse recovery.

H. Grüning (ABB Corp. Research, Baden)

How does the optical power required for the electrical trigger unit compare to that for a directly light-triggered thyristor ?

B. Danielsson

The length of the optical pulse is typically about 1 μs and the LED is driven with approximately 1 A for indirect instead of 3.5 A for direct light triggering. In the former case, there have never been any problems with LED's.

B.J. Baliga (NCSU, Raleigh, NC)

What is the progress in technology and what are the highest voltages you would expect in future ?

B. Danielsson

The device voltage is not mainly a question of technology but rather-more of a cost optimisation between voltage capability, losses, and in particular Q_{rr} that may lead to very high snubber losses.

P. Streit (ABB Semiconductors, Lenzburg CH)

Your triggering problem, caused by the reverse voltage, would probably vanish if you were using an asymmetrical device where forward and reverse blocking functions are separated. Would you propose to reconsider such a solution, as presented already 10 years ago ?

B. Danielsson

This AC triggering reluctance is not a major problem and can be overcome e.g. by using a longer gate pulse. The use of an asymmetric thyristor and a separate diode has always been a more expensive solution for HVDC compared to the conventional reverse blocking thyristor.

INTELLIGENT POWER INTEGRATED CIRCUITS

Jean-Marie Peter

SGS-Thomson Microelectronics
Rousset, France

ABSTRACT

The intelligent power IC, alternatively called SMARTPOWER, is a device, which integrates data processing and power functions in one chip.

For the next five years, we do not see any fundamental progress but several improvements (like cost reduction - design tools - less parasitic effects) and new technologies (like Silicon On Insulator, SOI - wafer bonding) opening new ways.

The major question concerning SMARTPOWER evolution is:

GIVE MUSCLES TO THE BRAIN ?
or
GIVE BRAIN TO THE MUSCLES ?

The first concept is the natural continuation of the IC evolution. The SMARTPOWER technology will allow to realize more and more complete systems on one chip.

The second concept, completely different, leads to realizing not a system on one chip, but an improved component. It is well known that the latest progress in power components is not due to the basic structures (bipolar and MOSFET transistors are close to their asymptotic limit), but is due to the combination of structures (example: GTO - IGBT - MCT).

With the SMARTPOWER technology, it will be possible to integrate in one chip all auxiliary functions the designer needs: drive - protection - alarm - status - sophisticated information etc.

Power Semiconductor Devices and Circuits, Edited by A.A. Jaecklin
Plenum Press, New York, 1992

These two evolutions, "complete systems", and "improved components" on one chip, correspond to an evolution in the design engineer's task: "More system and less hardware." The silicon technology will open many new ways to simplify their job.

But silicon is not able to perform "miracles". Especially, it will not be able to replace all functions taken care of by the magnetic components (transformers - inductances) and to suppress the parasitic phenomena. In order to utilise all silicon improvements, progress will also be needed in the field of electromagnetism (concepts - technology - teaching).

1. INTRODUCTION

The intelligent power Integrated Circuit (IC), often called "SMARTPOWER", is a monolithic device which integrates:
- data processing
- power processing

"Power" was defined some years ago by Dataquest as an output current greater than 1 A. In this paper, we will consider the following maximum ratings for SMARTPOWER:

V < 100 volts: I_{output} (total current if multi-outputs) > 4 A.

V > 100 volts: Switchable power V • I > 400 VA.

The aims of this paper are:

1) to give the state-of-the-art in the 1992 SMARTPOWER market.

2) to put into perspective some of the trends in the development of SMARTPOWER for the next five years, 1992/1996.

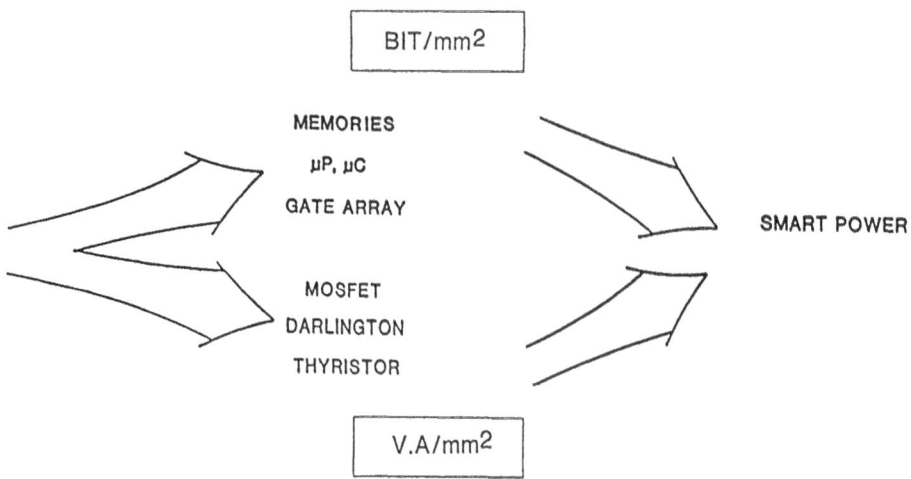

Fig. 1. Semiconductor evolution.

1.1 History

Six years ago, there were two main semiconductor areas (Fig. 1):
- The development of ICs, memories, microprocessors, op amps, gate arrays, etc., that focused on improving precision, speed and increasing the ratio "data processing/mm^2".
- The development of power discrete components: thyristors, bipolar transistors, Metal Oxide Semiconductor Field Effect Transistors (MOSFETs) and protection devices that focused on reduction of power losses, adapting semiconductor packages and increasing the ratio "switchable power/mm^2".

IC development and production had little or no connection with the development and production of power discrete devices. Each activity took place separately, generally in different factories with different management, in isolation from each other.

However, SMARTPOWER is the result of the union between these two different approaches, in a monolithic form.

1.2 An Example of SMARTPOWER

Figure 2 shows a typical SMARTPOWER IC. It is a 60 V self-protected switch, designed and manufactured for the automotive industry. The power stage is a 0.05 Ohm Double Diffused MOS (DMOS) transistor. The data processing function includes the electronics necessary to drive and protect the switch in the automotive environment and to send status signals to a central processor. This device will be manufactured in high volume for controlling lights, wiper motors and fan motors in cars.

2. SMARTPOWER IN 1992

2.1 The Technologies

Currently, several technologies can be used to integrate the two functions, signal processing and power stage, onto one chip. But overall, we have to consider only two major families, corresponding to the major direction of the current flow in the power stage.

The first family using "lateral" technology (Fig.3), is derived from the technologies used for integrated circuits. In this family, the power devices are made with a "horizontal" power transistor. This technology which uses the experience, gained in making ICs, offers many possibilities and allows the realisation of bridge or half bridge power stages, frequently used in power electronics. But the resistance per unit area is very high which leads to a high cost for the power stage.

50 mOhm - 60 V - SOLID STATE RELAY

- OUTPUT CURRENT (CONTINUOUS): 28A @ $T_c=25^{\circ}C$
- LOGIC LEVEL 5V COMPATIBLE INPUT
- THERMAL SHUT-DOWN
- UNDER VOLTAGE PROTECTION
- OPEN DRAIN DIAGNOSTIC OUTPUT
- VERY LOW STAND-BY POWER DISSIPATION

DESCRIPTION

The VN20 is a monolithic device made using SGS-THOMSON Vertical Intelligent Power Technology, intended for driving resistive or inductive loads with one side grounded.

Built-in thermal shut-down protects the chip from over temperature and short circuit.

The input control is 5V logic level compatible.

The diagnostic output indicates open circuit (no load) and over temperature status.

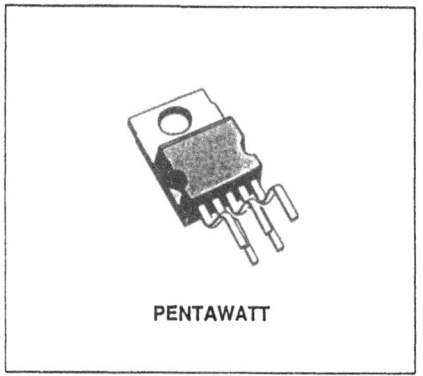

PENTAWATT

TYPE	V_{DSS}	$R_{DS(on)}$	I_{OUT}	V_{CC}
VN20	60 V	0.05 Ω	28 A	26 V

BLOCK DIAGRAM

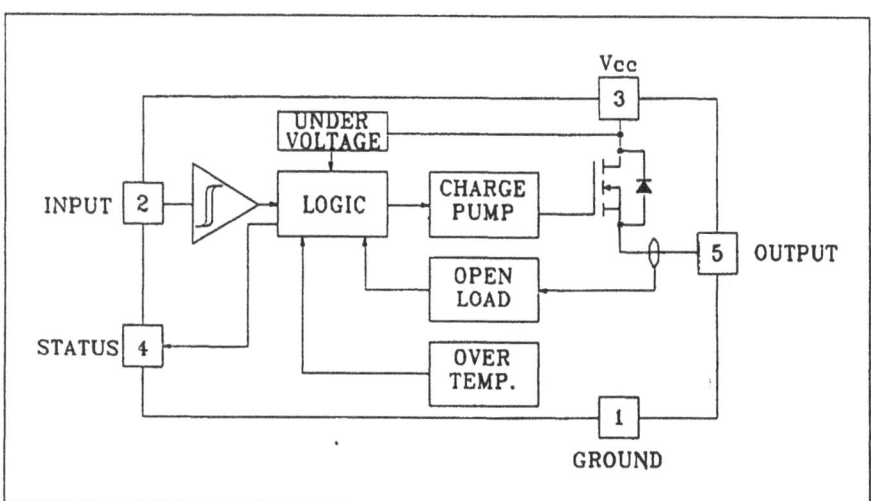

Fig. 2. Power switch for car application (60 V, 50 mΩ).

The second family called "vertical" technology is derived from that of discrete components. The power devices, MOSFETs and bipolar transistors, are designed and realised like a typical discrete component (the current flows vertically between collector/drain and the emitter/source). A signal island, insulated by a "well" formed in the silicon surface, contains the cells necessary for digital and/or analog control. The main advantage of this technology is that the power device has the same performance as a discrete component and has the same ratio of switchable power/mm^2. But this technology is less flexible than the horizontal technology and can only be used to realise a power stage with one power component (or several with a common collector/drain) and is not adaptable to bridge topologies. The quasi vertical technology - for example B.C.D.: Bipolar, Complementary MOS (CMOS), DMOS - combines some advantages of both technologies.

a   b

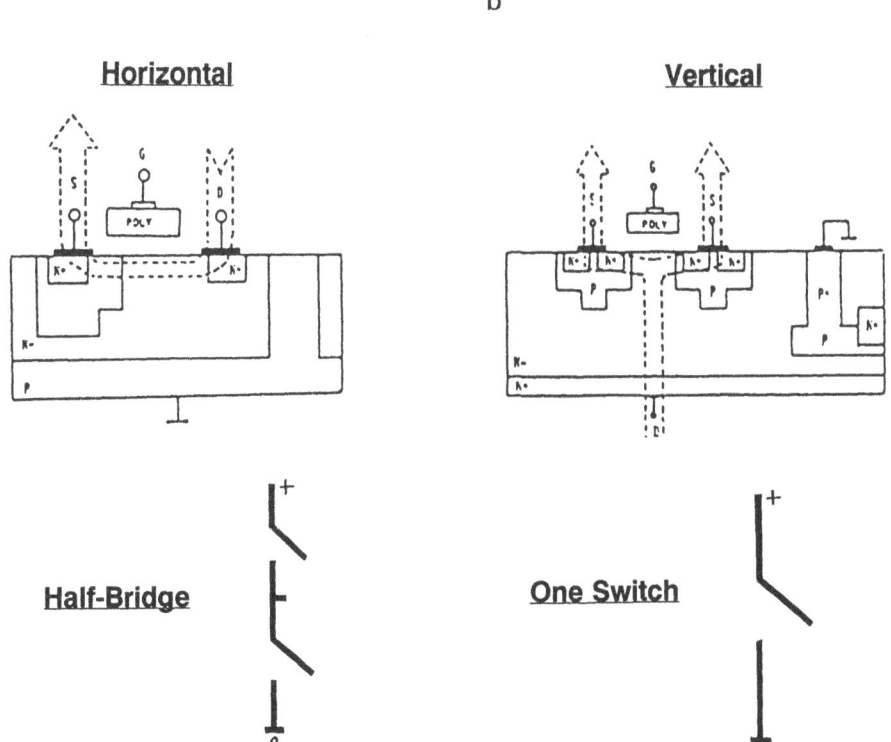

Fig. 3. Horizontal (a) and vertical (b) technologies.

2.2 Main Limits of SMARTPOWER

2.2.1 The Most Important Limit is Economics. When a manufacturer produces a SMARTPOWER device, the power component and the signal processing elements are created by the same processes, forming the complete device. Consequently, the process is a more expensive one than for just the discrete component. For example, the discrete component requires between one and five process steps while the SMARTPOWER requires either between six and nine (vertical technology) or between ten and sixteen (lateral technology) process steps.

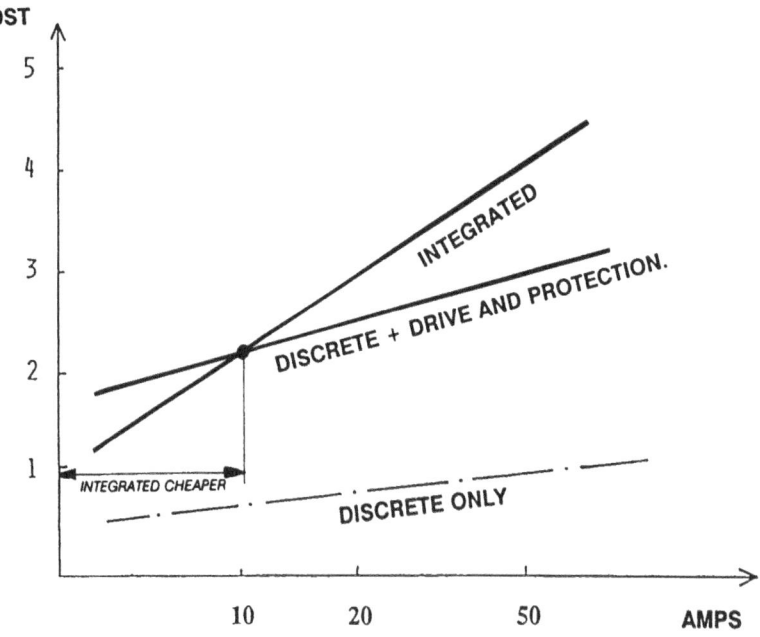

Fig. 4. Cost versus power for discrete and integrated solution.

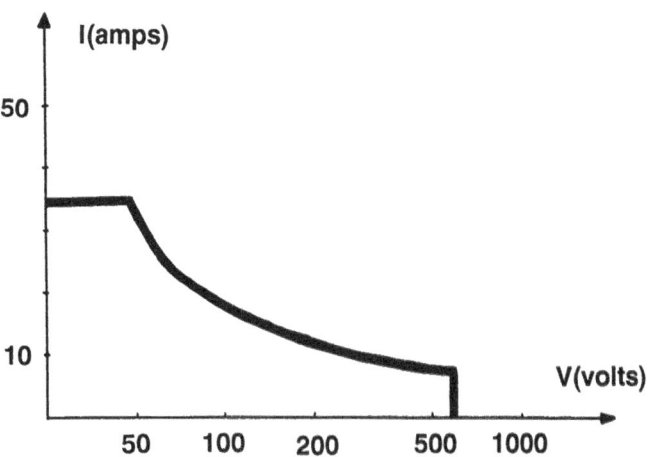

Fig. 5. Limit of switchable power for integrated circuits in 1992. This limit is essentially defined by economical considerations.

Hence the cost of a specific power capability in a SMARTPOWER device is always higher than the cost of the same power capability realised with a discrete power component.

In a SMARTPOWER device, the key factor to consider in order to be competitive with the discrete solution is to obtain a very low cost for the signal function. Figure 4 shows how to define the limit. It depends on the environment and on the complexity required for the signal and control function. Figure 5 shows an average limit for 1992.

2.2.2 Other Limitations Arise from the Technology itself. Today, in 1992, semiconductor manufacturers are not yet ready to integrate the following features or devices in a monolithic form:
- Fast rectifiers
- Thyristors and triacs
- Vertical Insulated Gate Bipolar Transistors (IGTBs)
- Galvanic insulation (5kV peak).

2.3 State of the Art 1992

2.3.1 Strong Points. Lateral and quasi vertical technologies are mature; a large number of products have already been made. (V < 200V, switchable power < 600 VA). Vertical technology is entering the market and is now able to provide, under specific economic conditions, a 3 kVA device with only one power component.

2.3.2 Some Difficulties. Vertical technology is still young.
    For both technologies there is:
- a lack of specialists (competent in both IC and power design).
- the influence of dV/dT.
- the cost of VA/mm^2.
- little flexibility.

2.4 Applications

The main applications, in terms of quantities, are:
- small motor controls (e.g. D.C., stepper, synchronous) for computer peripherals, robotics and automotive applications.
- electromechanical control for printers.
- switching regulators.
- audio amplifiers
- low voltage 2 A D.C. auto-protected switches for programmable logic computers.

These applications were started 4 to 6 years ago. The majority concerns small switchable power with low voltage (generally a voltage supply < 36 Volts).

The SMARTPOWER market is already an important semiconductor market; it represents (1991) 258 MUS$ (i.e. more than the total market of high power thyristors and GTOs).

For two years, new devices with higher switchable power have been in production.

The first example is the switch used in cars in order to manage the whole electrical main current. One device in a pentawatt package (TO-220 with 5 pins) is shown in Fig. 2. It is able to control 20 A under 60 V.

A 30 A device using the same package will be available in 1992. The majority of these switches are made, using vertical technology, in order to meet a cost compatible with the automotive market requirements.

The second example (Fig. 6) is a switching regulator, rated at 10 A, 40 V, able to operate up to 400 kHz.

The third example, a 30 A motor drive (Fig. 7), has a vertical power component, a power MOSFET with an $R_{DS(on)}$ of 0.036 Ohms.

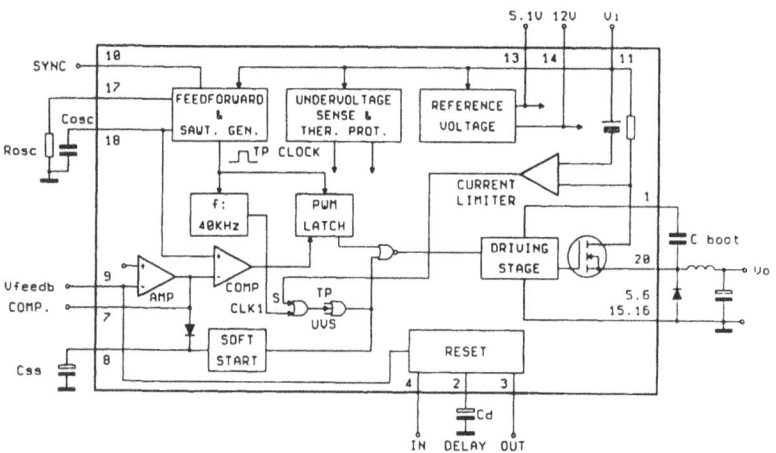

Fig. 6. Switching regulator 10 A, 40 V, 400 kHz.

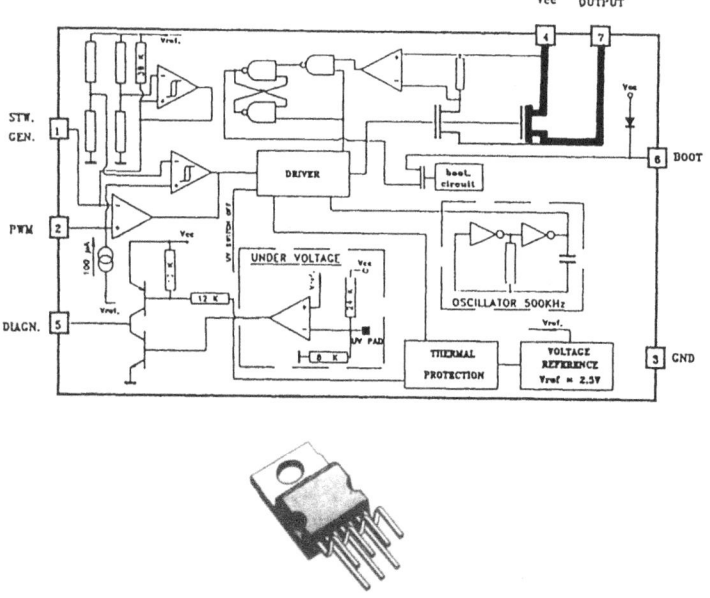

Fig. 7. Motor drive (30 V, 30 A) chopper circuit with current limitation.

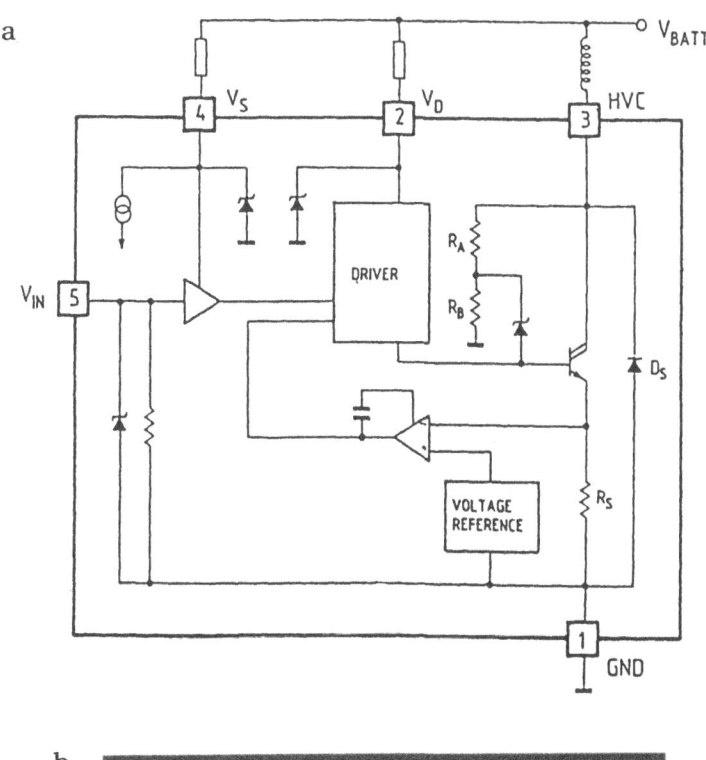

Fig. 8. Integrated Darlington transistor for car ignition (500 V, 6A): schematic (a) and chip (b).

Figure 8 shows a power system for car ignition. In the high voltage sector, two power components are competing: the MOSFET and the bipolar transistor. The choice depends on technical requirements, speed, drive consumption, and on economic considerations. For instance, a 22 mm^2 MOSFET has a R_{on} of 0.7 Ohms for V_{ds} = 500 Volts: the on-state voltage at 5 A (T_j = 100 °C) would be 5.2 V. With a smaller bipolar chip which would be much less expensive, it is possible to obtain 5 A with an on stage voltage of 2.5 V (V_{cbo} = 600 V , T_j = 100 °C). The ignition system for cars, shown in Fig. 8, is designed with a vertical bipolar power IC technology.

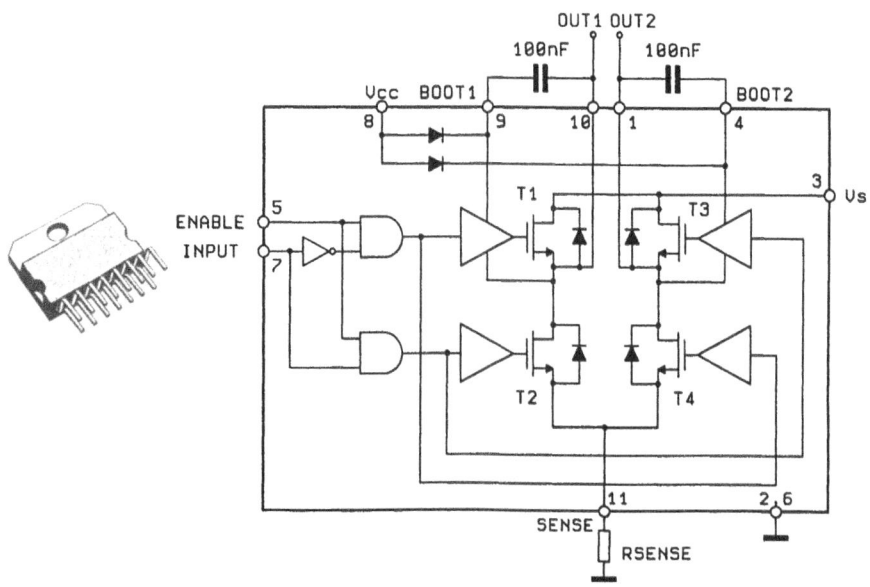

Fig. 9. Full bridge (MOSFET) with integrated drive (200 V, 2 A).

Switchable power is V • I = 500 V • 6 A = 3 kVA. In this voltage range, only the bipolar technology is suitable for such a device where the manufacturing cost must be compatible with the requirements of the automotive market.

A medium voltage, 200 V, full bridge with a complete drive has been made, using the horizontal technology (Fig. 9). Switchable power = 200 V • 2 A = 400 VA. This device is used for motor drivers and can support higher currents of short duration. Similar devices will have higher voltage ratings and will be used for electronic lamp ballasts.

The last example, Fig. 10, represents a complete stepper motor drive.

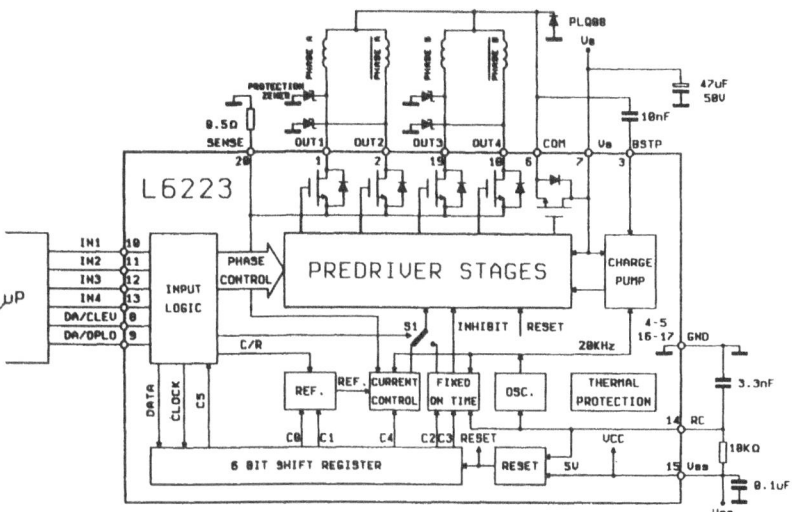

Fig. 10. Step motor drive.

3. EXPECTED EVOLUTION 1992/1996 OF SMARTPOWER

Two main parameters drive this evolution:
- The progress made in technology
- The market needs.

3.1 Technological Evolution

3.1.1 Maturation. Many products have already been made with the horizontal technology. Next year some progress is expected in:
- The field of C.A.D tools and pre-characterized cells, leading to a shorter design cycle and more flexibility.
- The voltage limits increasing up to 600 V, 1 A.
- New component integration, using the lateral IGBT.
- The control possibilities. With the B.C.D. technology it will be possible, in five years, to integrate several hundred thousand of digital components in addition to the power stage. That will lead to replacing analog control with digital control and dramatically increase the potential of the control circuit.

The vertical technology is still young in 1992. The first evolution will concern the optimisation of the design, achieving better control over the parasitic effects and the development of design tools.

The economic power range will be extended, especially in the field of high voltage, to 1200 V, 2 A with MOSFET power stages and 1200 V, 5 A with bipolar or IGBT power stages. In the very low voltage area, the progress already achieved with the discrete MOSFETs (high density) will allow a reduction of R_{on} from 0.5 Ω mm^2 to 0.2 Ω mm^2. At the same time, metallisation, bonding wire and packaging must be drastically improved in order to make full use of the potential offered by silicon.

3.1.2 New Technologies. SOI (Silicon On Insulator) will open up new possibilities in SMARTPOWER devices. The dielectric insulation between power and signal or between different power components will provide efficient separation, avoiding the majority of parasitic problems and lead to easier design. For example, the monolithic integration of several IGBTs in bridge configuration will be possible. However, this technology will also be more expensive.

3.1.3 Some Limitations. Should it prove possible to obtain monolithic integration of ultra fast rectifiers, thyristors and triacs in the laboratory, we do not think, that such devices will be introduced the next few years because the process would be more complex and hence too expensive for the market.

In another area, we do not see the possibility of obtaining monolithic integration of galvanic insulation according to the safety standards - 5kV peak.

3.2 Market Needs

The design of SMARTPOWER circuits will always be expensive. The first and most important consequence is that SMARTPOWER can not be considered for small quantities because the cost of the design must be fully compensated for by the quantity of devices sold.

In international competition, the cost of the function becomes more and more important. Despite the fact that SMARTPOWER represents some considerable technical progress - fewer components, smaller foot print, shorter connections between power and drive - many users want to opt of SMARTPOWER only if the cost for SMARTPOWER is less than the cost of the equivalent circuit with discrete components.

These two key points will define one aspect of the future of SMARTPOWER.

VOLUME MARKET - LOW COST OF SWITCHABLE POWER/mm^2.

One of the most important results of the maturation will be the economic aspect: reducing the ratio switchable power/mm^2, consequently we expect many new designs over the next few years, specially using the vertical technologies.

3.3 Two Different Directions

Considering the possibilities of integration and the market needs, two different ways are open:

provide   MUSCLES TO THE BRAIN ?
or give   BRAIN TO THE MUSCLES ?

The first way is the extrapolation of ICs toward the power. It will lead to the "SYSTEM" (several elements operating together for the same goal).

This evolution started several years ago and will continue with more sophisticated or higher power devices.

The second way is still new and will lead to completely different functions. Here, the goal is to improve the discrete components by integration, according to the market requirements:
- Less components
- Less hardware for drive and protection
- Automatic protection in case of danger
- More information about the component behaviour
- Improved switching behaviour.

3.4 Applications

Figure 11 shows:
a) an existing system to control a typewriter motor,
b) a circuit with the monolithic integration of the complete system.

The latter circuit includes the microprocessor interface, a small Switch Mode Power Supply (SMPS), two stepper motor drives, and a hammer control. This circuit, now entering the market, is very interesting because it gives an insight into the solution for a complete system integration.

Another application area (in development) is shown in Fig. 12 which represents a monolithic 40 W SMPS in a TO-218, 7 pin package. This SMPS operates from the 220/240 V AC mains. It is a flyback circuit and the power component is a bipolar Darlington transistor. The process, completely bipolar, is very cost effective. A 1000 V bipolar Darlington, driven without negative base current, is not fast enough for this kind of application (40/50 kHz). In order to obtain very fast switching with a low storage time, the Darlington is driven in the "cascode" configuration with a low voltage bipolar transistor in the emitter. This device is still in development.

The "SMARTMOSFET" shown in Fig. 13 is a new device (available in 1992). It is the first example of "brain to the muscle". The power component is a 500 V, 0.75 Ω MOSFET. This circuit includes:
- The drive (compatible TTL) with charge pump.
- The following protection functions:
     • a short circuit protection (Fig. 14)
     • a dV/dt limitation (Fig. 15) in order to limit the level of parasitics
     • a safety function which turns off the circuit if the supply voltage increases beyond its limits

a

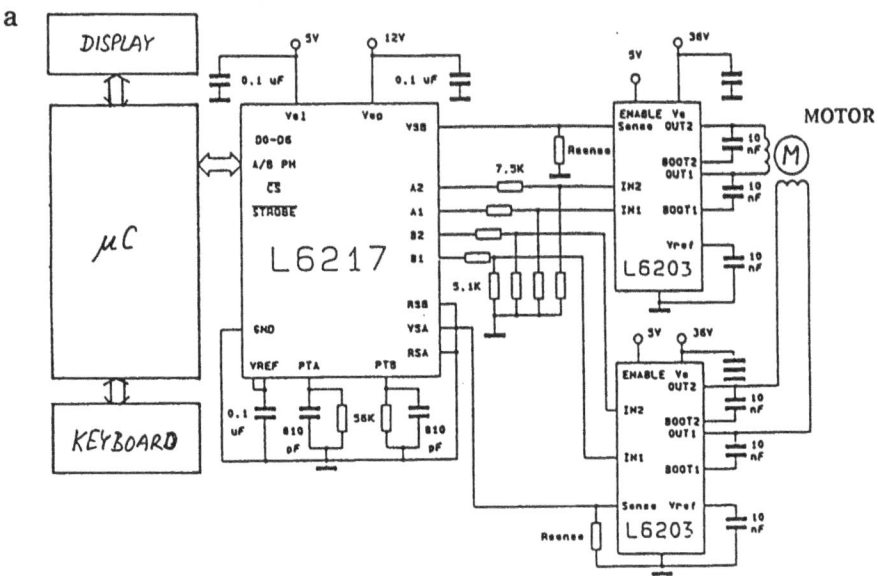

Step motor control for a typewriter

b

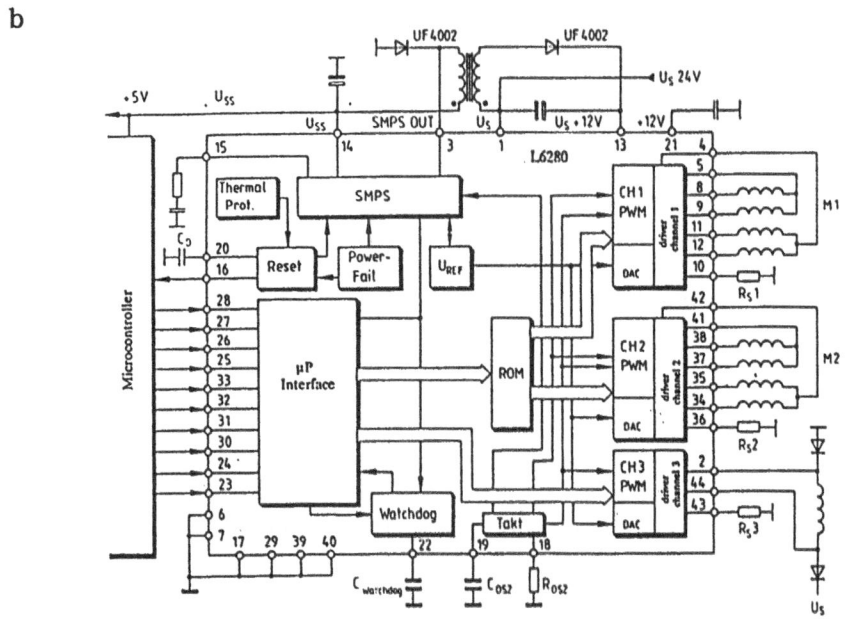

Complete integrated system for typewriter

Fig. 11. Step motor control for a typewriter: classical solution (a) and new solution (b).

• a temperature protection which stops the circuit if the junction temperature exceeds 150 °C.
- The following "information" functions:
   • an alarm, which delivers a signal if the junction temperature exceeds 150 °C
   • an output signal, giving an image of the drain current with a higher voltage level than the classical sense output in order to improve the signal/noise ratio.

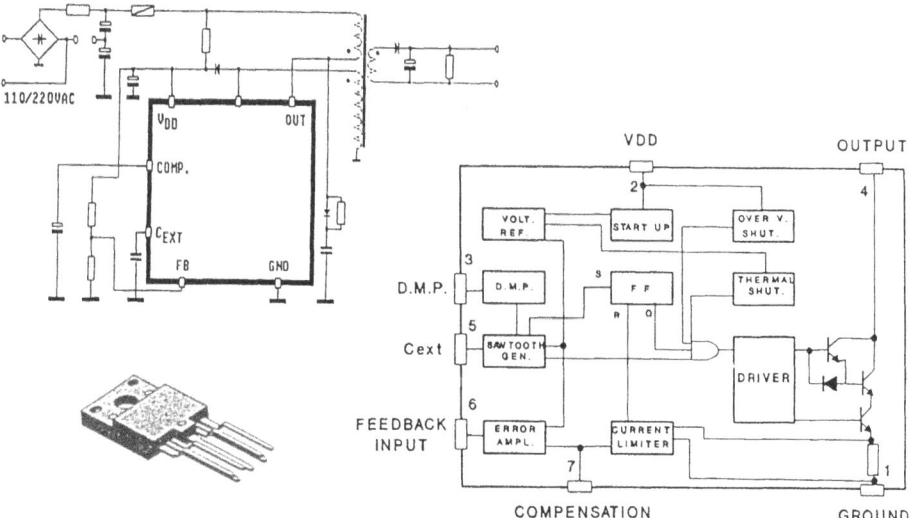

Fig. 12. Integrated flyback Switch Mode Power Supply (SMPS) in bipolar technology; device in development; the power stage is a 1000 V, 3 A Darlington.

This SMARTMOSFET is the first circuit which will replace a discrete MOSFET plus drive, protection and status circuits. It will be followed by similar MOSFETs, 200 V, 0.2 Ω and 1000 V, 4 Ω, and many others in the future.

S.O.I. technology will open new possibilities. Some complete ac motor drives (200 watts - 110 volts AC supply) are in development. Fig. 16 shows an experimental SMARTIGBT presented during the 1991 EPE-MADEP conference. Like the SMARTMOSFET, this device includes the power component and all protection and drive functions.

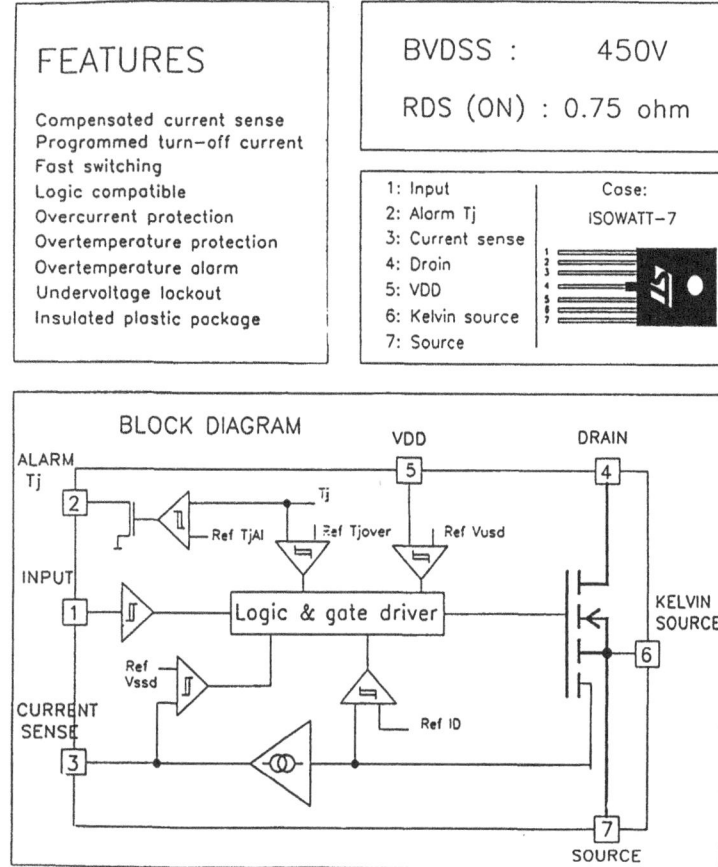

Fig. 13. SMARTPOWER 500 V, 0.75 Ω, with integrated drive, protection and status.

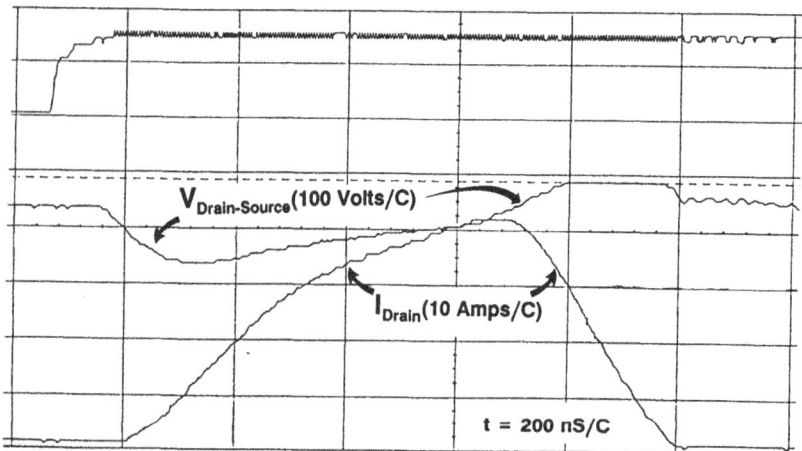

Fig. 14. Short circuit behaviour of a SMARTMOSFET without any limiting impedance. The switching dI/dt is limited in order to limit the turn-off overvoltage.

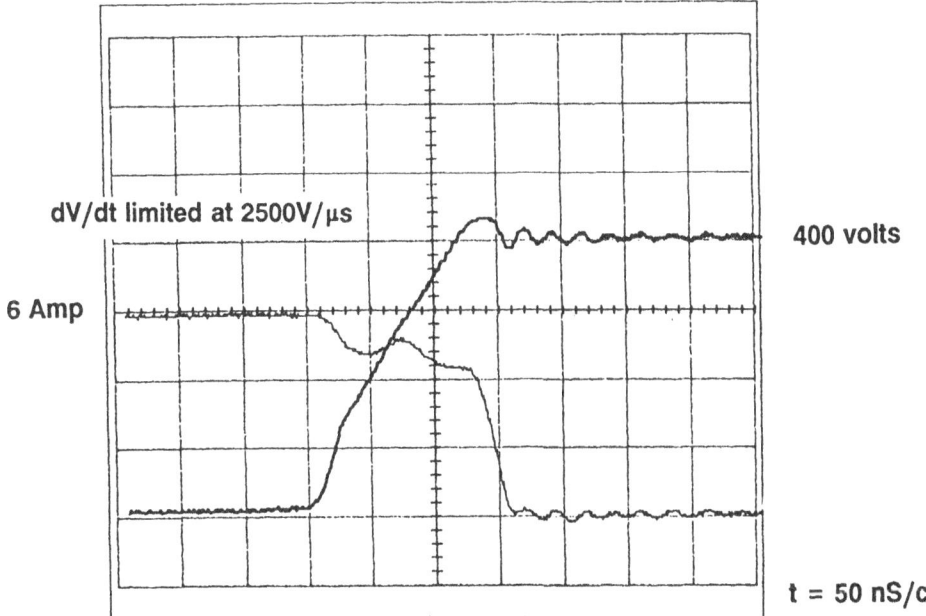

Fig. 15. Turn-off behaviour of the SMARTMOSFET. The reapplied dV/dt is limited by an internal circuit at 2.5 kV/microsecond.

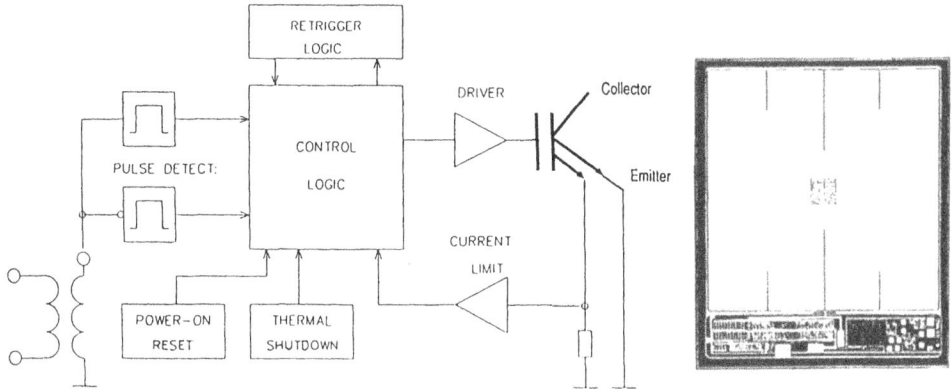

Fig. 16. SMARTIGBT (Prototype by R. Gabriel[8], © 1991 EPE-MADEP).

4. CONCLUSION

We have seen the essential characteristics of SMARTPOWER:
- The possibility of obtaining, by monolithic integration, power and signal functions on one chip.
- Some limitations:
  • Maximum economic switchable power, approximately 3 kVA today and a bit more in the next few years
  • Maximum voltage 600 Volts today, 1200 Volts in the future.
  • It will be possible to obtain (economically) monolithic integration of some components or some important functions such as galvanic isolation.

The evolution will go in two different directions:
- Complete systems on one chip.
- Improved discrete components. This important evolution is just starting and will lead to many new "improved discrete" components.

What can we see as consequences for power electronics ?
- Besides monolithic integration, we see an important development of the "hybrid integration" - multi-chips, or probably better multi-package assembly in order to extend the possibilities with higher power or with non-integrable functions like galvanic isolation.
- A part of the hardware will be transferred from the equipment manufacturer to the semiconductor manufacturer.

The job of the future power electronics engineer will be more system oriented. This evolution has already happened in the field of signal electronics, influencing the technical education in the engineering schools. In the field of power electronics, "electromagnetics" and thermal management cannot integrate.

So, contrary to what has happened in some areas in signal electronic education, even with the large expected progress in power integration, the general education in electromagnetics, thermal management and mechanics will always be mandatory to train a complete "system engineer".

REFERENCES

1   K. Rischmueller, "SMARTPOWER, quo vadis?" *EPE MADEP 1991 Proceeding*, pp. 420-427.

2   B. Murari, "SMARTPOWER technology evolves to higher level of complexity" *PCIM Europe*, Jan-Feb 91, pp. 27-31.

3   J. Mille, "A very high voltage technology for vertical SMARTPOWER IC", *Proceedings of the Symposium of High Voltage ICs*, May 7-11, 1989, Los Angeles.

4   J.M. Peter, B. Nadd, "SMARTPOWER, a way to improve discrete components". *IAS Conference*, Atlanta Oct. 1987.

5   R. Zambrano, M. Paparo, "High voltage IC with vertical current flow and buried emitter optimizes power handling capacity", *Proceeding of the Symposium of High Voltage ICs*, May 7-11, 1989.

6   A. Nakagawa, "Impact of dielectric isolation technology on power ICs." *CH 2986-6/91/0000-0016 1991 IEEE*, pp.16-21.

7   H.Myazaki et al., "A novel high voltage three phase monolithic inverter IC with two level current sensing", *CH 2987-6/91/0000-0248 1991 IEEE*, pp. 248-253.

8   R. Gabriel, "An intrinsic save SMARTPOWER IGBT" *EPE MADEP 1991 Proceedings*, pp. 179-182.

DISCUSSION

H. Grüning (ABB Corp. Research, Baden)

A very important issue is Electromagnetic Interference (EMI) and protection against high dv/dt, especially when using a smart power device for triggering high power devices. Where are the limits at about 1 kV isolation voltage ? What is the dv/dt limit today and what could we expect for the future ?

J.M. Peter

At SGS-Thomson, we have a vertical technology where one could reach up to 5 kV/μs. An additional difficulty in the case of bipolar technology is, however, the electrical field strength. Less problems are encountered in vertical technology.

B.J. Baliga (NCSU, Raleigh NC)

Do you include the packaging cost - which might be higher than the chip cost - in your cost projections ? What is the percentage cost of the package today and what will it be in future?

J.M. Peter

In the graph shown, the package cost was not included. The "multiwatt"-package shown is produced in large quantities and very competitive but not suitable for all applications. A "SMD"-like power package should be developed in the future, leading to a cheap solution. A figure for the percentage-cost of the package cannot be given now.

RESONANT LINKS: A NEW FAMILY OF CONVERTER TOPOLOGIES FOR SOLID STATE POWER CONVERSION

T.A. Lipo and D.M. Divan

University of Wisconsin
Madison, WI, U.S.A.

ABSTRACT

This paper summarizes recent development on the resonant link power conversion family of power converters. These converters utilize a high frequency link to introduce zero voltage or zero current intervals. By switching at the zero crossings of the voltage or current, the converter switch losses can be greatly reduced, permitting at least an order of magnitude increase in the converter switching frequency. The converters can synthesize nearly sinusoidal currents on both input and output converters and can also maintain unity fundamental power factor at the system input.

1. INTRODUCTION

The availability of high power gate turn-off devices such as the Bipolar Junction Transistor (BJT), the Insulated Gate Bipolar Transistor (IGBT) and the Gate Turn-Off thyristor (GTO) have contributed to remarkable advances in power frequency conversion in recent years. The most widely used and highly developed frequency changers are the variable amplitude (six step) and and fixed amplitude (pulse width modulated or PWM) DC voltage link inverter which synthesize variable frequency and variable voltage AC output from a DC voltage input. The second class of highly developed frequency changers is the variable amplitude DC current link converter. A third class of frequency converter, which is still in its infancy is the direct AC/AC converter which eliminates the need for a DC link but requires reverse blocking devices for satisfactory operation, a still elusive device property at the present time. These DC link based power conversion systems as well as the direct AC/AC converter have several inherent limitations. One important drawback is the excessive switching loss and device stress which occur during the switching intervals. As a result, the devices require a relatively large Safe Operating Area (SOA) and the reliability of the system may be compromised unless snubbers are employed. The typical switching frequency in medium size 10-50 kW PWM inverters is only about 5 kHz. Larger converters require lower switching frequencies or cannot even be contemplated because of the loss issue. Because of the relatively low switching frequency, it is difficult to realize dramatic gains in important system attributes such as faster system response, increased output frequency, improved power densities and reduction in audible and electrical noise, particularly when the motor is operating at high speeds.

In recent years, remarkable progress has been made in the development of high power density AC/AC converters which incorporate resonant-link schemes rather than the more conventional DC link. These converters, called "resonant link converters", utilize high speed devices such as fast recovery transistors, thyristors and GTOs to achieve a relatively high switching frequency and, thereby, markedly reduce the output current distortion compared to DC link schemes. In addition, these new converters also have high power capability made possible by very low switching losses. The purpose of this paper is to summarize the state of the art in this important new branch of high power electronics.

2. PRINCIPLE OF SOFT SWITCHING CONVERTERS

Soft switched converters are of two types. A Zero Current Switch (ZCS) refers to device turn-on and turn-off occurring with virtually no current in the device. This type of switching requires the use of purely inductive "snubbers" with no snubber reset mechanism. Turn-on losses are now dependent only on the size of the snubbing inductor which limits the rate of rise of current in the device. Turn-off losses, while device dependent, are relatively small since turn-off of the device uses only natural commutation (natural turn-off of the device when the anode current drops below the holding current).

Zero Voltage Switching (ZVS) converters use purely capacitive snubbers which requires that device turn-off occur in conjunction with an anti-parallel diode (usually in another branch of the circuit) which carries the current after turn-off. Turn-off losses are now dependent only on the size of the capacitance which limits the rate of rise of voltage across the device.

Turn-on losses, while device dependent, are also modest when the turn on of the device takes place under a zero voltage link condition. The two types of switching elements are shown in Fig. 1.

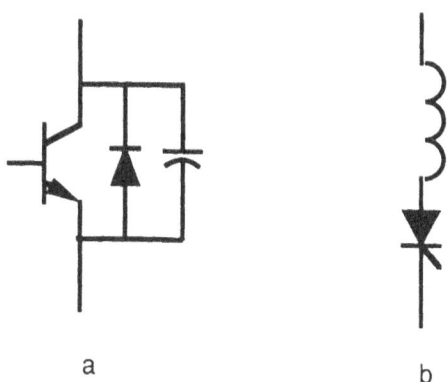

a   b

Fig. 1. Soft switching elements: (a) zero voltage switching, (b) zero current switching

3. SERIES RESONANT AC CURRENT LINK POWER CONVERSION

In general, switching schemes for high power resonant link converters can be classified according to whether they involve a resonant AC voltage or current impressed on the link or incorporate a pulsating DC component, i.e. the link is "DC resonant". Fig. 2 shows a schematic illustration of the simplest configuration, the series resonant AC current link.

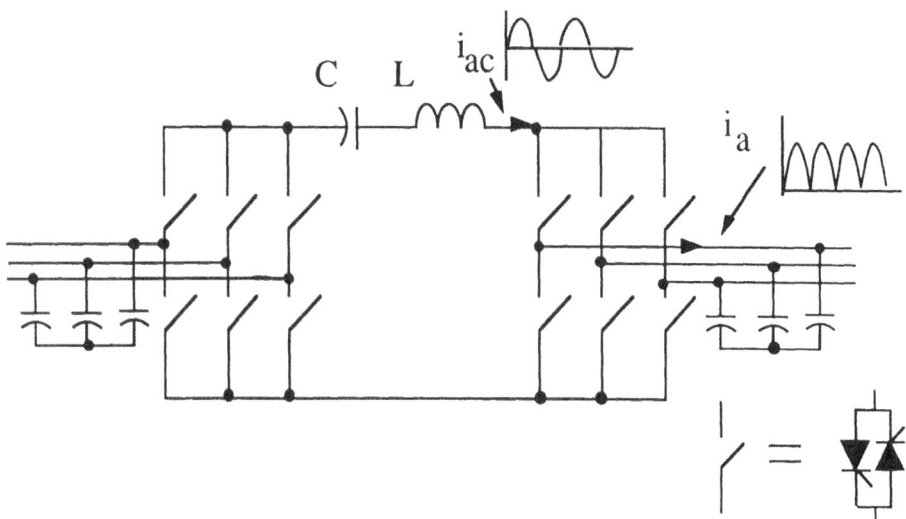

Fig. 2. Series resonant AC current link power conversion system.

Historically, this type converter was the first to be developed[1]. However, the initial conception involved a complicated magnetic structure which prevented application to high power ratings. In a more recent, and simplified realization, a single-phase AC current link operating at a fixed frequency of 20 kHz or higher serves as the interface between two six pulse bridges[2]. Shunt capacitors on the input and output serve to decouple the series resonant link from the source and load impedance. Since 4-layer devices turn off when the current becomes less than the holding current, this converter structure requires only thyristors in its implementation, an important advantage. Because the current in the link reverses, the switches are implemented with inverse parallel thyristors. The bridges operate from the bi-directional high-frequency, series resonant current of the link to synthesize low frequency (including DC or zero frequency) voltage or current source outputs as appropriate by means of suitable control of the bridges.

It is important to mention that the resonance of this converter is set solely by the link L and C and is essentially independent of the load parameters. This type of operation is in contrast to more traditional "resonant converters", used for example in induction heating, in which the load forms a portion of the resonant circuit.

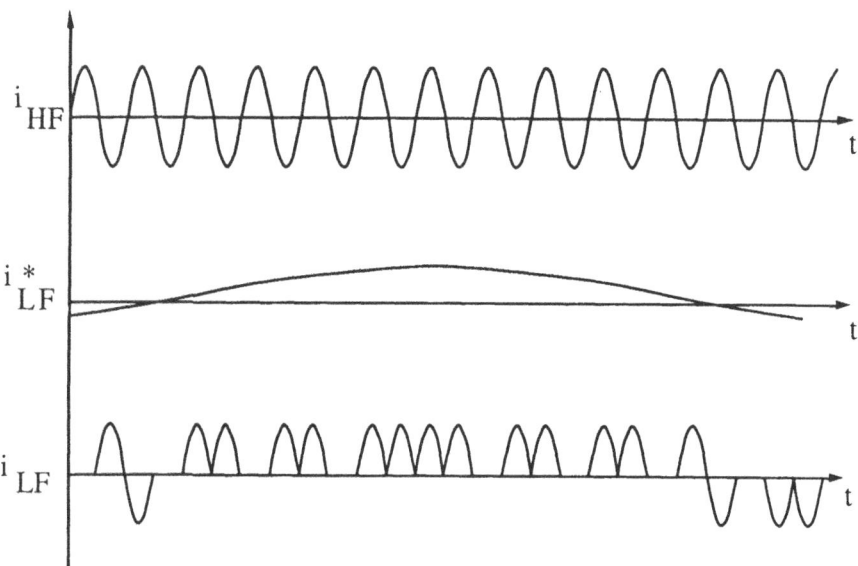

Fig. 3. Switching strateg. for series resonant AC current link power conversion.

With such constraints on the switching of the converter switches, one half cycle of the high frequency current becomes the basic unit of synthesis of the low frequency output signals. Fig. 3 illustrates the basic switching operation of the converter. In general, converter switching is restricted to the zero crossing points of the link current so that the switching losses, which dominate the converter losses at these high frequencies do not become excessive. Half cycles of the resonant link current are selected by the line side and load side converter so as to synthesize a sine wave of the appropriate frequency. Start up of the resonant link can readily be accomplished through proper control of the converter connected to the source. The major disadvantage of this converter is the need for 12 thyristors per converter to carry the bi-directional AC current of the link. Also, the voltage rating of the series resonant inductor and capacitor must be in the range of 2-3 times the voltage rating of the output, resulting in relatively high cost. Recently, the switch count of such a converter has been reduced to six as shown in Fig. 4, but at the cost of three additional, relatively large capacitors per bridge[3]. The concept continues to be pursued for variable speed power generation such as in small pumped hydro and wind turbines[2].

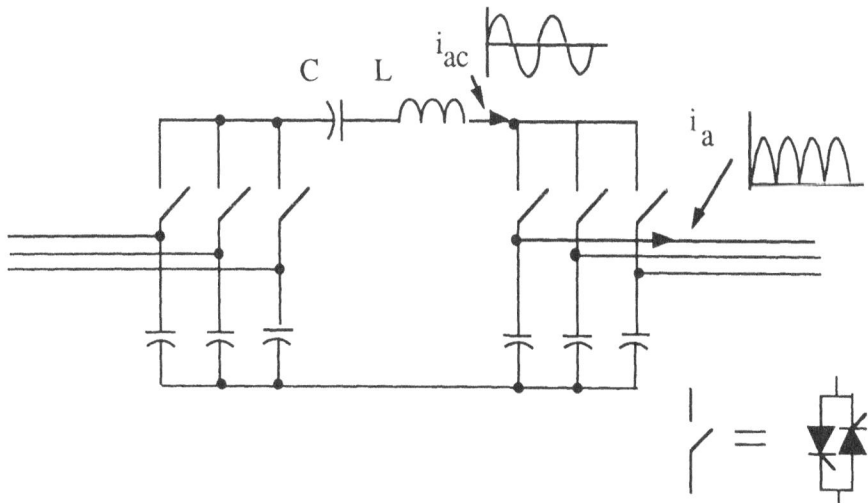

Fig. 4. Modification of series resonant AC current link converter, utilizing only six thyristors per converter bridge (© 1988 IEEE).

4. PARALLEL RESONANT AC VOLTAGE LINK POWER CONVERSION

Circuit duality plays an important role in power electronics and the duality of conventional DC voltage link and DC current link converters are well known. Current resonant links also have duals which possess similar properties as the conventional DC current link converters. Fig. 5 shows the circuit dual of the series resonant AC current link which has been investigated in detail at the University of Wisconsin for use in a high frequency link power distribution system such as for aircraft or aerospace applications[4-7]. In the case of this converter, the inductance L and capacitance C form a parallel resonant link. Switching of the devices of each converter occurs at zero voltage rather than zero current intervals. Fig. 6 shows the switching strategy for this converter. The waveforms are essentially identical to Fig. 3 except that the voltage variable is replaced by current.

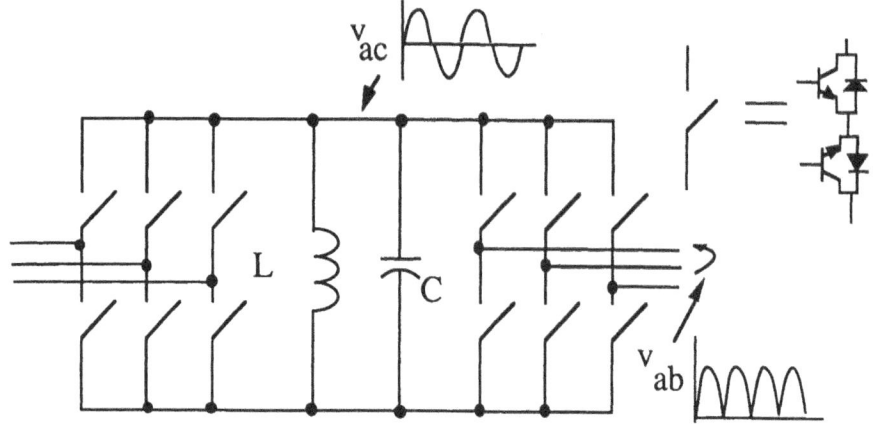

Fig. 5. Parallel resonant AC voltage link power conversion system.

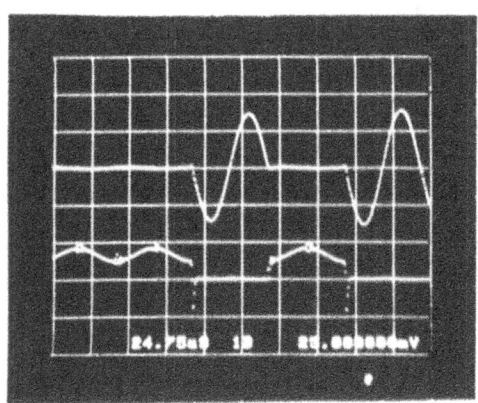

Fig. 6. Typical voltage and current waveforms observed across one converter switch of parallel resonant AC voltage link power conversion system. Upper trace: switch voltage, 125 V/div; lower trace: switch current, 5 A/div; time, 25 ms/div.

It can be noted that, since current is flowing through solid state switches when the voltage crosses through zero, the switches of this converter must be self commutated. Also, since the switches undergo voltage reversals while they are turned off due to the AC nature of the link, the switches must possess bi-directional voltage blocking capability.

Hence, each bridge must be implemented with switches consisting of 12 transistor or GTOs in series (to implement the bi-directional current conducting requirement) together with 12 diodes (to implement the bi-directional voltage blocking requirement). Alternatively, six transistors embedded in a diode bridge arrangement can be utilized.

Fig. 6 shows the typical voltage and current waveforms observed across one converter switch. The waveforms clearly demonstrate the zero voltage switching nature of the parallel resonant AC link converter. The high switch number together with the high current rating of the resonant link inductor and capacitor are the main disadvantages of this type of technology.

5. PARALLEL RESONANT DC VOLTAGE LINK POWER CONVERSION

An alternative to permitting the voltage link to reverse polarity is to bias the link with a DC voltage as shown in Fig. 7 [8]. This advancement in the resonant link technology was also developed at the University of Wisconsin, shortly after the introduction of the resonant AC voltage link. In this case, the link resonates at the tank frequency defined by L and C, in effect, a parallel resonant process. Capacitor C_0 is a conventional electrolytic capacitor used to provide the DC bias to the link. The link voltage now takes the approximate form of a biased cosine wave. Again, the switches of both converters operate only when the link voltage reaches zero. In order to reduce the effects of the stray capacitance, the link capacitor C can be distributed across the poles of both converters or even across the switches themselves. Because the voltage of the link does not actually reverse polarity, the switches need not block reverse voltage. Hence, the converter switches need be implemented only with a transistor and inverse parallel diode; the normal switch configuration for a self commutated voltage source inverter.

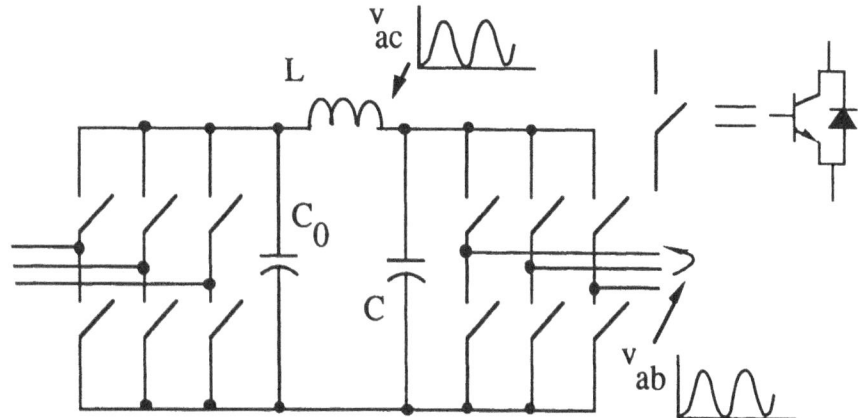

Fig. 7. Parallel resonant DC voltage link power conversion system.

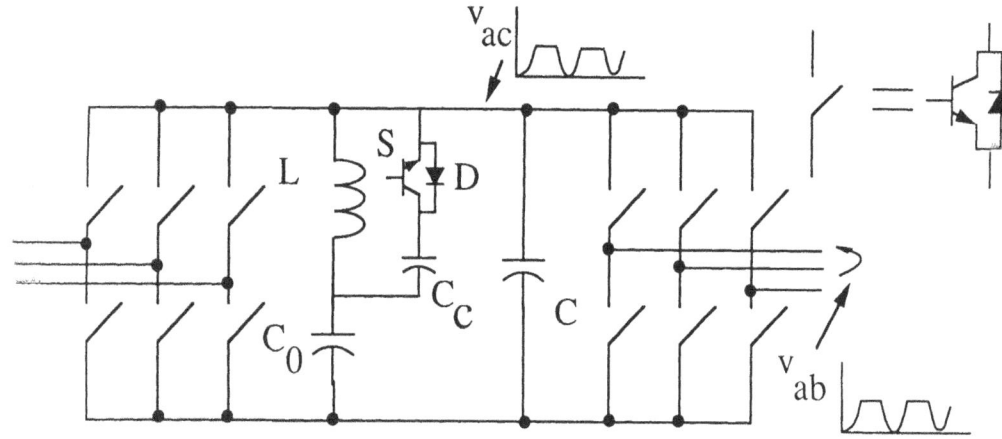

Fig. 8. Parallel resonant DC voltage link with active voltage clamp.

The major concern for successful operation of this scheme is to ensure that the link voltage reaches zero during each resonant pulse. It can
be shown that, if losses are included, the voltage across the capacitor C
is a damped sine wave which does not reach zero as predicted ideally. A
zero voltage interval can, however, be ensured by giving the link inductor
L an initial current condition. The initial condition can be accomplished
if the link resonant capacitor is briefly shorted by simultaneously triggering both top and bottom legs of one phase of the converter before the capacitor is released and an oscillation cycle commences. A potential disadvantage for such a converter is the relatively poor utilization of the
semiconductor switches since the RMS value of the output voltage is
roughly half the equivalent voltage of a DC voltage link PWM converter for
the same switch voltage rating. This difficulty can be overcome by incorporating an active clamp which limits the link voltage to a predetermined value as shown in Fig. 8 [9]. In this circuit, diode D turns on and
clamps the bus voltage at a predetermined value. With D conducting, the
device S is turned on in a lossless manner. The current eventually transfers from the diode to the device S. The charge transferred to the capacitor C_c with D conducting is recovered during the interval when S is on.
When the net charge transferred equals zero, S is turned off and the LC
circuit resonates until the DC bus voltage reaches zero and C is again
shorted. At this point, the resonant cycle is reinitiated. In this manner it
is possible to reduce the voltage stress from approximately 2.5 to 1.2-
1.4 times the link DC capacitor voltage. Hence, the switch elements become only marginally greater than the equivalent switches of a DC link
converter. Typical idealized waveforms for the case of voltage clamping
are shown in Fig. 9.

The DC offset in the link is a major advancement of the resonant AC
voltage link of Fig. 7 since the voltage bus is supported primarily by the
voltage of the DC capacitor C_0 so that the characteristic impedance of

the resonant L and C can be reduced to very small values compared to the corresponding elements of the resonant AC voltage link. Switching frequencies of the order of 60 kHz can be reached with devices losses not exceeding those of a conventional PWM inverter, switching on the order of 5 kHz. The primary disadvantage of this circuit is the need for the voltage clamp which adds significant cost to the inverter, compared with the normal DC voltage link type inverter. However, when issues other than cost, for example high efficiency and low weight are of concern, this new type of converter promises to be a very attractive new alternative in converter topology.

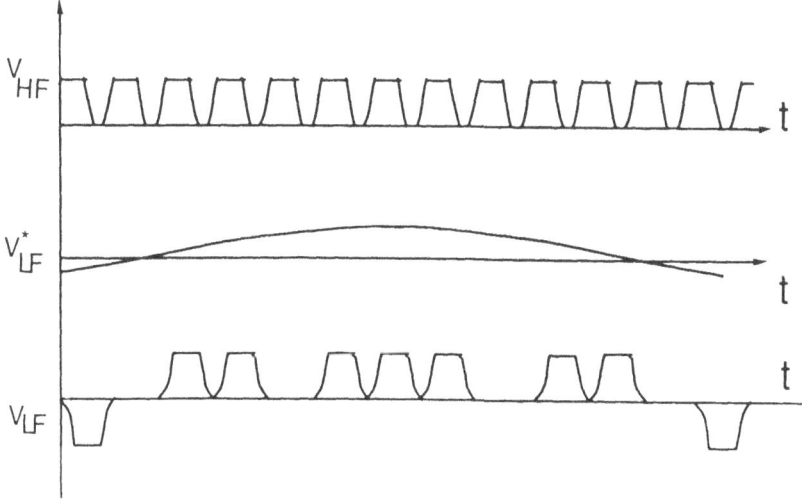

Fig. 9. DC link voltage V_{HF}, Command Voltage $V_{LF}{}^*$ and line to line output voltage V_{LF} waveforms for system of Fig.8.

6. SERIES RESONANT DC CURRENT LINK POWER CONVERSION

The dual of the Parallel Resonant DC Voltage Link is the Series Resonant DC Current Link converter shown in Fig. 10 [10]. This topology which also was developed at the University of Wisconsin utilizes a DC bias current which is added to the link by employing a DC inductor L_0. The link current now takes the form of a displaced cosine wave. Again, zero current intervals are used to turn off the conducting thyristors in much the same manner as the series resonant AC current link system. However, since the current now remains unidirectional, the converter switches can be implemented with single thyristors. Hence, each converter becomes no more complicated than the conventional thyristor bridge. In order to increase the switching frequency to as large a value as possible, fast turn off devices must be employed so that the cost of the bridge is somewhat greater than a simple phase controlled bridge. The link current now again becomes resonant at a high frequency (say 20 kHz) and the switching instants of the bridge are selected to produce a sinusoidal distribution of current pulses on each line of the output. The capacitors on the output serve to filter the high frequency content in the output (resonant frequency).

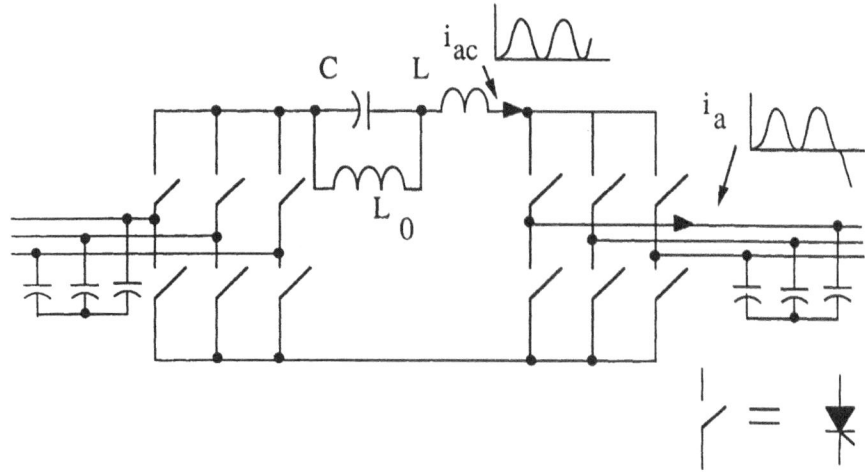

Fig. 10. Series resonant DC current link power conversion system.

When the losses in the system are considered, the link current does not return to zero after each oscillation interval, in much the same manner as the DC resonant voltage link. The problem can be solved in this case by properly regulating the current in inductor L_0 and the instant of turn-on of the next thyristor to ensure that the link current reaches zero after each cycle of oscillation. It is apparent that this converter suffers from the dual of the disadvantage of the DC resonant voltage link. That is, the RMS AC output current is relatively small compared to the current switch rating. However, in this case the problem is less serious since thyristors are used for the switch elements so that sufficient current carrying capability can easily be accommodated with relatively modest cost.

If necessary, the problem of switch utilization can be eased by incorporating a current clamp as shown in Fig. 11 [11]. In this case, a DC biased inductor ("swinging choke") can be utilized as the resonant inductor. The DC bias current is conveniently implemented by using the DC current in the bias inductor L_0. Waveforms obtained from measurements on of the DC link current for a typical motor load are shown in Fig. 12 for two different resonant switching frequencies.

Since the series resonant DC current link converter does require only simple thyristors, high power applications for this strategy are particularly attractive, including DC motor drives [12], rectifiers for superconducting energy storage [13] and large AC motor drives [14]. The future of this circuit depends heavily on availability of suitable thyristors with fast turn-off capability (less than 20 μs). The major disadvantage of the circuit is the voltage rating of the resonant capacitor and consequently the converter switches which must be on the order of 1.5 per unit of the input line to line voltage.

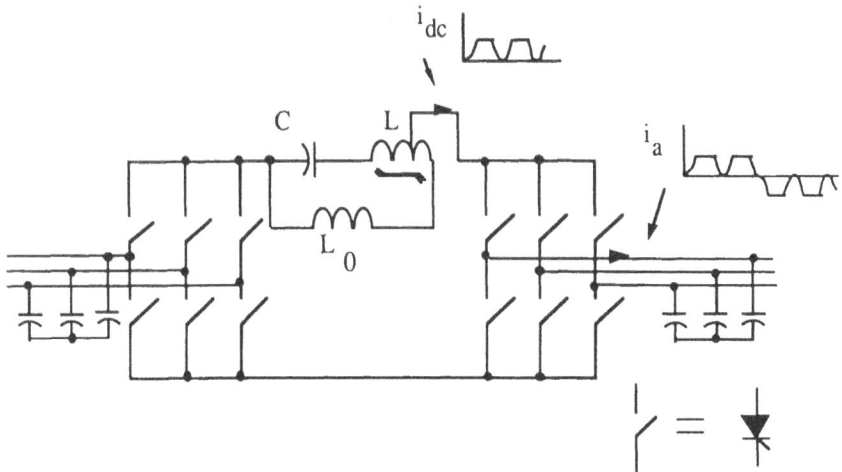

Fig. 11. Series resonant DC current link with current clamping inductor.

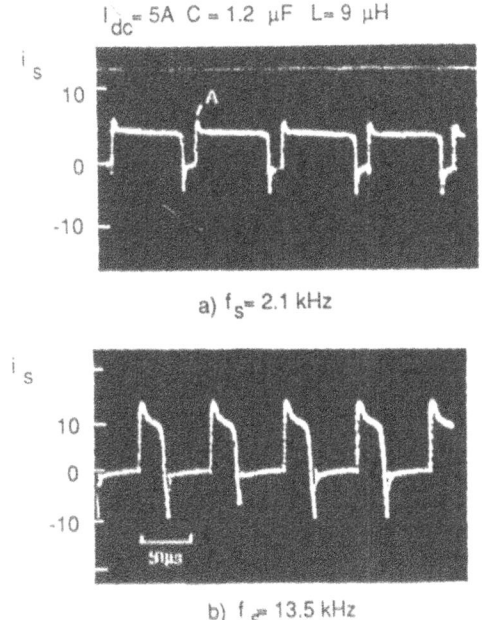

a) f_s= 2.1 kHz

b) f_s= 13.5 kHz

Fig. 12. Typical link current waveforms with current clamping: (a) link frequency f_s = 2.1 kHz, (b) link frequency f_s = 13.5 kHz.

7. VARIATIONS IN CIRCUIT TOPOLOGY

Over the past several years, activity in resonant link converter has increased exponentially and circuit topology variations have appeared for both the DC voltage resonant and DC current resonant topologies. Fig. 13 shows a typical modification of the basic DC resonant voltage link (Fig. 7) in which the resonant capacitor is shorted by an additional switch [15]. In this manner, the link voltage can be limited to safer values than for the basic circuit and required zero crossing of the voltage can be ensured without the need for "precharging" the resonant link inductor.

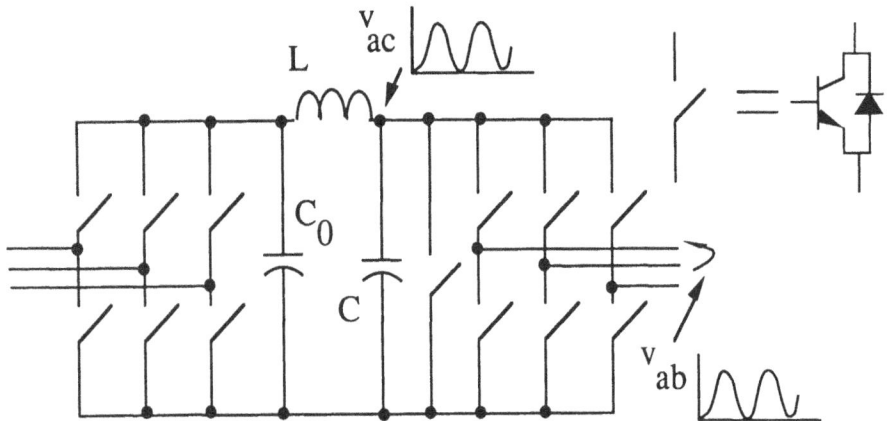

Fig. 13. Modification of DC voltage resonant link by utilizing an additional switch to ensure zero crossing of resonant capacitor voltage (© 1988 IEEE).

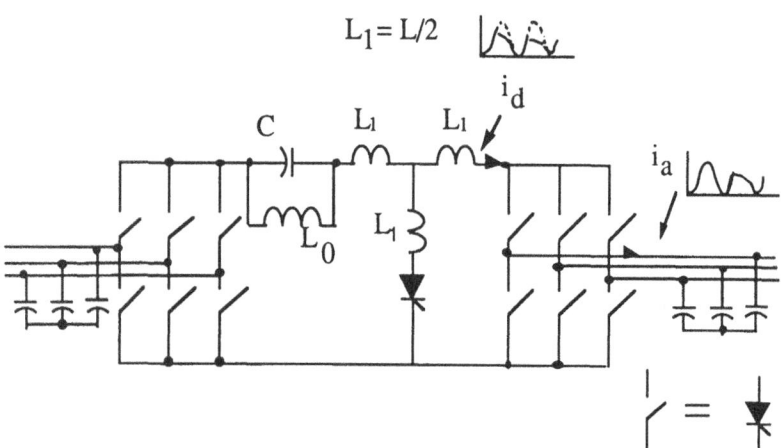

Fig. 14. Pulse splitting topology with bypass thyristor.

In Fig. 14, a modification of the basis DC current resonant link is shown in which the current flow to the load is diverted by means of a by-pass thyristor [16]. Use of the by pass thyristor allows the current pulses to be "trimmed", i.e. varied in size so that the current fundamental component can be more accurately synthesizes without the "lumpiness" created by fixed amplitude current pulses. The principle of pulse splitting is illustrated by Fig. 15. It is shown in [17] that the pulse splitting concept can be realized without the need of an additional thyristor. This modification is illustrated in Fig. 16. In this case half the resonant inductance is placed in the legs of the converter bridge. Current is diverted from the load when both top and bottom switches of the same leg are turned on. A circuit employing both pulse splitting (Fig. 14) and clamping (Fig. 11) has been reported in [18]. Fig. 17 shows typical high fidelity waveforms which can be achieved with such circuits.

It has been shown by Park and Cho that capacitors on the ac side of the converter rather than the DC (link) side of the converter can be used to form a series resonance with the DC link inductor[19]. In this case, the bias inductor and DC link capacitor can be eliminated. An extension of this principle which resonates both the line and load side capacitors with the link inductor is shown in Fig. 18. Figure 19 shows a typical line

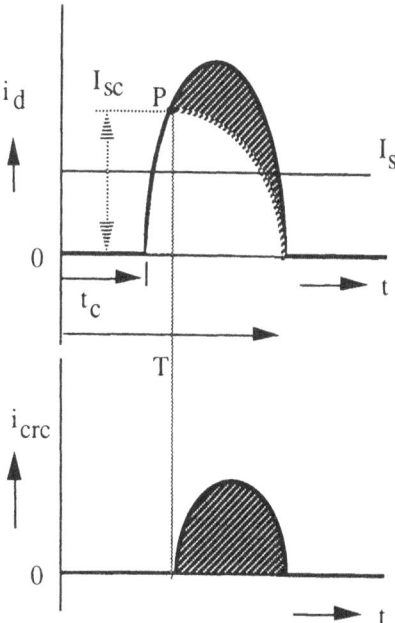

Fig. 15. Illustrating control method of pulse splitting.

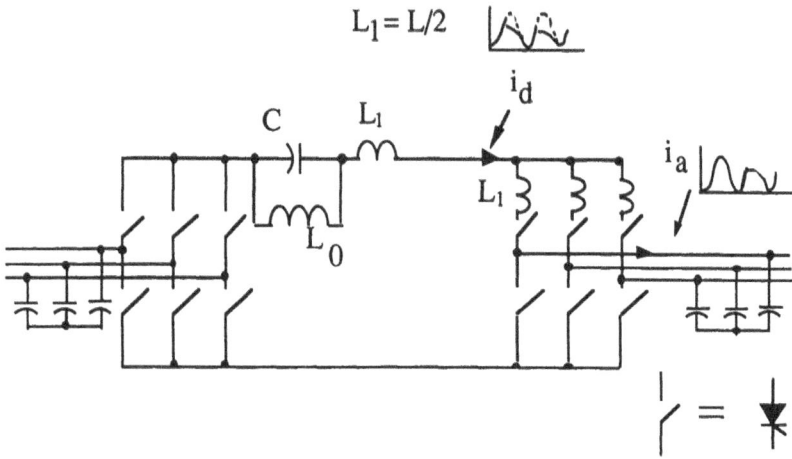

Fig. 16. Pulse split circuit obtained by eliminating the additional thyristor of Fig. 13.

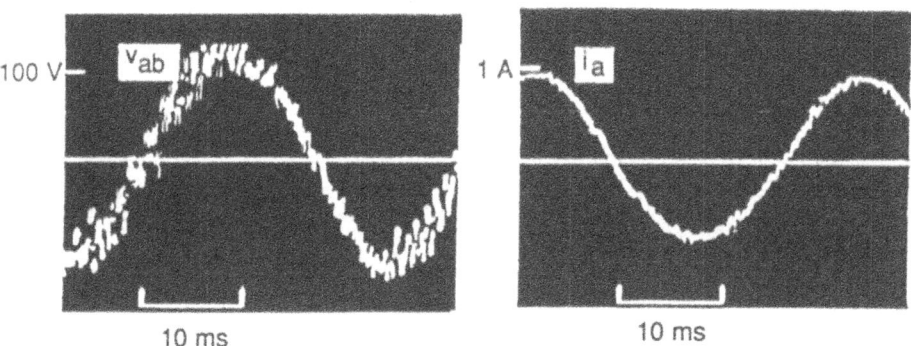

Fig. 17. Output voltage and current waveforms obtained by pulse splitting technique.

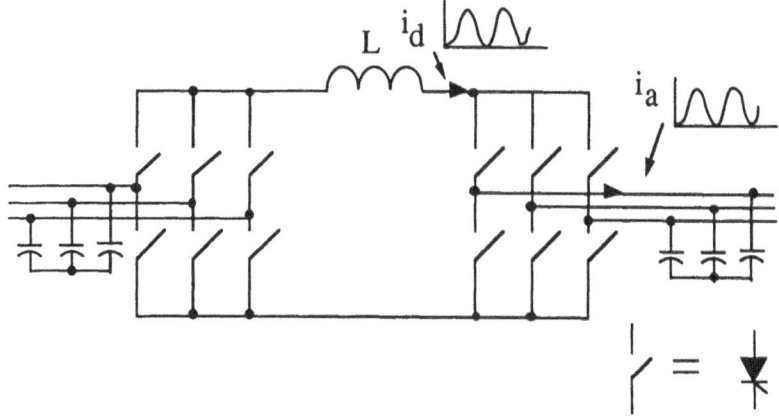

Fig. 18. Series resonant converter with AC side resonant capacitors.

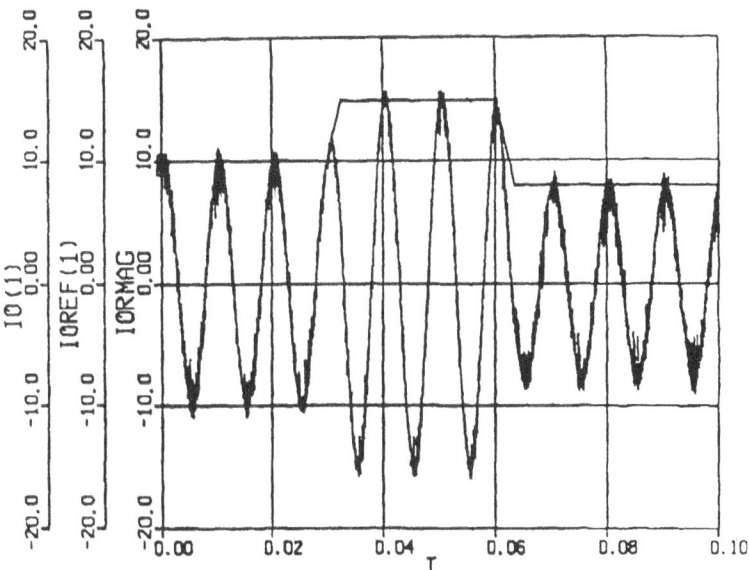

Fig. 19. Typical line side current waveform obtained with circuit of Fig. 18.

side current waveform which can be achieved. Research on this topology is actively being pursued at the University of Wisconsin.

8. MODULATION STRATEGIES

The quantum nature of resonant link power conversion has necessitated the development of specialized schemes for regulation of the output current or voltage. Fig. 20 shows the block schematic of a typical modulation scheme used for resonant voltage link systems, based on the concept of sigma modulation which relies upon integration of the current error [5]. In this scheme, the area under the reference signal is compared with the area of the synthesized voltage signal. If the comparison indicates that the integral of the synthesized signal is more (less) than the desired value, then the controller causes the next half cycle pulse to be applied so that the integral is increased (decreased). In this manner, voltages having a fundamental component of DC, sinusoidal AC or any other smooth waveform may be synthesized using a single integrator, a comparator and a few logic gates. This simple implementation results in the density of the half cycle pulses in the synthesized voltage to be modulated in close accordance with the amplitude of the reference signal.

Figure 21 shows a typical spectrum of the voltage obtained by such a modulation strategy using, in this case, the resonant AC voltage link. It can be noted that very low level harmonics distribute evenly over the spectrum from the fundamental to twice the link frequency (in this case

40 kHz). This spectrum is in marked contrast to many types of DC link converters in which the harmonic spectrum has characteristic "blips" at well defined multiples of the fundamental. Similar results can be shown to be possible for the DC resonant AC voltage link configuration except that the basic switching frequency is one full cycle rather than a half cycle resulting in a "blip" at 20 rather than 40 kHz.

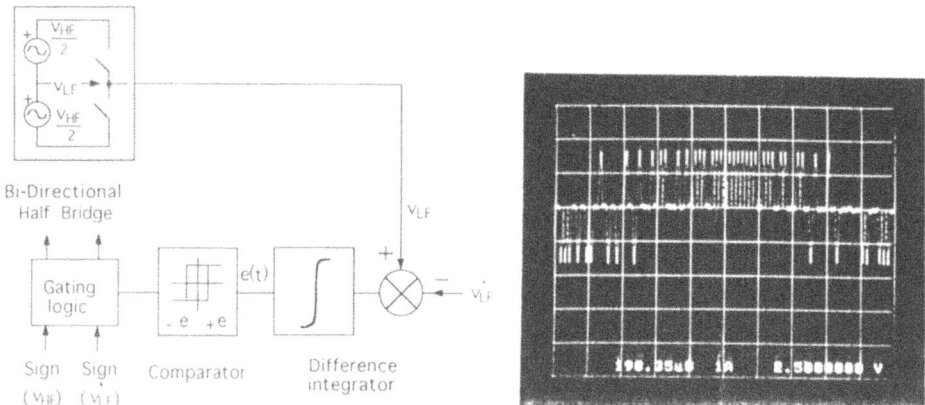

Fig. 20. Control block diagram for area comparison pulse density modulation (delta modulation) and resulting line-to-line voltage waveform; voltage: 250 V/div, time: 198.35 ms/div.

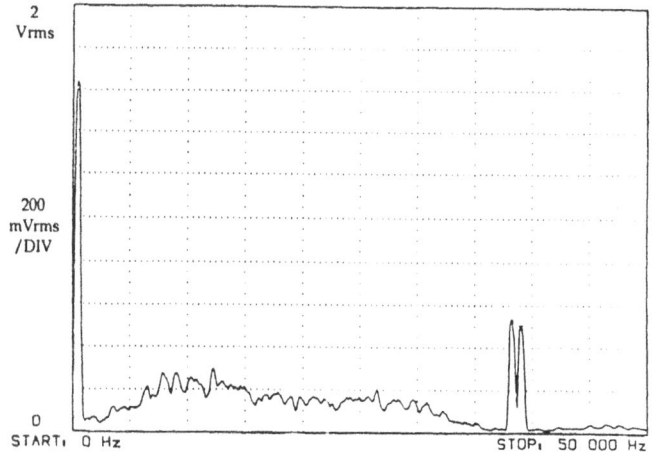

Fig. 21. Harmonic spectra associated with the line voltage waveform of Fig. 13; amplitude: 20 V/div, frequency range: 50 kHz.

The importance of current regulation in AC motor control has prompted a close look at current regulation in such drives [20,21]. Fig. 22 shows the current waveform for the AC resonant voltage link system, using a simple bang-bang controller. While this simple type of current controller is very adequate at low frequencies, the need for more sophisticated current regulators increases as the motor frequency rises.

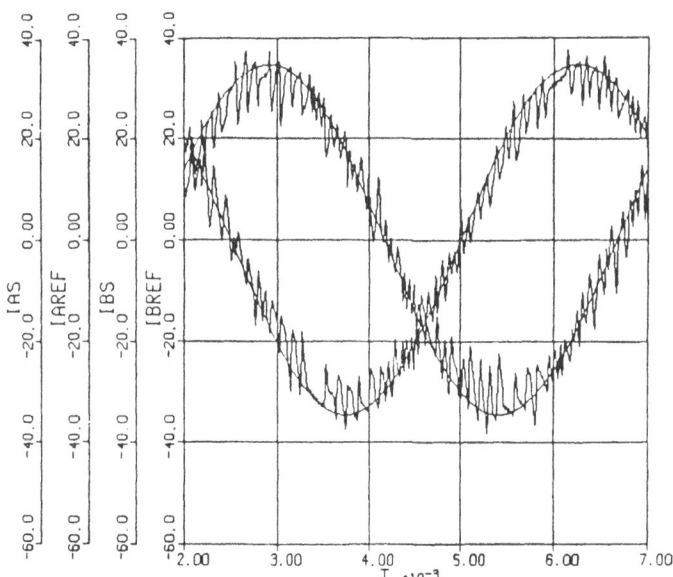

Fig. 22. Response of current regulator employing bang-bang current regulation. Traces: A phase and B phase line currents and corresponding reference currents in amperes, frequency: 143 Hz.

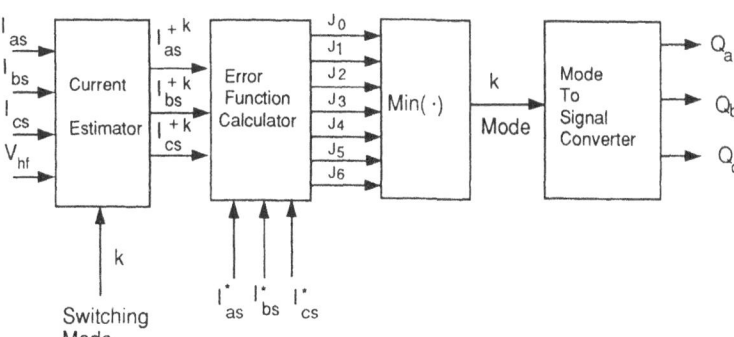

Fig. 23. Block diagram of mode controller for current regulation. Min (·) selects the minimum value of the error functions J_0 to J_6.

An alternative method is evident when it can be noted that in zero voltage or zero current switching schemes, the switching instant and pulse duration is specified. Hence, the problem of current modulation reduces to finding the next optimal combination of switch states at each switching instant, a switching strategy impractical for conventional pulse width modulated DC link converters. If the load current for the next switching instant can be predicted for all possible switching states before the switching instant, the switching pattern can be selected which minimizes a specified error function. If only current regulation is required, the error function may be simply the sum of the absolute current regulation errors of each phase or, alternatively, the square of the individual errors. Fig. 23 shows the block diagram of a mode selection controller based on this principle which realizes substantial improvement in the current waveform at high frequencies. A typical waveform for this controller is shown in Fig. 24 and can be compared with Fig. 22.

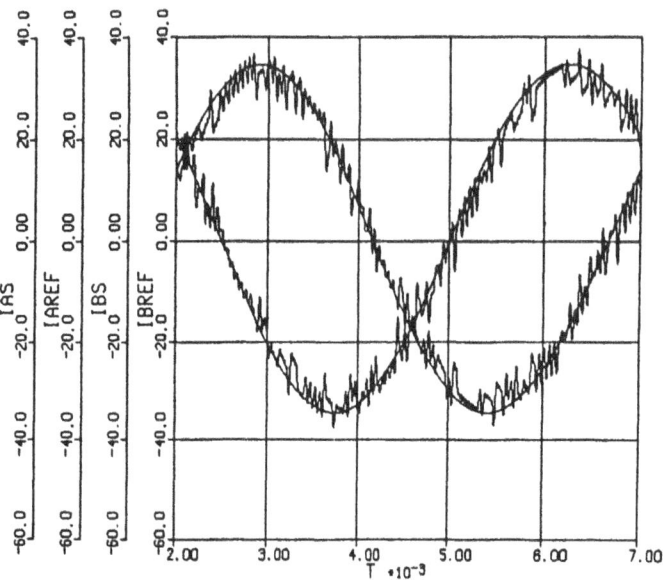

Fig. 24. Response of current regulator, employing mode selection current regulation. Traces: A phase and B phase line currents and corresponding reference currents in amperes, frequency: 143 Hz.

9. FIELD ORIENTED CONTROLLER

A complete system diagram for a field oriented induction motor controller, utilizing high frequency link technology, in this case a resonant DC current link topology, is shown in Fig. 25 and illustrates the complexity which must be dealt with when applying resonant link technology. The current of the load side converter (outer loop of Fig. 25) is controlled by a field oriented controller in which the speed error is used to command a stator current component orthogonal to the rotor flux while the rotor flux itself is adjusted by the in phase component (I_{qs}^* and I_{ds}^* respectively in Fig. 25). The output of the inner current regulator is used to command a voltage by an inner link voltage regulator. This inner voltage loop regulator is needed to ensure stability of the overall control.

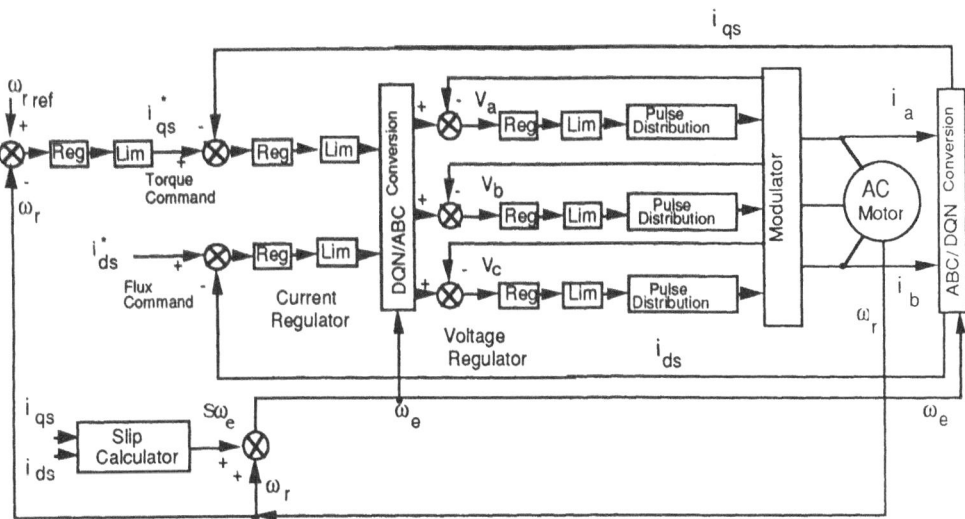

Fig. 25. Diagram of controller of the output converter of a resonant DC current link converter feeding and induction motor load.

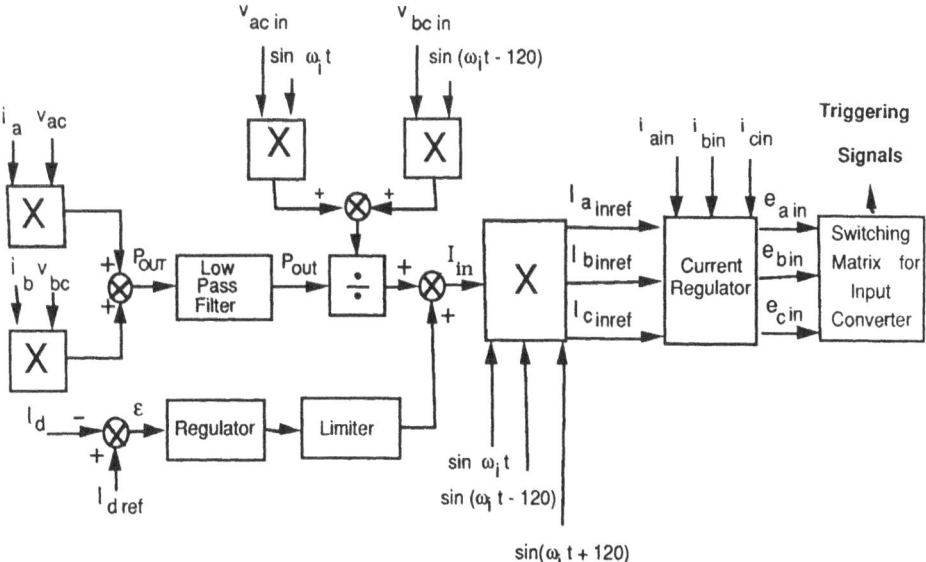

Fig. 26. Complete diagram of control ot the input converter of a resonant DC current link power conversion system.

The input power converter is controlled as shown in Fig. 26. The power estimator P_{out} provides an estimate of the current value of active power by calculating the average load side power and the system losses, based upon measurement of the current operating conditions. The output power is divided by the amplitude of the input side AC voltage to form a current amplitude command for the source side converter. Small errors in this calculation are compensated for by the regulator which regulates the current in the DC link side inductor to a prescribed value. The sum of the current to support the load side power plus the current to maintain the DC inductor current form the current amplitude command I_{in}. The current amplitude command is then converted to equivalent sinusoidal line frequency current commands by means of templates sin $\omega_i t$, etc. which are obtained from taking algebraic combinations of the line to line voltage divided by its amplitude.

A typical trace showing the transient characteristics of the drive is shown in Fig. 27. In particular, the trace shows system performance for a step change in torque command from 0 to 8 Nm and then back to 0 Nm. Again, the DC resonant current link converter of Fig. 7 is modeled. It is apparent that the high effective switching frequency of the converter permits very rapid changes in motor current results, in turn, in an extremely fast acting torque controller. Figure 20 shows an experimental trace of the system for the condition where the motor is operating under a steady load condition and demonstrates the feasibility of operation with the supply current having a nearly sinusoidal waveform at unity power factor.

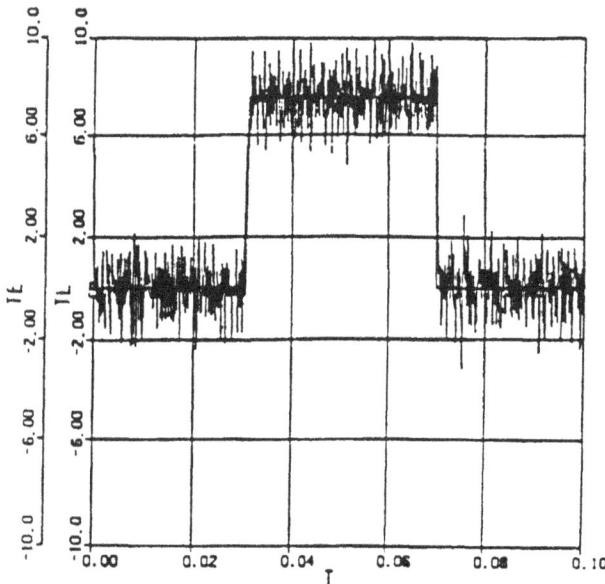

Fig. 27. Torque response of system, employing resonant DC current link. Traces: electromagnetic torque in Nm, speed in RPM and torque command.

Figure 28 shows an experimental trace of the input voltage as well as the input and output current for the case of a resonant AC voltage link conversion system. Note that the input current is in phase with the input voltage. This important property is common to all of the circuit topologies that have been discussed in the paper and points to a major advantage of this technology compared to conventional thyristor bridge rectifiers.

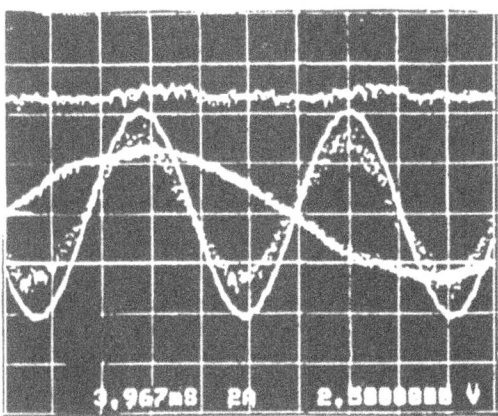

Fig. 28. Illustrating unity power factor operation of line side converter during motoring operation. Top trace: peak of the link voltage: 130 V/div; low frequency sinusoidal trace: phase A induction machine current (30 Hz): 5 A/div; high frequency sinusoidal trace: phase A source voltage (60 Hz), 40 V/div; remaining trace: phase A source current (60 Hz): 5 A/div; all grounds 4 div from the bottom; time: 3.97ms/div.

10. CONCLUSIONS

This paper has summarized recent work on a new class of power converters, the resonant link power converter. These converters utilize the zero crossings of the link voltage or current to realize nearly zero loss switching. Of particular importance is the fact that these new converters do not necessarily require extremely fast turn-off devices (i.e. IGBTs or MCTs) for satisfactory operation but can, in fact, utilize low cost bipolar transistors and thyristors as switching elements. Since the effective switching frequency is high, the currents at the input and output of the converter are nearly sinusoidal. These new converters promise to play an important role in the next generation of power conversion equipment.

REFERENCES

1   F.C. Schwartz, "An improved method of resonant current pulse modulation for power converters", *IEEE Trans on Proc. Ind. Electronics and Control Instrumentation*, vol. IECI-23, May 1976, pp. 133-141.

2   H.K. Lauw, J.B. Klassens, N.G. Butler and D.B. Seely, "Variable speed generation with the series-resonant converter", *IEEE Trans. on Energy Conversion*, vol. 3, no. 4, December 1988, pp. 755-764.

3   J.B. Klassens, F. de Beer, "Three-phase AC-to-AC series-resonant power converter with a reduced number of thyristors", *1989 IEEE Power Electronics Specialist's Conference*, Milwaukee, WI, 26-29 June 1989, pp. 376-384.

4   P.K. Sood and T.A. Lipo, "Power conversion distribution system using a resonant high-frequency AC link", *Conf. Rec. 1986 IEEE-IAS Annual Meeting*, pp. 533-541.

5   P.K. Sood, T.A. Lipo and I.G. Hansen, "A versatile power converter for high frequency link systems", *IEEE Trans. on Power Electronics*, vol. 3, no. 4, Oct. 1988, pp. 383-390.

6   S.K. Sul and T.A. Lipo, "Field oriented control of an induction machine in a high frequency link power system", *IEEE Trans. on Power Electronics*, vol. 5, no. 4, October 1990, pp. 436-445.

7   S.K. Sul, I. Alan and T.A. Lipo, "Performance testing of a high frequency link converter for space station power distribution system", *1989 Intersociety Energy Conversion Engineering Conference*, Washington DC, August 1989, vol. 1, *Aerospace Power Systems and Power Conditioning*, pp. 617-624.

8   D.M. Divan, "The resonant DC link converter - a new concept in static power conversion", *Conf. Rec. 1986 IEEE IAS Annual Meeting*, pp. 648-656.

9   D.M. Divan and G. Skibinski, "Zero switching loss inverters for high power applications", *Conf. Rec. IEEE IAS Annual Meeting*, pp. 627-634, 1987.

10   Y. Murai and T.A. Lipo, "High frequency series resonant DC link power conversion", *1988 IAS Annual Meeting*, Oct. 2-7, Pittsburg PA, pp. 772-779. (Accepted for publication in *IEEE Trans. on Industry Applications*).

11   Y. Murai, S.G! Abeyratne, T.A. Lipo and P. Caldeira, "Current peak
      limiting for a series resonant DC link power conversion using a
      saturable core", *1991 European Power Electronics Conference*,
      Florence Italy, Sept. 1991, vol. 2, pp. 8-12.

12   P. Caldeira, K.W. Marschke, T.A. Lipo and Y. Murai, "Utilization of
      the series resonant DC link as a DC motor drive", *1990 IEEE IAS
      Annual Meeting*, Oct 1990, pp. 266-272.

13   K.W. Marschke, P.Caldeira, T.A. Lipo, "Utilization of the series
      resonant DC link as a conditioning system for SMES", *IEEE Power
      Electronics Specialist's Conference*, June 1990, pp. 266-272.
      (Accepted for publication in IEEE Trans. on Power Electronics).

14   P. Caldeira, T.A. Lipo, Y. Murai and S. Mochizuki, "Design and
      control of a series resonant DC link power converter drive", 1990
      International Power Electronics Conference (IPEC), Tokyo Japan,
      April 1990, pp. 397-404.

15   J.S. Lai and B.K. Bose, "An improved resonant DC link inverter for
      induction motor drives", *IAS Annual Meeting*, Oct. 1988, pp. 742-
      748.

16   Y. Murai, S. Mochizuki, P. Caldeira and T.A. Lipo, "Current pulse
      control of high frequency series resonant DC link power
      converter", *1989 Industry Application Society Annual Meeting*, San
      Diego, CA, Sept. 1989, pp. 1023-1030.

17   Y. Murai, S.G. Abeyratne, T.A. Lipo, P. Caldeira, "Dual-flow pulse
      trimming concept for a series resonant DC link power
      conversion", *1991 IEEE Power Electronics Specialists Conference*,
      June 1991, pp. 254-260.

18   Y. Murai, H. Nakamura, T.A. Lipo, M.T. Aydemir, "Pulse-split
      concept in series resonant DC link power conversion for
      induction motor drives", *IAS Annual Meeting*, Oct. 1991, pp. 776-
      781.

19   S.-S. Park and G.H. Cho, "A current regulated pulse width
      modulation method with new series resonant inverter", *IAS
      Annual Meeting Conference Record*, 1989, pp. 1045-1051.

20   M. Kheraluwala and D.M. Divan, "Delta modulation strategies for
      resonant link inverters", *Conf. Rec. of 1987 IEEE Power Electronics
      Specialist's Conference*, Blacksburg, Virginia, pp. 271-278.

21   R.D. Lorenz and D.M. Divan," Dynamic analysis and experimental
      evaluation of delta modulators for field oriented AC machine
      current regulation", *Conf. Rec. of 1987 Annual Meeting IEEE Ind.
      Appl. Soc.*, pp. 196-201.

DISCUSSION

P. Tenti (Univ. Padua, Padua I)

The main advantage of the solutions you propose seems to be in
higher frequencies, beyond the inherent limitation of the di/dt and
dV/dt. On the other side, you have a penalty in terms of switch stresses
and circuit complexity that also affects the reliability of the system, and
there is essentially no possibility to integrate this kind of system. What is
the trade off in industrial applications ?

T.A. Lipo

By far the most interesting circuit to the manufacturers is a resonant
DC voltage link system. To be economical, it requires a clamp, typically,
a rather sizeable capacitor and an additional transistor which is working
at 6 times the frequency of the other switches (e.g. at 20 kHz). So far,
most manufacturers have decided not to pursue this concept in the market place, discouraged by the additional cost of one transistor. Only in
special applications, particularly in areospace, aircraft and certain specialized consumer applications, people are very seriously going ahead.

In the area of the series resonant DC link, there is a lot of potential
if we can get a good device which will turn off quickly. A thyristor with a
turn-off time of 20 or 30 μs would allow for a 20 kHz link and, therefore,
make a very cheap drive. If we take just half of this link, the rectifier
part or the AC to DC part of the series resonant link, and connect it to a
DC motor, we can have a DC motor drive with a very nice sinusodial input
current at unity power factor with only 6 thyristors. Of course, these
thyristors will be more expensive than normal ones. The logical application of a resonant current link would be high power, just like the application of a DC current link converter is high power. But that presumes
that we have a very fast turn-off thyristor or perhaps a fast GTO with reverse blocking capability. It is up to the device people to say if that is
possible on the megawatt scale. So far, I have not seen any devices that
would allow us to go beyond a few 100 kW.

A. Mertens (RWTH Aachen, Aachen D)

In our laboratories, we have found that the field controlled thyristor,
used as a zero current switch, might be a good alternative for your application. May I have your opinion on the comparison between resonant

link topologies and resonant switching converters using PWM techniques, as introduced first by DeDoncker ?

T.A. Lipo

I have limited my presentation to resonant links rather than trying
to cover other areas of zero voltage switching which are being worked
on, e.g. at GE. Although I cannot give a quantitative answer, I do believe
that DeDoncker proposes basically a circuit just used in a different way. I
think that this circuit is very attractive and may be suitable for higher
power than the resonant DC voltage link. It seems that we have to go
through a phase of looking at all the commutation circuits with the interest of not turning off a device or not shifting the current from one
device to another but actually going through a zero voltage switching
mode.

R. Dutton (Stanford U., Stanford CA)

You have mentioned the problems of the non-ideality of the
switches, their temperatures sensitivity, etc. What is the status of the
switch modelling and how do you actually handle it to try and translate
this from a theoretical switch capacitor view of to a practical circuit ? If
the real limitations to switches are so important, what kind of modelling
tools are you actually using to treat the switches?

T.A. Lipo

We basically rely upon SPICE circuit models. Some of the phenomena that I talked about, the storage time, and the resonant circuit, are
very complex issues and I do not know if it is worth studying them on a
computer. Basically, you have to build up a circuit, e.g. to properly
account for the temperature. We prefer to work on hardware rather than
rely upon simulation.

ADVANCED POWER MODULES FOR DRIVES AND AUTOMATION

H.-J. Krokoszinski and S. Kjellnäs[*]

Asea Brown Boveri Corporate Research
Heidelberg, Germany

[*]ABB Control AB, Västerås, Sweden

ABSTRACT

For power switches with high repetition rates (solid state relays) and for motor starters with ramped speed (softstarters), power modules, i.e. the combination of several power semiconductor devices, are widely used. In contrast to electro-mechanical switches, they provide the required long-term reliability and the control of load current, respectively. Today, the drive circuits for the power devices are mostly attached to the switching devices inside of or closely adjacent to the module housings. In the former case, the integration of more micro-electronic functions is easily accomplished by adding simple logic circuitry or even complex microprocessor capability. This enables the design of versatile and flexible product systems since characteristics and parameters can be adjusted by software modifications. In this work, we try to elucidate the common trend in power electronics of replacing components more and more by integrated devices and show some examples of advanced power modules, developed in ABB laboratories, which demonstrate these system-technical aspects. Special emphasis will be put on the features needed for automated production, testability, thermal management, and those which enable self-protection of the devices as well as digital communication.

Power Semiconductor Devices and Circuits, Edited by A.A. Jaecklin
Plenum Press, New York, 1992

1. INTRODUCTION

Cost effective production of power electronic devices for drives and automation requires the integration of an increasing number of components into preproduced subsystems which can easily be combined to higher-level systems in various different versions ("system technique"); thus, engineering cost can be reduced significantly. By adding the benefits of system technique and of modern production technologies, the overall production cost can be reduced which is the most prominent goal in industrial manufacturing today. In the case of power electronic circuits like rectifier bridges, power switches, inverters, converters etc., the first step of integration comprises the assembly of all power semiconductor chips involved on one substrate leading to a "power module". These are available on the market from many suppliers (e.g. Toshiba, Mitsubishi, Powerex, SanRex, Eupec, Semicron, ABB-IXYS etc.) in a variety of types (Schottky diode modules, bipolar transistor or Darlington modules, PowerMOSFET modules, IGBT modules, thyristor modules)[1,2]. In most cases, all free-wheeling diodes required are included in the package. However, only some of the module suppliers provide the next step of integration in terms of insulated gate drive circuits and self-protection ability. Then, conventionally, the term "smart power module" is justified. Since this often leads to confusion with the "smart power devices", describing the monolithic integration of power device and gate drive circuit[3], the expression "intelligent" instead of "smart" is sometimes preferred[4,5]. However, this type of intelligence by electronic means does not compare to the versatility of a microprocessor which should define the level where intelligence of electronics begins. To bypass this open discussion, we are dealing in this paper with "advanced" power modules, which can belong to either definition, depending on their level of system-technical integration and technology.

2. ELECTRONIC LEVELS

In Fig. 1a, the three electronic levels ("power", "control" and "communication") are depicted, which are successively integrated in the device when system-technical benefits represent the driving force. However, the electronic levels do not necessarily coincide completely with the geometrical levels inside of a package which, of course, have to be stacked in the third dimension in order to minimize area and space (Fig. 1b). In most cases, the communication level will be combined with the control level on a common hybrid circuit or Printed Circuit Board (PCB), mounted in an intermediate position inside of the housing. On the other hand, those parts of the control circuitry which are on high potential or dissipate power are conveniently placed on the power level represented by an electrically insulated but thermally conducting module substrate, attached to a heat sink.

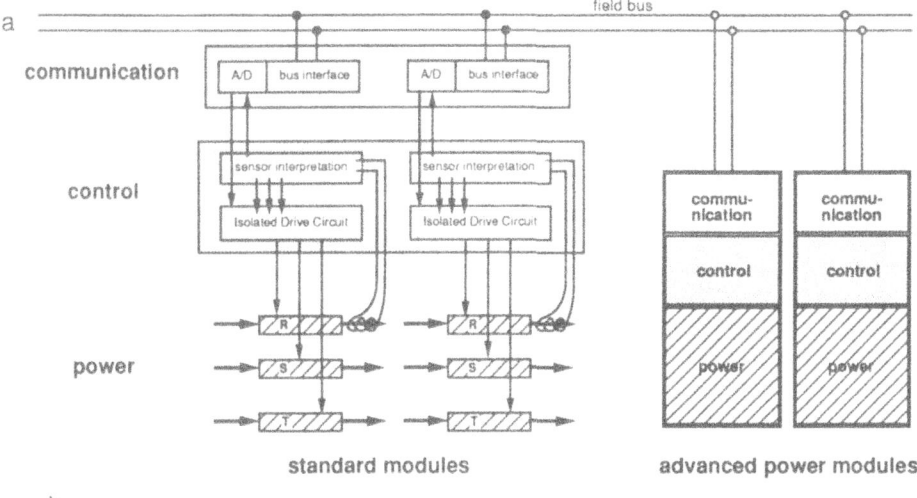

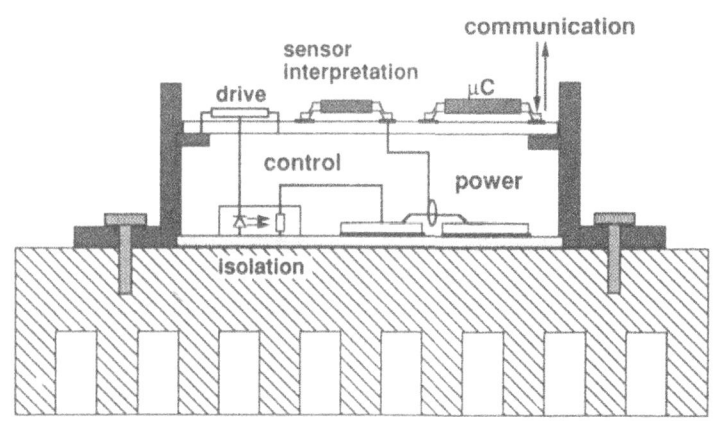

Fig. 1. System technical integration of communication and control electronics into power electronic devices; a) electronic levels b) schematic of possible arrangement in a module housing.

2.1 The Power Level

Standard modules, comprising, for instance, half-bridges (2-pack), full-bridges (6-pack) or 2 antiparallel thyristors (1-phase AC-switch), represent the most simple type of power level (Fig. 2). They are widely used for the construction of drives by screwing them down to a heat sink, side by side and interconnecting them to each other or to the control electronics by conventional cables or boards. The big advantage of using standard modules is certainly the lower price due to high production numbers compared to any kind of custom designed module with special features. However, when the high-level device, which is constructed from these parts, reaches larger numbers, then the assembly cost basing on the number of components to be mounted is likely to exceed the benefit of using standard modules. Hence, the increasing batch numbers will, at some instant, make it worth while designing a custom specific power module which can be equipped with properly located test pads and with components required for control and self-protection features, especially voltage, phase, current and temperature sensing devices. Furthermore, the insulation barrier between power level and drive circuitry is conveniently integrated on such an advanced power module substrate.

Fig. 2. Examples of "standard modules" (courtesy of ABB-IXYS).

2.2 The Control Level

The control level comprises the gate drive circuit and the sensor signal interpretation circuits. In most cases, it is insulated from the power level by either pulse transformers or by optical means, using optocouplers or optically fired triacs (Fig. 3). The latter devices, as realized for instance by Siemens ("SITAC"[6]), are able to directly switch the ignition pulse current of thyristors (typically 300 mA) without the need of an auxiliary power supply. For signal transmission and a Transistor-Transistor-Logic (TTL)-voltage supply, small ferrite core transformers, providing an isolation barrier of typically 2.5 kV are used as for example in the "smart power Input/Output (I/O)"-chip set (Fig. 4) of ABB-IXYS, intended to drive Metal Oxide Semiconductor Field Effect Transistors (MOSFET) or Insulated Gate Bipolar Transistors (IGBT). It consists of a "low"-side interface chip, IXD7100, a "high"-side driver Integrated Circuit (IC), IXD7200, and a dual transformer for power transmission and bidirectional manchester-coded digital communication between the two chips[7]. Since all three components are packaged in a Dual In-line Package (DIP), they have to be mounted on a PCB and connected to the semiconductor switches.

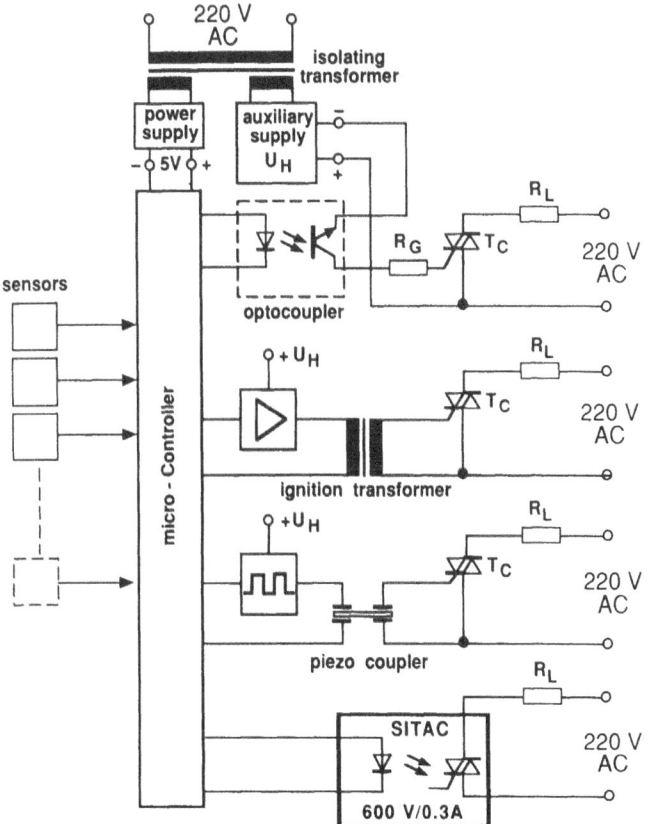

Fig. 3. Schematic representation of different realizations of a potential barrier between microelectronic control (e.g. microcontroller) and the drive circuit of power electronic devices[6] (SITAC® : Registered Trademark of Siemens).

A *purely analog* low-cost version of an isolated gate drive for IGBTs was developed in the ABB-labs in cooperation with ABB-IXYS[8]. It contains a pulse transformer with two primary windings (Fig. 5a) for gate drive signal transmission (prim. => sec.) and fault signal transmission (sec. => prim.), respectively. The complete drive for one IGBT consists of 10 ordinary transistors, 20 resistors, 2 capacitors and 1 diode. Of course, in a half-bridge these numbers have to be doubled but both legs share a common auxiliary power supply, fed by an additional transformer. The drive voltage for the gate of the IGBT is generated by a pulse in the primary winding W1 (caused by switching on U1) and transferred by the secondary winding W3 to the flip-flop T5+T6. There, it is stored and fed to the bases of the amplifiers T10+T11 which turn the IGBT on. The diode, connected to the collector, is used for V_{CEsat}-detection: when the anode potential of the diode sharply rises, then the base potential of transistor T8 exceeds its emitter potential and T8 is turned off. This over-current condition, which can be preset by the resistor R17, is instantaneously used to shut down the gate of the IGBT by turning off transistors T5+T6 and, thus, extinguishing the gate voltage of T10+T11. Simultaneously, due to turn-off of T5+T6, the constant current through the secondary winding is interrupted; this causes an induction peak in the primary winding W2 which is stored in the flip-flop T2+T3 and interpreted as a fault signal.

Due to handling the V_{CEsat}-condition only on the secondary, the reaction time of this circuitry to short-circuit induced current rises is outstanding (Fig. 5b): for the nominal current of 50 A, a short-circuit is detected and turned off within 1 µs; with the given line inductances, the current could only rise up to 150 A within the first 500 ns.

The complete circuit for two legs of a half-bridge, including auxiliary power supply, was integrated on a small PCB by conventional double-side Surface Mounted Device (SMD)-technique. Thus, the control electronics could be included as a second plane in the normal module housing (Fig. 5c).

In contrast to this, the drive circuitry for a 1 kVA hybrid converter (Fig. 6) of HANNING[5] is monolithically integrated in a *central control IC* which is mounted on the same ceramic substrate (100 mm x 60 mm) that carries the power electronic switches.

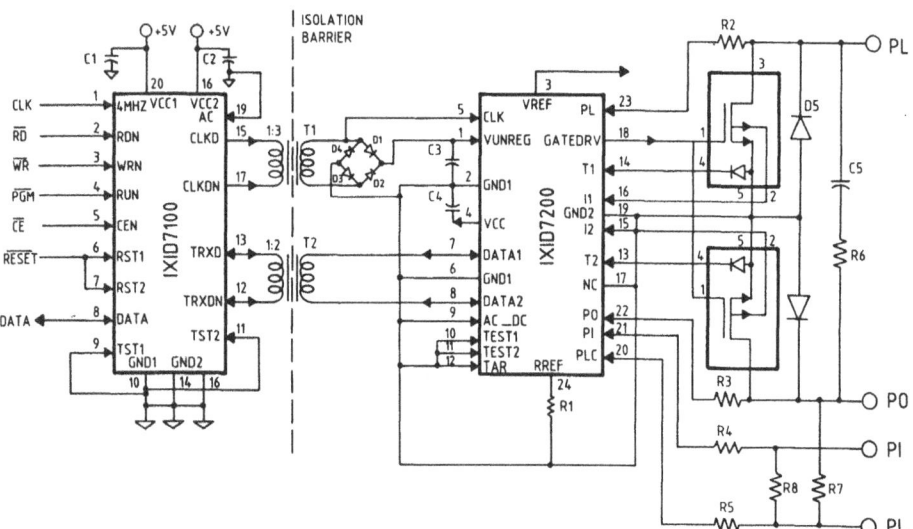

Fig. 4. Block diagram and wiring scheme of the "smart power I/O chip set" (courtesy of ABB-IXYS); IXID7100 is the low-side interface chip which is separated by a dual transformer device (T1+T2) from the high-side driver chip IXID7200, driving a half-bridge of Power MOSFETs or IGBT's with integrated current mirror.

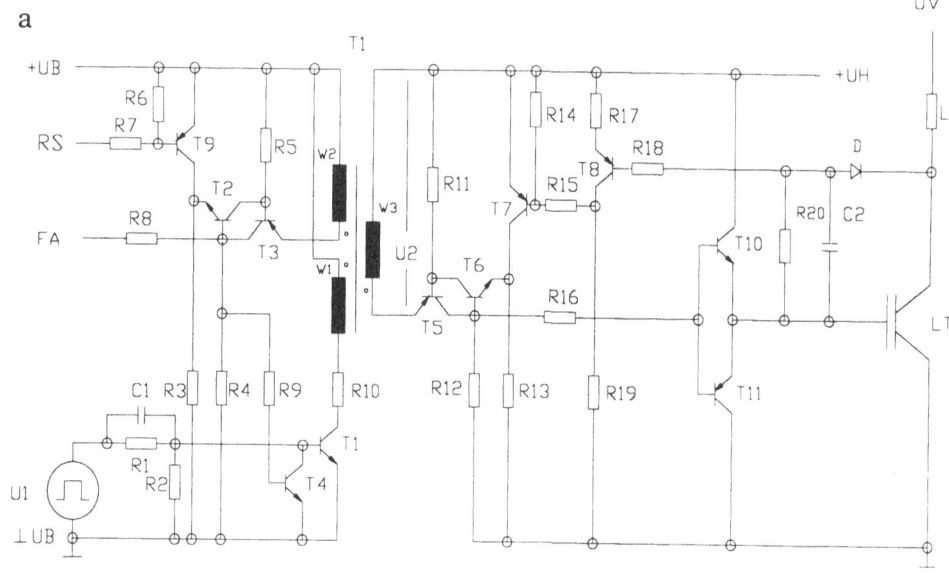

a

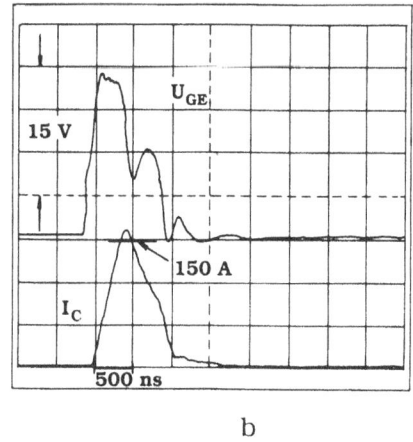

b

c

Fig. 5. a) Low-cost circuitry of an insulated gate drive circuit for IGBT's[8], performing V_{CEsat}-detection, gate voltage shut-down and fault signal generation on the "high" side; trigger signals and fault signals are transferred from and, respectively, to the "low" via a ferrite core transformer with two primary windings;

b) IGBT over-current shut-down characteristic, traced on an oscilloscope; the short-circuit is detected by the circuitry in Fig. 5a within 500 ns, and the current is reduced from its peak value (150 A, nominal current 50 A) down to zero within another 500 ns;

c) IGBT half-bridge module (ABB-IXYS), supplemented (inside of the standard housing) by a SMD-type of PCB, carrying two circuitries of the type shown in Fig. 5a plus an auxiliary power supply, coupled to the drive circuits by the bigger ferrite core transformer in the center of the PCB; the primary sides of all three transformers are on top and the secondaries are on the rear side of the PCB.

An even more "intelligent" solution is the integration of a *complete microcontroller* which is able to interpret all sensor signals attached to its A/D-converters and to set the required control voltages at the primary of the galvanic separation devices (Fig. 3). Such a controller, e.g. on the basis of the 8051-processor from INTEL, is already available as a complete credit-card-sized PCB. In the 80535 version[9], it includes 24 parallel bidirectional I/O-ports, one serial port, 8 analog inputs for 8 bit-A/D-conversion, different timer functions, 64 kByte Random Access Memory (RAM) and 64 kByte Erasable Programmable Read Only Memory (EPROM) or, optionally, Electrically EPROM (EEPROM). Since these microcontrollers are extremely versatile and, in large numbers, are going to be price-competitive with any custom designed Application Specific IC (ASIC), they are likely to be more wide-spread in the future. By mounting them in a second plane in the housing, the major advantage of system technique can be exploited: the same control and communication hardware can be combined with various power parts, differing e.g. in their current or voltage ratings. Then, simply by changing the program in the ROM of the Microcontroller (µC), the control functions can be adapted to different ratings and load characteristics, leading to completely different product characteristics.

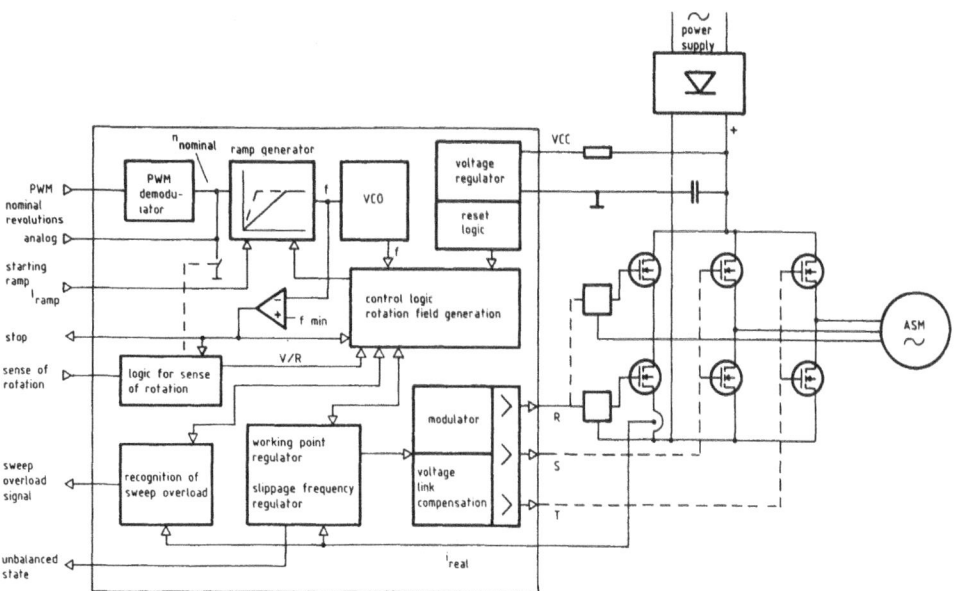

Fig. 6. Hybrid converter in thick film technology (HANNING) according to [5]; the complete control electronics is integrated in a central IC (solid frame).

2.3 The Communication Level

The combination of many different intelligent devices, like sensors and actuators in an industrial process system, requires an I/O-system which is sufficiently failsafe, maintenance-free and easy to run. In addition to that, there are, for different types of processes, many different requirements in terms of I/O-speed (real-time behaviour) and granularity (number of I/O-spots per subsystem)[10]. Due to the diversity of applications, no commonly accepted field bus conception has been developed up to now, matching all requirements. In Table 1, the properties of the currently available field bus systems and their special applications are collected. For example, high speed local area connection of sensors and actuators in automobiles is especially met by the Control Area Network (CAN)-bus[11]. For drive systems, both the SErial Read Time COmmunication System (SERCOS) and the PROcess Field BUS (PROFIBUS) are suitable candidates. The former was specially designed for the drive applications by the German Electrical and Electronic Manufacturers' Association (ZVEI)[12]. Up to 254 slaves can be connected to the master in a ring configuration by a single optical fibre. With the highest number of users, the slowest transmission rate of 1 ms for each slave is achieved, which is still fast. Due to optical transmission and the CRC-check of data, the resistance to electrical interference is good. Up to now, this bus has only been used between numerical controls and drives, but it has the impact to become a standard for drive systems[13].

The PROFIBUS is used with standard 2-wire cable (RS485) for the interconnection of up to 32 participants. Since only cycle times of 2-4 ms per user are achieved, this cannot be considered "real-time" where $\leq$ 1 ms is required. However, data transmission security is attained by a special hand-shake mechanism and fault correction. Special performance management functions and features, like exchange of participants during operation, are of large interest in drive systems since they help in commissioning, configuration, and service[13]. Up to now, only software solutions of PROFIBUS-interfaces are available, but chip solutions are likely to arise in future. However, when μC-control is used in advanced power modules, the bus coupling is easily incorporated by software means.

3. POWER MODULE TECHNOLOGIES

According to the power ratings of the semiconductor devices to be combined in a module, different technologies are available. The first decision is done with the choice of *substrate material*, the second is introduced by different methods of *copper deposition* used for the production of current leads on the substrate, and an additional increase of variety is induced by various *chip mounting* and *interconnection techniques*.

3.1 Substrates and Copper Deposition

The basic choice with respect to substrates is between Insulated Metal Substrates (IMS) or insulating (ceramic) materials; all of them have to be attached to some kind of heat sink, using a thermal grease and considerable pressure. There is only one exception, the Compact Power Circuit (C.PC)-technique[14], which is able to produce alumina and copper layers directly on massive aluminum heat sinks by means of plasma and high-velocity-powder spraying (see sec. 3.1.2).

Table 1. Properties of available field bus systems [10,13]

property	bitbus	Interbus	Profibus	SERCOS	CAN
			PROcess FIeld Bus	SErial Real-time COmmunic. Syst.	Control Area Network
target application	process control, installation	process control, installation	process control, installation	drives	automotive
trans-mission medium	2 pairs of wires RS485, ser. trans. direct or isolated	field bus 9x2 wires, ser. trans.	shielded 2-wire cable, RS485, ser. trans.	optical fibre ring	twisted pair of wires
access	master-slave-slave	serial polling	Token-passing ("Flying Masters")	allocated time-slot per slave	bytewise
number of stations	255 repeater after 32	32 cluster, 8 I/O-modules /cluster	122 repeater after 32	254	16
implemen-tation	microcontroller (Intel, Inrec)	protocol chip (Phoenix)	software, later on: microcontroller master-slave-slave	spec. interface	controller chip (Motorola, Intel)
installation distance	600 m up to 10 km (with 10 repeaters)	32x400 m (12 lm)	4.8 km, 3 repeaters requ.	synth. fibers; above 40 m: glass fibers	40 m
baud-rate	62.5 kBit/s, 375 kBit/s 2.4 MBit/s synchr.	300 kBit/s	kBit/s: 9.6, 19.2 90, 187, 500	125 kBit/s 1 MBit/s planned	1 MBit/s
granularity	16...64 I/O-spots	128	16...1000		1...4
communic. level	short messages (1...50 bytes)	serial polling (telegrams)	Token passing Token control	time slots	Message passing
software-interfaces	RAC-task direct read/write of remote registers	process imaging high voltage drivers	MMS-Subset ("small" MAP), LLI (Lower Layer Interf.)		communic. object layer
special properties	local fito-coupling to CPU	emergency stop release	message passing slow betw. intelli-gent and fast betw. simple partners	CRC-check ring check	max. message length: 8 bytes, isolation of faulty nodes

3.1.1 Insulating Substrates. The most commonly used insulating substrate material up to now is aluminum oxide (alumina, 96 %). Beryllium oxide (beryllia, BeO) is certainly the best material concerning thermal conductivity (Table 2), but it is in most cases avoided in power electronics due its toxic dust and its high price. In contrast to this, aluminum nitride (AlN) is non-toxic and provides a similar (but still varying) thermal conductivity for a (still high) price which is certainly decreasing the more users require outstanding thermal properties of module substrates.

Table 2. Characteristic properties of module substrates [15,16,17]

property	unit	96 %-Al_2O_3	BeO	AlN
thermal conductivity	W/mK	20-25	220-250	150-240
thermal expansion	ppm/K	6.7-7.5	8-8.5	3.4-4.5
dielectric strength	kV/mm	8-8.3	18-19.7	14
cost factor (rel. to Al_2O_3)		1	10	15

Alumina substrates and aluminum nitride substrates can be coated with copper layers by the same techniques; however, on AlN it is considerably more difficult to produce well adhering coatings, using the following well-known copper deposition methods:

- screen printing (*Copper Thick Film technology*, CTF)
- electroless plating (*Copper On Ceramic*, COC)
- eutectic oxide bonding (*Direct Copper Bonding*, DCB).

Different module suppliers prefer different copper deposition techniques; PowerCompact, for example, only offers screen printed layers, both on alumina and aluminum nitride substrates[15], whereas ABB-IXYS has confined itself up to now to DCB on alumina substrates. This may be partly due to historically grown special know-how and partly to an economically motivated restriction to certain power ranges.

In Table 3, the various copper layers are compared with respect to their specific resistivity and typical thickness achieved by the individual deposition technique. Due to its extremely inhomogeneous structure (sintered grains), copper thick film interconnects exhibit (for single printing of Cu, d ≈ 28 μm) typical sheet resistances around 1.1 mΩ/sq. In contrast to this, the massive copper sheets of 300 μm thickness used on DCB-substrates have 0.07 mΩ/sq. This explains why CTF is only used in cases of small currents (single print: < 20 A) whereas, for example, DCB-modules can carry currents of more than 200 A, provided that the total thermal resistance, R_{th}, of the multilayer "chip to heat sink" is sufficiently low to handle the according losses in the semiconductor devices on the module (Fig. 7);

$$R_{th,total} = [T(Chip) - T \text{ (heat sink surface)}] / P_{el} \qquad (1)$$

where T is the temperature and P_{el} the electrically dissipated power.

Table 3. Comparison of copper resistivity values

type of copper	typ.thickness [μm]	spec.resistivity [μΩcm]	typ. sheet resistance [mΩ|sq.]
bulk copper	arbitrary	1.673 [18]	-
copper thick films	28	3.07	1.096
electroless plated copper	75	2.53	0.337
laminated copper sheets	60	2.05	0.342
direct copper bonding	300	2.12	0.071
hv-powder-sprayed Cu	200	4.9-5.9	0.245-0.295

Table 4. Calculated 1-dim. thermal resistance of a 7 x 7 mm^2 silicon die soldered on a DCB-substrate (see Fig. 7); no heat transfer through the top lead.

layer i	material	area F_i [cm^2]	thickness d_i [μm]	th. cond. λ_i [W/mK]	trans. coef. a_i [W/m^{2}K]	th. resist. $R_{th,i}$ [K/W]
1	silicon chip	0.49	170	145[*]	0.876E6	0.0239
2	solder	0.49	60	33	0.58E6	0.0371
3	copper	0.49	300	330	1.1E6	0.0186
4	alumina	0.49	630	24[**]	0.0333E6	0.5357
5	copper	0.49	300	330	1.1E6	0.0186
6	thermal grease	0.49	20	0.42	0.021E6	0.9718
1-2	junction-substrate[**]					0.061
3-6	DCB-substrate (incl. thermal grease)					1.545
1-6	junction-heat sink[***]					**1.606**

[*]   at T = 50 °C, whereas λ = 149 W/mK at 25 °C [18]

[**]   Rubalit 708 (Hoechst), T = 42 °C

Assuming 1-dimensional heat conduction, i.e. excluding heat spreading in the individual layers, then the total thermal resistance given by eq. (1) can be calculated from the sum of the thermal resistances of all involved layers including the thermal grease between module and heat sink:

$$R_{th,total} = \sum_i \frac{1}{\alpha_i F_i} = \sum_i \frac{d_i}{\lambda_i F_i} \qquad (2)$$

where α_i = thermal transfer coefficient of layer i
         F_i = area of layer i
         λ_i = thermal conductivity of layer i
         d_i = thickness of layer i

This calculation is often used to gain a rough estimate of the chip temperature which can be expected for a given multilayer of substrate materials. Table 4 gives as an example the calculation for the 7 x 7 mm^2 silicon die soldered to a DCB substrate as shown in Fig. 7. The resulting value of 1.606 K/W for the thermal resistance is certainly too high, since lateral heat flow in good thermal conductors enlarges their effective heat transfer areas F_i, resulting in smaller thermal resistances $(\alpha_i F_i)^{-1}$. In order to include thermal spreading, a computer program is used which is able to calculate (for radial symmetry) the temperatures of all layers i for the case of a heat source (hot chip, e.g. 7 x 7 mm^2) on a large substrate. In Table 5, the results for different types of substrates are compared using the same chip size and power as well as a heat sink surface temperature of 25 °C in all cases.

Thermal resistance: R_{th} = [T(Chip) - T(heat sink surface)] / P_{el}

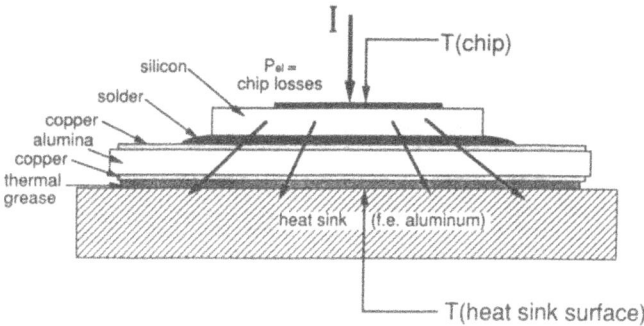

Fig. 7. Schematic representation of a "junction-to-heat sink" layer sequence used for thermal resistance determination; in this case, the chip is soldered onto a DCB (Direct Copper Bonding) substrate which is attached to a massive aluminium heat sink by using thermal grease.

Table 5. Calculated thermal resistances of a 7x7 mm² silicon die on:
   A) Cu-thick film on alumina + Cu-base plate
   B) DCB-substrate
   C) Polyimide (PI) insulated Cu-Mo-Cu
   D) Cu-Mo-Cu with high thermal conductivity PI including heat spreading effects
   in the layers (power: 30 W)

	A				**B**		
layer	d(μm)	λ(W/mK)	T(°C)	layer	d(μm)	λ(W/mK)	T(°C)
silicon	170	145[*)]	48.70	silicon	170	145[(*)]	48.54
solder	100	33	48.1	solder	100	33	47.94
TFT-Cu	30	200	46.58	DCB-Cu	300	330	46.45
alumina	630	24[**)]	46.5	alumina	630	24[**)]	46.09
Cu-plate[***)]	3000	399	33.87	DCB-Cu	300	330	38.10
th. grease	20	0.42	30.35	th. grease	20	0.42	37.88
Al-heat sink	10 000	237	25/26.8	heat sink	10 000	237	25/28.5
Rth(junction-heat sink)		**0.730 K/W**		Rth(junction-heat sink)		**0.668 K/W**	

*) at 48.5 °C

**) Rubalit 708 (Hoechst), T = 40 °C

***) including 20 μm of Ag-filled adhesive between ceramic and base plate

	C				**D**		
layer	d(μm)	λ(W/mK)	T(°C)	layer	d(μm)	λ(W/mK)	T(°C)
silicon	170	142.5[*)]	92.06	"	170	144[**)]	69.74
solder	100	33	91.50	"	100	33	69.16
copper	70	330	90.32	"	70	330	67.83
glue	12	0.4	90.26		12	0.4	67.74
PI I	25	0.155[***)]	82.69	PI II	25	0.9[****)]	57.21
glue	12	0.4	42.02	"	12	0.4	47.47
Cu-Mo-Cu	1000	237	34.46	"	1000	237	36.94
th. grease	20	0.42	33.64	"	20	0.42	35.82
heat sink	10 000	237	25/27.7	"	10 000	237	25/28.1
Rth(junction-heat sink)		**2.148 K/W**		Rth(junction-heat sink)		**1.388 K/W**	

*) at 92 °C   **) at 70 °C

***) see [19]

****) see [20]

Part B of Table 5 recalculates the example of Table 4. Heat spreading decreases the thermal resistance to 0.668 K/W which is less than 50 % of the estimated value! For comparison, the thermal resistance of copper thick film power circuits on alumina are calculated in Table 5A for the case that the alumina substrate is glued or soldered to a massive copper base plate. This is shown to almost compensate the smaller heat spreading and heat conducting capability of thick film copper compared to DCB-copper: the temperature drop across the thermal grease is less because of the base plate, whereas the drop across the alumina is larger due to the smaller heat spreading in the printed copper layer. However, as was shown in Table 3, the electrical sheet resistance of a TFT-circuit is 15 times larger than that of DCB interconnects. Therefore, their good thermal performance can be best exploited in high voltage/low current applications.

3.1.2 Insulated Metal Substrates. The mechanical stress applied to large area ceramic substrates, especially when they are bonded to copper layers and covered by some sort of mold inside of a module housing, leads to a restriction in size for DCB-modules. In contrast to this, metal based substrates are not principally limited in size and mass but here an insulation of sufficient dielectric strength and thermal conductivity has to be deposited on the surface to form an "Insulated Metal Substrate" (IMS). There are three basically different solutions of this problem available on the market:

1) Plasma sprayed alumina films on massive aluminum substrates which then are covered with high-velocity powder sprayed copper layers, patterned during deposition by means of laser cut copper masks. This technology has been developed by the small Norwegian company C.PC [14] The alumina is additionally impregnated by an organic filler or scalant which provides outstanding dielectric strength to the porous ceramic layer. Although its thermal conductivity is only about 1/4 of sintered alumina, the thermal resistance of the complete sprayed layer sequence down to the base plate (which may be a finned heat sink, Fig. 8) is of the same order of magnitude as DCB shown in Table 4, since thermal grease is not required. A considerable advantage of this technology is certainly due to the fact that multilayer strip line designs of modules, including high speed switching devices (IGBT's), are possible with a multiple masking and spraying process.

2) Porcelain Enamelled Metal Substrates (PEMS) which consist of a metal core (low-carbon steel or Cu-clad Invar, "Cu-Invar-Cu", or Cu-clad Molybdenum, "Cu-Mo-Cu"), coated with some sort of enamel, depending on the supplier. Ferro Electronic, for example, developed an enamel named ELPOR which softens between 580 °C (on steel) and 670 °C (on Cu-clad Invar) whereas Fujikura offers a PEMS named "Fujimetax" which can be refired at normal thick film burning temperatures (850 °C). Both kinds of PEMS are supposed to be subsequently processed in thick film multilayer printing and burning sequences (provided that adapted paste systems are available). This, however, represents a limitation of current handled by the module since the sheet resistance of Cu-thick film interconnects is high (Table 3) and its heat spreading effect under hot chips is poor. Thus, Cu-plates as heat spreader will certainly be required. Moreover, the breakdown voltage of the enamel might be too low for

Fig. 8. Samples of power circuits, produced by plasma spraying of alumina directly on massive aluminum heat sinks and by subsequent high velocity power spraying of copper through massive copper masks (courtesy of C.PC [14]).

some applications; Fujimetax is specified with > 1 kV for its 150 μm thick enamel.

3) Polymer Insulated Metal Substrates (PIMS) are thin plates of metal, typically aluminum, steel or Cu-clad Invar or Molybdenum which are coated with a thin layer of polymer like epoxy or polyimide (typical thickness 25 μm). On top of this insulation layer, a sheet of copper is laminated (i.e. glued) which is 35 μm or 60 μm thick (see Table 3), similar to that of conventional PCB-material; accordingly, the patterning of the single layer circuit is done by a simple etching process. The insulation by polyimide is excellent (280 kV/mm = 7 kV/25 μm[19]) but the thermal conductivity is poor (0.155 W/mK [18]). When the thermal resistance (including heat spreading) of a complete PIMS is calculated on that basis (2.15 K/W, Table 5C), it is a factor of 3.2 higher than that of DCB (0.668 K/W, Table 5B). If, however, a high thermal conductivity polyimide is available (as stated by Redpoint for their aluminum based product CLEApi[20]: λ = 0.9 W/mK), then the total resistance is considerably decreased (1.39 K/W; Table 5D) and the temperature drop across the substrate multilayer (including thermal grease, 30 W losses in a 7 x 7 mm²-chip) comes down from 62.6 °C (Table 5C) to 39.7 °C.

The choice of the most appropriate substrate and the suitable copper deposition technique has to be based on the special requirements in a given application. For small area modules, insulating substrates will certainly be advantageous, whereas for large area power hybrids, metal based substrates are an interesting alternative.

3.2 Chip Mounting

The most widely used technique for large area die attach is the soldering of the back plane of a silicon die to a landing area on the module substrate, using fluxless and high melting point solder like Pb95Sn5 or Pb92.5Sn5Ag2.5 (melting point 307-315 °C).

In most cases, solder preforms are used which are placed in the milled holes of a graphite form together with all dice and clips required for the complete module. Then, all parts together are passed through a hydrogen flushed belt oven and soldered simultaneously. An alternative is the use of an "eutectic die bonder" which acts as a pick-and-place machine positioning one chip after the other onto the substrate which is presoldered on all landing areas and pre-heated by the supporting table to a temperature closely below the melting point of the solder (Fig. 9). During deposition of the die (which is additionally preheated in the die tool), a low frequency scrub movement of the die tool is applied in order to scratch through the oxide layer of the solder on the substrate. In spite of this, an inert or even reducing atmosphere around the die position is strongly recommended to minimize oxide formation during the bonding process.

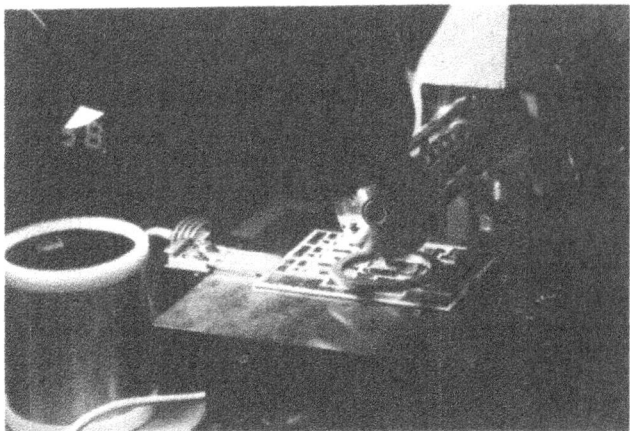

Fig. 9. Die bonding of power chips by using a "eutectic die bonder" which picks the die from a stack (circular plane on the left-hand side) and places it onto liquid solder on a module substrate; a low-frequency scrub movement of the chip, produced by the pre-heated die tool (in the center of the ring, representing the inert gas supply), is necessary to penetrate the oxide layer on the solder.

Since fluxless high temperature soldering is a complicated technical problem, it is worthwhile looking into the performance of alternative die attach technologies.

Up to now, silver filled adhesives have been used mostly in low power hybrid technology due to the ease of deposition by dispenser or screen printing and the simple die bonding process. Now, several suppliers[20,21] claim the suitability of adhesives, even in the higher power ranges where soldering has been unevitable ever since. The problem is easily recognized using the comparison of thickness and thermal conductivity numbers. A solder with a thermal conductivity of 35 W/mK and a typical thickness of 100 μm can only be replaced by an adhesive without deterioration of thermal performance if it has a λ of at least 3.5 W/mK and if it can be deposited with a mean thickness of 10 μm (disregarding roughness of substrate layers). In Table 6, the properties of adhesive materials are summarized, comparing them with a high melting point lead-rich solder (95Pb5Sn) and a gold-silicon eutectic die-bond. From the thermal conductivity data, it seems possible to reach similar thermal performance with different silver filled materials. Comparative measurements[23] showed that the thermal resistances of these alternative die attach materials come indeed close to that of solder material (see Table 7) but do not really reach it. However, thermal resistance is not the only property which has to be considered when a particular application in envisaged: the elastic modulus, stresses, electrical stability under current load, degradation of materials, and processing requirements may conflict with each other. Moreover, the void content in the bond layer is varying for different commercially available expoxy adhesives whereas for polyimide paste adhesives the removal of solvent is the most critical process step which affects the quality of the bond.

Table 6. Properties of materials used for large area die attach[23]

material	melting point/ max. exposure temperature[°C]	thermal conductivity [W/mK]	modulus of elasticity [kN/mm²]
Au-Si eutectic	363	295	83
Pb95-Sn5 solder	310	28	18
Epoxy Ag-filled	≈ 300	1-6	1-10
Silicone Ag-filled	≈ 250	1-6	0.005
Polyimide Ag-filled	≈ 350	1-6	1- <10
Ag-glass	≈ 400	≈ 80	38
Thermoplastic ribbon	≈ 500	≈1	1-10

A very important feature of a chip mounting technique is the automatization capability. If in a highly automated fabrication line the only manual step is the die attach by soldering in order to meet thermal management requirements, then the specifications should be critically reconsidered in terms of, for example, lower current density by bigger chips in order to enable computer controlled pick-and-placing, using some sort of Ag-filled adhesive. The benefits of lower assembly cost will in most cases exceed the higher silicon cost. As proven by Japanese module manufacturers, fully automated fabrication, using screen-printing of solder or adhesive for die bonding and heavy wire bonding for interconnections, can indeed yield high-quality products.

Table 7. Summary of thermal resistance measurements on a range of die attach materials used to bond Si to Al_2O_3-substrates [23]

material	thermal resistance [K/W]
solder - preform	4.5
Ag/glass - paste	4.6
Epoxy paste (Ag filled)	5.6-8.2
Polyimide paste (Ag filled)	7.2-8.7
Acrylic ribbon (Ag filled)	6.7

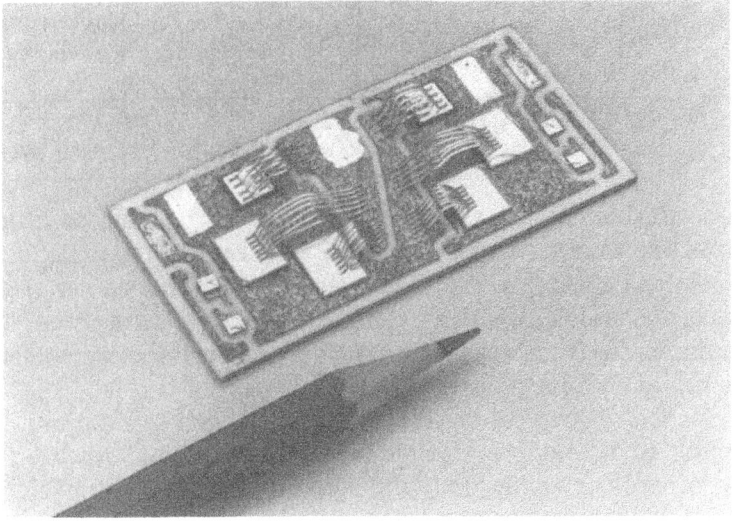

Fig. 10. Half-bridge module (DCB substrate, electroless plated with nickel) comprising two IGBT chips per leg and a free-wheeling diode; note the interconnection of the die top to the substrate by multiple wire bonding, required to carry the load current.

3.3 Interconnections

3.3.1 Conventional Techniques. In many applications, the bare dice like IGBT´s are supplied with solderable back side metallization to be soldered to the substrate and with aluminum bonding pads on the top surfaces. In these cases the chip-to-substrate and chip-to-chip interconnections are done with heavy wire wedge-wedge bonds which have to be parallelled several times, depending on the current level (Fig. 10). A pure (99.999) aluminum wire of, for example, 175 µm diameter has a burn-out current of 4.9 A [24]. Typically, multiple bonds consist of 5 parallel wires, leading to burn-out currents of 25 A for 175 µm dia. and at least 100 A for twice the diameter of 350 µm, but, of course, only about 1/3 of this burn-out current is applied.

Table 8. Types of connections in modules

component 1	component 2	connector	connection
substrate	bare die	direct or solder/Mo/solder	soldering
substrate	terminal screws	Cu-straps	soldering
substrate	SMD	direct	soldering
substrate	multipin-connector	Cu-litz wire	soldering
die 1	die 2	Al-heavy wire	us* wire bond
die (Al-Pad)	substrate	Al-heavy wire	us* wire bond
die (solderable)	substrate	Cu-clip	soldering
PCB or hybrid (2nd level)	substrate	Cu-litz wire or Cu-stud	soldering soldering
SMD	SMD	PCB or hybrid (2nd level)	soldering

* us = ultrasonic

Since wire bonding can be fully automated, the process cost can be kept within reasonable limits. However, soldering is required for various interconnections inside of a module (see Table 8) and sometimes for the top surface of the dice, too (thyristors, for example, have to be soldered). The techniques used today for the assembly of parts before soldering still comprise many manual process steps:
- the *production of graphite preforms* which are needed to adjust all chips, copper clips and solder preforms in their predefined position for the soldering process is costly and unflexible, since the slightest change of layout or dimensions of components induces replacement of the preform by a new one;

- the *assembly of all (sometimes dozens of) parts* to be simultaneously soldered is done by inserting piece per piece with a pincette into the holes of the graphite form.

In order to achieve automatization of soldering processes, the substrate has to be presoldered on the chip mounting pads. This can be accomplished a) by a screen-printing process or b) by a wave soldering process. Both of these alternatives require certain compromises in terms of modified specifications:

a) Screen printing: solder creams, of course, contain volatile constituents which have to be outgassed during reflow. This can lead to a certain amount of bubble voids in the solder layer after die bonding which might be acceptable when known and taken into account during thermal design. We calculated that the thermal resistance of a solder layer increases by only 11.6 % when a volume share of 25 % voids is introduced in the solder[25].

b) Wave Soldering: prior to presoldering on a solder wave, a solder stop mask has to be applied to the module surface in order to deposit the liquid solder selectively. The masking can be accomplished, for example, by a magnetically attached stencil[26] or by screen-printing of a high-temperature resistant polyimide or thick film glass layer. However, according to the local wettability of the surface, the thickness of the deposited solder layer can vary from pad to pad which introduces a limited uncertainty in the thermal and electrical resistances of the solder bonds. The scatter has to be determined whether or not it is acceptable. If, on the other hand, it is necessary to improve the wetting behaviour of the substrate surface, the contact to a liquid flux is a possible solution, but then flux residues are likely to be incorporated in the bond layer which can lead to corrosion problems. In any case, long-term quality of a product has to be proven for any of these production processes.

3.3.2 The Foil-Clip Technology. As an interesting alternative to the described interconnection technologies, we developed a novel wireless type of interconnection which resembles the well-known Tape Automated Bonding (TAB)-technique[27], used in microelectronic packages. It consists of a polyimide foil (Kapton, thickness 25 µm) glued to a 70 µm thick copper sheet. Double-sided Cu-clad Kapton foil has already been used for low-inductance packaging of fast switching power semiconductors [28]. In our case, this "foil-clip" is used to provide all interconnections between all chips and the substrate simultaneously by just one piece of conductor.

The preparation of a foil-clip is shown in Fig. 11. The copper sheet is patterned, using conventional photolithography and etching. When mounted into the module, the copper is always on top of the clip. The connections to the underlying pads are achieved by via-holes which are produced by masked UV-ablation[29] (Fig. 12a). These openings in the polyimide (see Fig. 12b) are subsequently filled with solder by passing a solder wave or with a silver-filled adhesive using screen-printing. As can be inferred from Fig. 12c, the polyimide used for insulation between the contact patterns on a die simultaneously acts as a solder stop layer. The results of the main production steps are shown in Fig. 13.

Using a foil-clip, the assembly of the module reduces to one of the chip mounting process described above, including pick-and-place of additonal SMD-components and a single "thermode" soldering step on the top side of the foil-clip which is adjusted to the module via registration holes fitting to studs on the module (Fig. 17). A possible alternative to thermode soldering is an additional (low temperature) reflow solder step in a belt oven; then, the foil-clip has to be loaded or pressed down by an appropriately shaped body which provides the required contact pressure to the solder joints during reflow and cooling.

The following advantages are provided by the foil-clip technology:
1. Since the connection to a second-level PCB, carrying control and communication electronics can be achieved by the flexible Cu/Kapton-material, too, the number of different connectors in a module is considerably decreased (Table 9).

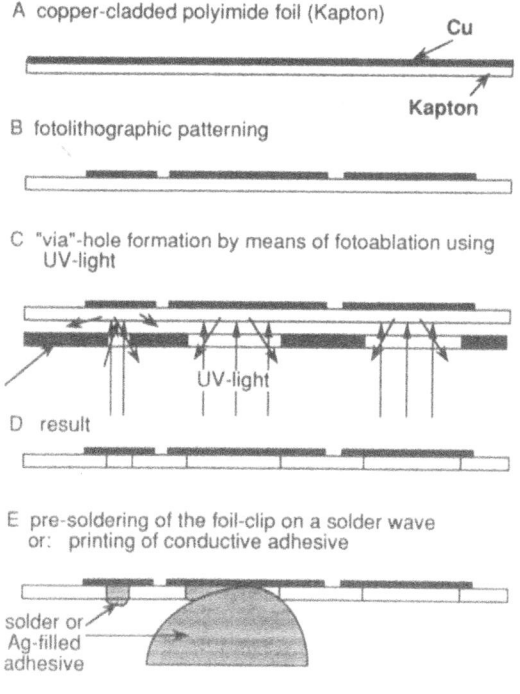

A copper-cladded polyimide foil (Kapton)
                                                                    Cu
                                                                Kapton
B fotolithographic patterning

C "via"-hole formation by means of fotoablation using
    UV-light
                            UV-light
D result

E pre-soldering of the foil-clip on a solder wave
    or: printing of conductive adhesive
solder or
Ag-filled
adhesive

F shaping of the foil-clips by cutting and stamping

Fig. 11. Schematic representation of the foil-clip production process (steps A-F) as developed in the ABB-laboratories[29].

a

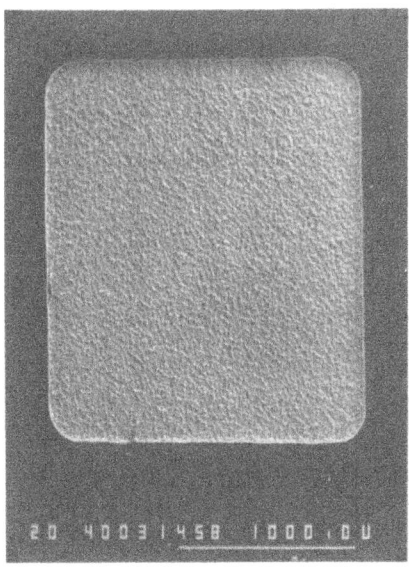

b

c

Fig. 12. a) Realization of process step C of Fig. 11 (via hole formation by UV-light ablation): the sample (polyimide layer) is covered by a metal mask containing the via hole patterns; the substrate/mask-combination is moved (by means of stepping motors) in front of a scanning excimer laser beam in order to illuminate the whole surface; alternatively, a UV-lamp [30] can be used for simultaneous illumination of the complete surface.

b) SEM-picture of a via hole produced as described above; note the sharp edges of the polyimide layer;

c) Solder bump produced by moving the polyimide side of the foil-clip through the tip of a solder wave (process step E, Fig. 11).

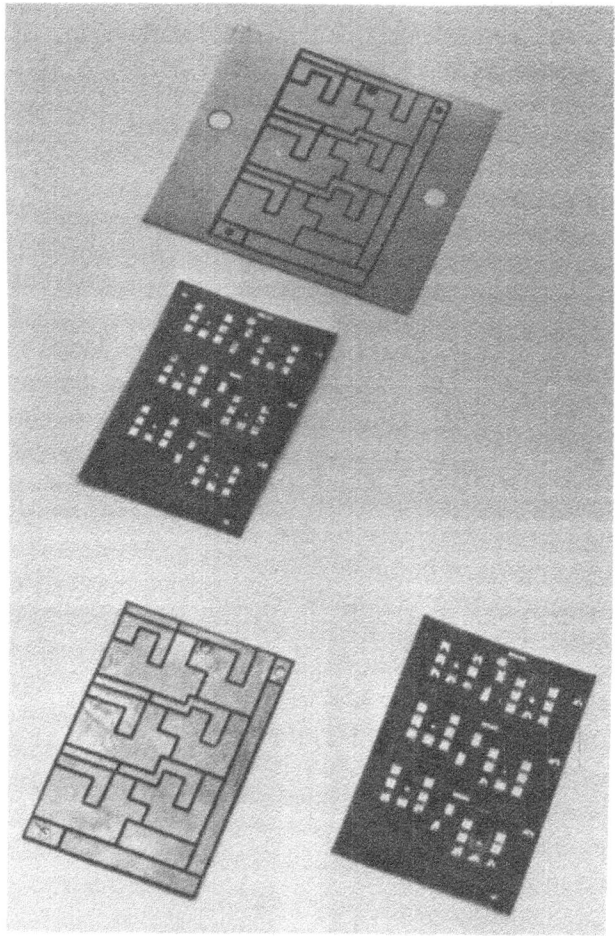

Fig. 13. Examples of foil-clips in different stages of production; from top to bottom: etched copper pattern on the top side; via holes in the polyimide layer on the rear; nickel plated copper layer; via holes with solder bumps.

Table 9. Types of connections in "Advanced Power Modules"

component 1	component 2	connector	connection
substrate	die	direct or solder/Mo/solder	die-bonding (soldering or glueing)
substrate	terminal	Cu-clips	soldering (reflow)
substrate	SMD	direct	soldering (reflow)
all other	all other	**Foil-clip**	soldering (thermode or reflow) or: Ag-filled adhesive

2. For high current conduction, the cross-section per millimeter width of the track is equivalent to one Al-wire bond of 350 μm dia. (specific resistivity relation of 2:3 for Cu:Al taken into account).

3. The production of foil-clips can be rationalized and automated by
   a) large area photolithography and etching similar to PCB production,
   b) large area photoablation (dry process!) using stepper motor driven laser scanning or large area illumination by a UV-lamp [30],
   c) automated presoldering by wave soldering,
   d) automated deposition of conductive adhesive by screen-printing.

4. The assembly of the modules is considerably simplified by the reduction of parts.

5. All electrical interconnections can be contacted from top (by soldering studs of wires on the copper tracks) and connected, for example, to test pins.

4. A PRACTICAL EXAMPLE:
SOFTSTARTER FOR 3-PHASE AC-MOTORS

4.1 The Function

In order to reduce the cost and volume of existing softstarters, using sophisticated digital control without loosing the main functions of softstart and softstop, we developed a purely analog circuitry for ramp generation and phase control (Fig. 14), leading to a limitation of the inrush current to 4.5 x I_e during motor start. Thus, the size of the thyristors can be considerably decreased compared with hard switching. The thermal resistance of the aluminum heat sink takes into account that at e.g. 85 °C heat sink temperature, the full inrush starting current is available. At higher temperatures, an integrated Positive Temperature Coefficient (PTC)-sensor derates this overcurrent limit to values which keep the motor running but which would not allow for a restart (for example 1 x I_e at 100 °C).

The thyristors are separated from the gate control by 3 x 2 optically fired triac devices (SITAC), described in sec. 2.2. In order to account for line voltages as high as 500 V, two of them per phase are coupled in series. The control circuit for one phase is depicted in Fig. 15. A phase control IC (TCA 785, Siemens) is connected to the primary of the SITAC in order to generate the ignition current for the thyristor after zero-crossing at a phase angle which is proportional to a DC voltage supplied by a ramp generator. The maximum allowed motor current during start and normal operation is set by a potentiometer. If the measured current exceeds this setpoint, the phase angle for the firing point is increased, leading to a reduced motor voltage and current.

The current measurement is performed by a low cost current transformer with a low-permeability core (COROVAC, Vakuumschmelze) which does not saturate up to about 200 A load current. Accordingly, the induced voltage which is proportional to the time derivative of current is low. After it is amplified, the rectified and integrated signal is compared with the preset maximum value (see Fig. 14). As long as the actual current is lower than this, the phase angle is constantly decreased from 120° to zero or, for softbrake, increased to 120°.

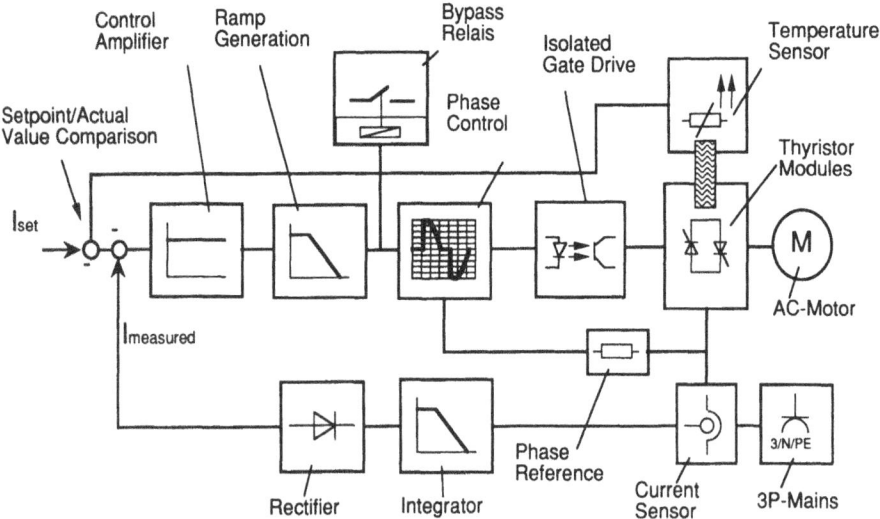

Fig. 14. Schematic block diagram of the analog softstarter (for details see text).

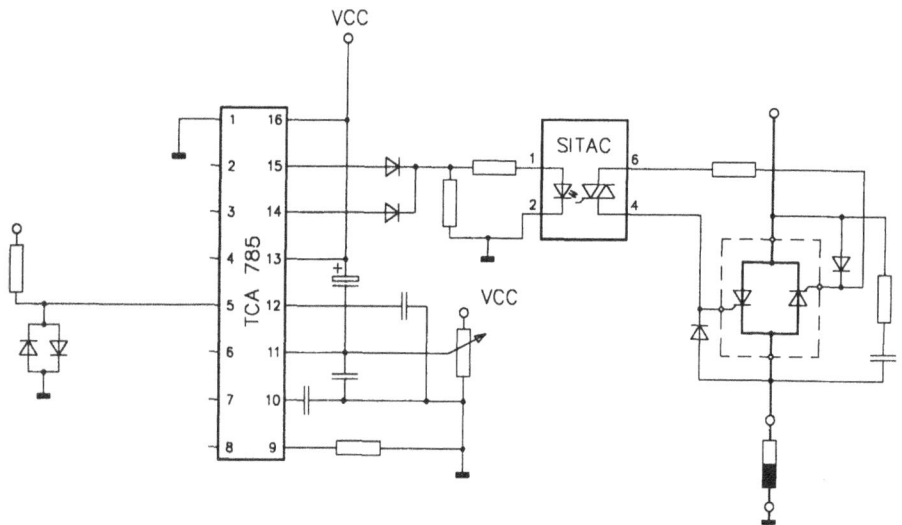

Fig. 15. One phase of wiring scheme of the analog softstarter from phase control chip (Siemens TCA 785) including HV-insulation chip (SITAC® : Registered Trademark of Siemens), antiparallel thyristors, diodes and RC-network.

At a presettable lowest angle of about 15°, a bypass relay (400 V, 4 A) switches in order to bridge the thyristors during normal operation.

4.2 Technical Realization

The first (conventional) prototype of the softstarter for motors up to 11 kW is shown in Fig. 16. The power level on the aluminum heat sink only consists of the three standard thyrister modules (MMO-75 from ABB-IXYS) and the R-, S-, T-wiring, including a 1-phase current measurement, performed by the small reddish coil on the black wire. The RC-network for the thyristors is integrated on a 2nd-level PCB of standard Europe-format which also contains the complete control circuitry with isolation barrier (SITACs), phase control chips, ramp generation and interpretation circuitry for the current sensor signal. The small potentiometers on the left foreground side are used for the setting of ramp slope (up and down) and current limit.

Fig. 16. Prototype of the analog softstarter with standard thyristor modules on an aluminum heat sink and a second-level PCB, carrying the RC-network, insulation barrier, phase control and auxiliary power supply.

The main goal with this standard solution was the reduction of oversized heat sink mass, thyristor chip size and control PCB area as well as the minimization of cost and number of components. This could all be achieved by the simple analog circuitry confining strictly to the major functions of a softstarter.

The next step towards "advanced power modules" comprises the integration of all thyristors on one module substrate. The production steps for the 3-phase AC-switch module of 11 kW on an area of 75 x 58 mm^2 is shown in Fig. 17. After etching of the copper layer, the substrate is electroless plated with Ni. In parallel, a foil-clip is produced as shown in Fig. 13. From Fig. 18, the simple module assembly can clearly be inferred. The DCB-substrate also carries the 6 SITACs and the ferrite core for 1-phase current measurement. Hence, the only high-side devices which could not be integrated on the ceramic is the RC-network because of the large capacitor volume. In an appropriate plastic housing it can be located on a small separate PCB which is conveniently mounted in a vertical slot on one side (Fig. 19). Then, the low-side electronics should be placed into the slot on the opposite side of the housing.

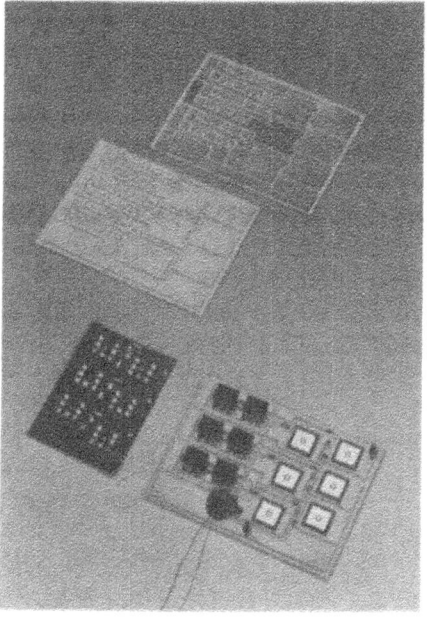

Fig. 17. Production stages of a three phase thyristor module, comprising a DCB-substrate of size 75 x 58mm^2 where 6 thyristors, 6 SITACs and a current transfomer are integrated; the interconnection of the thyristors is accomplished by a foil-clip (bottom, left-hand side), produced as described in Fig. 11-13.

Fig. 18. Three-phase thyristor module with foil-clip interconnection as described in Fig. 17.

Fig. 19. Digital softstarter, containing the three-phase thyristor module of Fig. 18, mounted into a plastic housing and equipped with the RC-network; the small PCB in front of the power module carries a microcontroller (80535, PHYTEC[9]), performing both phase control and bus interfacing of the "advanced power module".

In a last step of intergration, the control circuitry can be supplemented by a small microcontroller like the 80535 with ROM as shown in Fig. 19. This unit is able to generate the ramp voltage and the phase control signal as well as to handle the current and temperature sensor signals. Then, significant parts of the analog circuitry of the first prototype are replaced by digital means. The main advantages of this final version of an advanced power module is obvious: differences in the characteristics of all involved devices in general and of the attached motor in particular can be accounted for by software adaptation. By a software interface to a field bus the condition of the device can be monitored by a central survey intelligence. Furthermore, the same control circuitry can be used for the integration of special additional functions (kick-start, energy saving, slow-speed etc.), which are not required in all cases. Thus, the number of PCBs produced increases significantly due to the addition of those for all power ranges and special feature devices, leading to a more economical production.

5. CONCLUSIONS

Cost reduction in the production of power electronic devices can not only be achieved by using high volume products like standard power modules which are assembled on a heat sink and wired to remote control electronics. It has to be taken into account that the higher cost for the production of more complicated "advanced power modules" can be compensated by automated production techniques which lead to smaller numbers of parts to be assembled, i.e. less mounting, assembly and wiring cost during installation of the devices. In addition, the versatility in terms of system technical combinations of subsystems and software adaptation to particular requirements in different applications reduce the variety of components assembled to higher order systems and, hence, reduce engineering cost. Since, moreover, the self-protection ability of advanced power electronic modules reduces their failure rate and since field bus connection can be used for remote condition, monitoring both maintenance and repair are facilitated.

Consequently, it has to be concluded that the *overall cost* for advanced drive and automation systems might even decrease when all benefits of system technique and automated production techniques are appropriately evaluated.

ACKNOWLEDGEMENTS

The authors wish to thank Dr. K. Langer for substantial work in the softstarter development and for performing thermal resistance calculations. Moreover, the important experimental contributions of W. Eichhorn (electronics), Dr. H. Esrom and Mrs. Zhang (laser ablation) and D. Gilbers (module technology) are gratefully acknowledged.

REFERENCES

1   Glyn GmbH: *TOSHIBA Leistungselektronik*, 1991.

2   ABB-IXYS: *Thyristor/ Diode Modules*, 1991.

3   a) J.G. Melbert, "Smart power: Technologien für integrierte Leistungschaltungen", *Siemens Zeitschrift*, 6/1990, pp.32-36
      b) G. von Treek, R. Ferrari and S. Storti, "Hochstrom-Motortreiber-ICs", Elektronik, 14/1991, pp. 58-61.

4   "Intelligent Power IC Module", *PCIM Europe*, July 1989, pp.131-133.

5   B. Büchau and D. Silber, *"Intelligente anwenderspezifische Leistungs-Halbleitermodule"*, ed. by VDI/VDE Technologiezentrum Informationstechnik GmbH, Berlin 1990, pp 1-117.

6   W. Schott, "SITAC - ein neuer, mikrocomputerkompatibler AC-Schalter mit galvanischer Trennung", *Siemens Components*, vol. 23, no. 2, pp. 77-81, 1985.

7   R. Bayerer, "New concepts of power modules", *PCIM Europe*, July/Aug. 1990, p.223-224.

8   W. Eichhorn, R. Bayerer and H.-J. Krokoszinski, ABB patent pending *(DE PA 40 07 539.7).*

9   PHYTEC document L-018/89.

10   F. Furrer, "Ein-/Ausgabe-Konzepte für die Computer-Prozess-Steuerung", *Elektroniker*, vol. 29, no. 12, pp. 87-95, 1990.

11   P. Zimmermann, "CAN - Serielle Datenübertragung für Echtzeitanforderungen", *Elektronik*, vol. 40, no. 5, pp. 76-78, 1991.

12   A. Scharf, "Digital drive control with personal computers", *PCIM Europe*, Sept. 1989, pp. 180-181.

13   A. Meisinger, "The suitability of computer buses and networks for drive systems", *PCIM*, May/June 1991, pp. 128-131.

14   C.P.C. = Compact Power Circuits, P.O.Box 305, N-3191 Horten, Norway.

15   P. Dupin and P. Massiot, "Aluminium nitride ceramic in power hybrid technology", *PCIM Europe*, July/Aug. 1990, pp. 221-222.

16   P. Lenk, "Substrate und Pasten" in *Hybridintegration*, ed. H. Reichl, Hüthig-Verlag, Heidelberg, 1986, pp. 28-65.

17   C. Zardini, S. Bontemps, P. Despagne and P. Dupin, Seminar "Power Hybrid" during the *PCIM 1989*, München, June 1989.

18   *Handbook of Chemistry and Physics*, 56th edition, CRC-Press, Cleveland, Ohio, 1975/76, p. F166.

19   A. Krempel Söhne GmbH, Katalog: *Flexible Basismaterialien für gedruckte Schaltungen.*

20   Redpoint Ltd, Redpoint CLEApi.

21   R. H. Estes, R.F. Pernice, "Die attach adhesives: evaluation of V_{ceSAT} and θ_{jc} performance in power devices", *Proceedings of the ISHM'89*, Baltimore, Maryland, USA, 1989, pp. 664-667.

22   J.P. Mollie, "Silicone adhesives and encapsulants", presented at the *C.PC-Workshop*, Dec. 3.-4., 1990, Horten, Norway.

23   N. Stockham, "Large area die attach (adhesives and soldering)",
      paper presented at the *C.PC-Workshop*, Dec. 3.-4., 1990, Horten,
      Norway.

24   A. Kolbeck, "*Aufbau- und Verbindungstechniken*", ed. VDI,
      Düsseldorf, 1984, table p. 75.

25   H. Bogs, ABB Corporate Research, Germany, internal
      communication.

26   H.-J. Krokoszinski , K. Langer, A. Neidig, H. Oetzmann, ABB-
      patent pending *(PCT/EP 91/00272)*.

27   A. Kolbeck "Montage und Kontaktierung ungehäuster
      Halbleiterbauelemente" in *Hybridintegration*, ed. H. Reichl, Hüthig-
      Verlag, Heidelberg, 1986, pp.183-221.

28   F. Kloucek, *ABB-patent, EP 0 381 849*.

29   H.-J. Krokoszinski, K. Langer, H. Oetzmann, ABB-patent pending
      (DE PA   41 30 637.6).

30   U. Kogelschatz et al., *ABB-patent, EP-OS 0 254 11*.

DISCUSSION

G. Crawshaw (Brush, Loughborough UK)

Could you comment on applications for low volumes - in contrast to
the high volume applications discussed - such as might be found for
locomotives where you have a requirement for high reliability and more
difficult environmental conditions ? An example might be an integrated
gate drive for GTOs.

H.J. Krokoszinski

For locomotives, the volume may be a few hundred pieces, and it will
be very hard to compete with the decreasing cost for standard
components on a base plate. Unless you require real sophisticated
additional functions, this will not pay. However, if you have several such
advanced modules in your locomotive to be connected centrally, and cost
is not the driving force, then it may be worthwhile. The low volume
production can only be economical if you exploit all the additional
functions and the versatility of such an advanced module.

H. Langer (RWTH Aachen, Aachen D)

You have shown a very fast overcurrent protection within approximately 500 ns. Turning on a device in an inverter leg, you have to wait a certain time until the voltage is coming down and then you have a dynamic saturation of I think at least 1 μs, even in the case of an IGBT. Could you comment on how you measure this short circuit condition ?

H.J. Krokoszinski

These short circuit conditions were produced in two different ways. One was under normal operating condition, just short circuiting the load with a screwdriver. The other involved turning on under short circuit condition. Both cases were leading to such a result.

J. Gobrecht (ABB Corp. Research, Baden)

You have shown a nice comparison of the thermal resistivities of various substrates like direct bonded copper, polyimide etc. Could you give us an idea how the costs compare ?

H.J. Krokoszinski

Despite a very hard competition, the base material prices of copper-moly-copper amount to about 10 US\$ per dm^2. I did not find any pre-produced polyimide insulated metal substrate but you can purchase copper on top of polyimide and glue it down to this copper-moly-copper base. The costs are certainly high, but the main advantage of those metal based substrates is the lack of any restriction in size of the substrate. With ceramic substrates, you cannot produce modules of more than 7, 8 or 10 cm edge length because of warpage when pressing them down to the heat sink. Since this is no problem for thick metal base substrates, you are bound apply these for large area modules and pay the high price.

APPLICATION OF HIGH POWER ELECTRONICS IN ELECTRICAL

POWER TRANSMISSION SYSTEMS

Åke Ekström

The Royal Institute of Technology
Stockholm, Sweden

ABSTRACT

This paper gives a summary of the existing and expected future applications of high power electronics for electrical power transmission
systems. Power electronics has been used already during the 1950's in
high voltage electrical power transmission in the form of High Voltage
Direct Current, HVDC, links and was introduced during the 1970's for
reactive power control in the form of Static Var Compensators, SVC. A
number of new concepts, mainly based on the usage of forced-commutated voltage-source converters, are now being studied and some pilot installations have also been built. This will be further treated in this paper.
The importance of further development of the semiconductor devices
will also be discussed.

1. INTRODUCTION

The successful commissioning of the first HVDC link between the
mainland of Sweden and the island of Gotland in 1954 started the application of power electronics, directly connected to the main circuits of
electrical power transmission systems. We will in the following use the
term "high power electronics" for the treated applications as the power
ratings for these applications are of the order of MVA or MVar rather
than of kW or Watts.

Power Semiconductor Devices and Circuits, Edited by A.A. Jaecklin
Plenum Press, New York, 1992

The Gotland project was later followed by other HVDC systems for higher power. During the 1950's and 1960's, mercury arc valves were used as controlled semiconductor elements. However, the invention of the thyristor towards the end of the 1950's resulted in replacing the mercury-arc valves by thyristors for motor drives and later, towards the end of the 1960's, also for HVDC applications.

Since the 1950's, the development of thyristors has been very substantial, especially with regard to maximum current and voltage capabilities as illustrated in Fig. 1. This development is very closely linked to the development of the technique for the manufacturing of large high quality silicon wafers. One of the major driving forces for this development has been the use of silicon for microelectronic circuits.

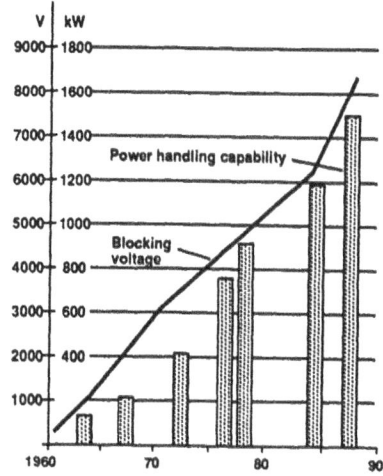

Fig.1. Development of the power handling capability of ABB thyristors for HVDC.

The majority of the thyristors used today are Electrically Triggered Thyristors or ETT. When used for HVDC, requiring series connection of a large number of thyristors, the disadvantage with ETT thyristors is that each thyristor has to be supplied with a special gate unit for the conversion of the optical triggering pulse to an electrical triggering pulse. For many reasons it is advantageous to use an optical triggering system for the complete valves. The first tests with Light Triggered Thyristors, LTT, were performed already in the end of 1960's. However, it was not until twenty years later, in the beginning of 1980's, when light triggered thyristors were designed, requiring much lower light intensities for triggering and also including protective firing capabilities that the light triggered thyristors began to offer an interesting alternative to electrically triggered thyristors.

In the beginning of 1970's, the technique of using pairs of anti-parallel thyristors as switching elements for the connection of capacitors and reactors was also developed. As the thyristors can be switched twice per cycle, at each zero crossing of the current, it is possible to obtain fast control of reactive power. Earlier, such a control of reactive power had usually been performed by rotating synchronous machines. Stressing that no rotating parts were used in the new reactive power devices, based on thyristors, it was termed "static compensators"

When capacitors or reactors are used for reactive power control, the reactive power generation or absorption corresponds to stored energies in the capacitors or reactors. However, reactive power can also be generated or absorbed in three-phase converters with no requirement of capacitors or reactors for storing the corresponding amount of energies. Instead, the energies are cyclically exchanged between the phases in the converter. In the early 1970's, it was investigated if converters could be used for reactive power control. However, since a line-commutated converter, using thyristors only, can consume but not generate reactive power, it was found that this solution was more expensive than using antiparallel thyristors as switching elements in combination with capacitors and reactors for reactive power control.

The limitation of line-commutated converters, always consuming reactive power, which usually had to be compensated for by installation of capacitor banks, stimulated investigations of other types of converters, commutating the current independent of the line-voltage, already in the 1930's. These types of converters, the so called forced-commutated or self-commutated converters, were rather complicated, as long as they were based on valves, which required the assistance from extra commutation circuits to commutate the current. Despite of that, during the 1970's, they were used for special applications, e.g. motor drives and traction. However, during the 1980's, the fast development of thyristors which can turn off the current controlled from the gate, the so called GTO thyristors, have drastically increased the interest and usage of forced-commutated converters, especially for motor-drives in traction applications. As a forced-commutated converter can both, consume and generate reactive power, it is also of interest to investigate the usage of such converters for static Var compensators. Some pilot installations have already been built. However, the costs of forced-commutated converters are still substantially higher than for line-commutated converters and also higher than for a combination of thyristor controlled reactors and thyristor switched reactors. Because of that they have to offer other additional advantages in order to be used instead of conventional static Var compensators.

As forced-commutated converters can operate on very weak AC-systems, they might also be of interest to be used for high voltage direct current transmission systems. This will be further treated later.

The interest in using different concepts of power electronics for electrical power transmission systems have during the last years been further stimulated by EPRI, introducing the Flexible AC Transmission Systems concepts or FACTS. The basic idea is that it should be possible to increase the power transmission capacity of the existing AC power transmission systems by fast control of the active and reactive power flow. This might be achieved by using different concepts based on the usage of power electronics. It should be noted that most of the different Flexible AC Transmission Systems concepts, to be further described in this paper, had been presented elsewhere before EPRI introduced the FACTS concepts. However, credit has to be given to EPRI for presenting a more complete overview and stimulating the interest among utilities to further investigate the merits these concepts could offer to their systems.

As alternatives to improving the performance of existing AC-systems by further usage of Flexible AC Transmission Systems concepts, the conversion of AC-lines to DC-lines as well as more extensive use of HVDC back-to-back stations should also be considered.

2. THE EXISTING USAGE OF HIGH-POWER ELECTRONICS IN ELECTRICAL POWER-TRANSMISSION SYSTEMS

As mentioned above, power-electronics, directly connected into the main circuits of electrical power transmission systems, are today mainly used in the form of high voltage direct current transmissions, HVDC, and Static Var Compensators, SVC.

The major advantages with HVDC transmissions are, that they give:
- reduced costs of cables with no problems concerning reactive power generation of the cables and no limitation of transmission distance
- reduced costs of overhead lines with increased power transmission capability and no distance limitation
- asynchronous connection of AC-networks
- fast and accurate control of the power flow.

The major HVDC applications are:
- cable transmissions
- overhead-line transmissions
- back-to-back converter stations.

Most of the HVDC schemes are two-terminal transmission schemes, although a few multi-terminal schemes have also been built.

The HVDC technique can now be considered as a widely accepted valuable complement to the AC-transmission technique. HVDC schemes are installed all over the world and the installed HVDC transmission capacity is steadily increasing and is now well above 30 000 MW.

The converter stations are usually bipolar with each pole consisting of two current-source line-commutated six-pulse converter bridges connected in series on the DC-side as shown in Fig. 2. As shown in the figure, AC-filters and capacitor shunt banks are usually connected on the AC-side to take care of the AC-harmonic currents generated by the converters and to compensate for the reactive power consumed by the converters. The major drawback of the HVDC technique, preventing a more extensive usage, is the fairly high costs of the converter stations. In some cases, the inherent weakness of line-commutated converters, preventing the converters from supporting the voltage control of the AC-network, might also have certain influence.

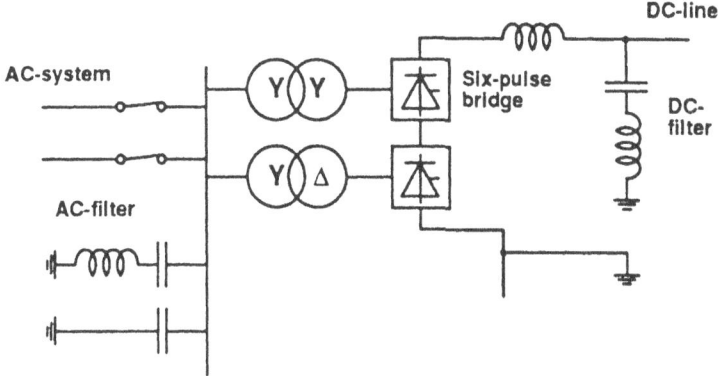

Fig. 2. One pole of a bipolar HVDC converter station.

The major applications of static Var compensators for electrical power transmissions are:
- voltage control including limitation of temporary overvoltages
- improvement of transient stability
- balancing of unsymmetrical loads.

The previously used synchronous compensators for reactive power control have today to a great extent been replaced by static Var compensators, as static compensators will be less expensive, give lower losses and require less maintenance.

Today, synchronous compensators are only installed for special applications, e.g. for converter stations for which they can offer an increase in short-circuit capacity. When the reactive power is generated by capacitors, as in a conventional static Var compensator, the installation of the static Var compensators will further weaken the AC-network from an impedance point of view.

The installed total capacity of static Var compensators using thyristors shows a similar trend as for HVDC stations .

The conventional static Var compensators based on pairs of anti-parallel thyristors as switching elements are usually built-up of a combination of Thyristor Switched Capacitors (TSC), Fixed Capacitors (FC), Thyristor Switched Reactors (TSR) and/or Thyristor Controlled Reactors (TCR) as illustrated in Fig. 3. The continuous control of reactive power is obtained by the TCR. The static Var compensation for electrical power transmission applications is usually asymmetrical with regards to generation and absorption capacity of reactive power as shown in the circuit diagram in Fig. 4 for an SVC station in USA.

The thyristor valve design for HVDC and SVC applications have many similarities. For both applications, the valves are usually air insulated and installed indoors. A number of thyristors have to be connected in series, as the operating voltage across the valve might vary from a few tens of

kilovolts to some hundred kilovolts. The thyristors are usually cooled by
water and triggered by electrical signals, although recently also valves
with light triggered thyristors have been built.

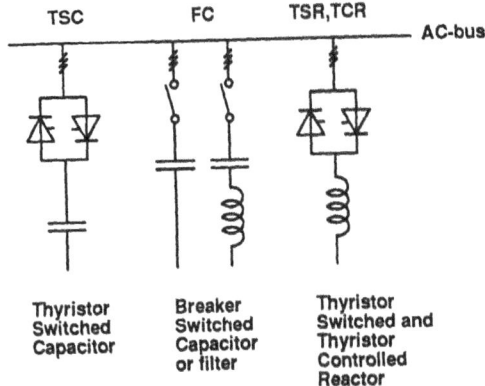

Fig. 3. Thyristor switched capacitor, fixed capacitor, thyristor switched and thyristor
controlled reactor.

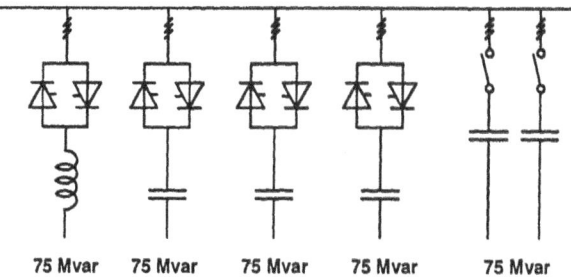

Fig. 4. Circuit diagram of a typical SVC-station.

The thyristors are usually rather symmetrical with regard to maximum blocking voltage in forward and reverse direction, with the maximum voltage being in the range of seven kilovolts. As the thyristors can
be designed for wafer-areas up to 60 cm^2, single thyristors without parallel connection can handle the required maximum current.

The development of the HVDC and SVC techniques during the past
twenty years can best be characterized as an evolutionary process with a
continuous development of the technique. It is to be expected that the
development will continue in the same way in the future with the objective of decreasing costs and losses and improving performance.

The item in an HVDC or SVC station responsible for the largest share
of the total station costs is the transformer. The cost of the transformer
for a HVDC station is about 25 % of the total station costs as compared to
15-20 % for indoor valves excluding the costs of valve hall and cooling
equipment. During the past twenty years, it has been possible to substantially reduce the costs of the thyristor valves, mainly due to the fur-

ther development of the thyristors as illustrated in Fig. 1. The potential for further development is probably smaller but is still substantial. However, one limiting factor is that increased thyristor voltage must not result in increased losses determined by the on-state losses, thyristor turn-on and turn-off losses and losses in the snubber circuits. It should also be considered that the evaluated costs of the losses in a thyristor valve for HVDC often is of the same order of magnitude as the cost of the valves. In order to increase the blocking voltage capability of the thyristor, the thyristor wafer has to be made thicker which will result in increased on-state voltage drop and/or increased recovery charge depending upon the life-time of the charge carriers. This also indicates that it might be easier to design thyristors for higher voltage for TSC than for HVDC as the switching losses are of less importance for TSC applications.

3. EXPECTED FUTURE REQUIREMENTS ON ELECTRICAL POWER TRANSMISSION SYSTEMS OF IMPORTANCE FOR THE USE OF HIGH POWER ELECTRONICS

The following factors are expected to be decisive for the further usage of high power electronics for electrical power transmission systems:
- It is to be expected, that in the future, the increase of electricity consumption will be rather limited in many industrialized countries. However, the increase can be substantial in certain regions. It will also be more difficult to get permission to build new overhead lines. This will make it necessary to increase the transmission capacity on the existing lines, which might be achieved by applying different Flexible AC Transmission Systems concepts or by conversion of the AC-lines to DC-lines. Installation of energy storage devices, e.g. in the form of Superconducting Magnetic Energy Storages, SMES, is also expected to be used to cut off the peaks in required transmission capacity and to improve the transient stability.

-The tendency in many countries to deregulate electrical power generation and transmission in order to stimulate competition is also expected to increase the need for improved control of the power flow in the AC-systems. However, in most cases the slow control offered by conventional phase-shifters or series capacitors might be sufficient to provide this control.

- Finally, it is also expected that the requirement of the quality of the electrical power supply will increase in the future. By quality, we understand both, minimizing the risk for short interruption of the power supply as well as reducing distortion and variation of the AC-voltage. One way to avoid short interruptions could be to install energy storage devices e.g. in the form of superconducting magnetic energy storages, SMES.

The distortion and fluctuation in the AC-voltage can be minimized by installation of active filters.

4. LINE-COMMUTATED CURRENT-SOURCE CONVERTERS AND FORCED-COMMUTATED VOLTAGE-SOURCE CONVERTERS

The use of high power power electronics in electrical power transmission systems has until now mainly been based on the use of thyristors. They can only conduct current in one direction but are rather symmetrical with regard to blocking voltage capability, as the maximum forward blocking voltage is of the same order of magnitude as the maximum reverse blocking voltage. The maximum blocking voltage is of the order of 7 kV which requires a silicon wafer thickness of the order of 1 mm of which the major part consists of a weakly n-doped layer.

When a forward-blocking voltage is applied across the thyristor, it will start to conduct current as soon as a triggering pulse is fed to the gate. This triggering pulse can either be in the form of a short electrical pulse for ETT thyristors or in the form of a short, high intensity light pulse for LTT thyristors. As soon as the thyristor has started to conduct current, it will latch in the conducting state and continue to conduct current until the current is decreased to zero.

The type of converter that has been used for HVDC applications is the so called line-commutated current-source six-pulse converter, as shown in Fig. 5. It is called line-commutated as a positive voltage determined by the line voltage is required for the commutation of the current from one valve to the succeeding valve. The converter is called current-source or current-stiff converter as the smoothing reactor on the DC-side prevents fast changes of the DC-current. This also means that the DC-current is commutated between the three valves connected to the same DC-terminal. The duration of the commutation is determined by the reactance on the AC-side and the AC-line voltage. The basic relationships for the current-source converter is derived from the fact that the fundamental of the AC-current in each phase, $I_{L(1)}$, is proportional to the DC-current of the bridge, I_d. This gives

$$I_{L(1)} = k_I \cdot I_d \tag{1}$$

where k_I is a proportionality factor.

If we neglect the losses in the converter, we can see that the active power on both sides is equal. This gives the following relationship between the voltage on the DC-side, U_d, and the line-to-line voltage on the AC-side, U_L, and the phase-displacement angle, j,

$$U_d = \sqrt{3} \cdot k_I \cdot U_{L(1)} \cdot \cos j \quad . \tag{2}$$

As the valves only can conduct current in one direction, the change of power direction has to be obtained by change of the direct voltage polarity. This is obtained by changing the polarity of cos j which is controlled by the timing for firing the valves or the firing angle. For valves where only the firing but not the turn-off of the valves can be controlled, the phase-displacement angle has to be limited to the inverval 0 - 180°. This means that the converter will always consume reactive power.

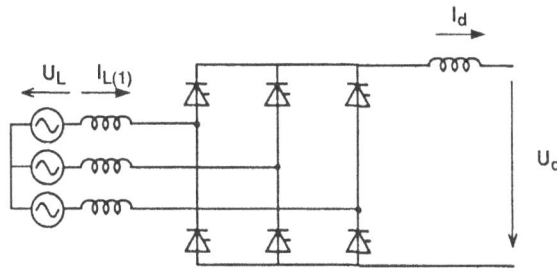

Fig. 5. Line-commutated current-source six-pulse converter.

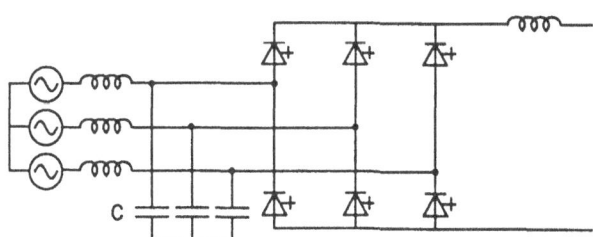

Fig. 6. Forced-commutated current-source converter.

To get around the limitation that the converter will always consume reactive power and that a sufficiently stiff line-voltage has to be supplied by some other source, valves with current turn-off capabilities have to be used. Figure 6 shows the configuration of a converter which still is of the current-source type but which is forced-commutated. The symbol for the valve has here been supplied with a cross at the gate to indicate that the valve also can turn-off the current by the generation of a high counter-voltage. The commutation of the current has now to be very fast to avoid high energy generation in the valves. Because of that, the impedance in the commutation circuit has to be very low. For this reason, capacitors are connected between the AC-terminals of the bridge. The valves have to sustain both forward and reverse blocking voltage in the same way as for a line-commutated current-source converter according to Fig. 5.

However, it has turned out that forced-commutated voltage-source converters according to Fig. 7 are of greater interest for high power applications than forced-commutated current-source converters according to Fig. 6. The possible applications of this type of converter for HVDC and for Flexible AC Transmission Systems will be further treated in the following. Forced-commutated voltage-source converters with GTO (Gate Turn Off) thyristors are today commonly used for drive systems with asynchronous motors. As can be seen from Fig. 7, each converter arm consists of one thyristor valve with current turn-off capability which we will call GTO-valve in the following and one diode valve connected anti-parallel to the GTO-valve. A large capacitor is connected directly across the DC-terminal of the bridge, preventing fast voltage fluctuations. Because of the diodes, the voltage polarity will always be unidirectional. This also means that the GTO-valves only have to be designed for for-ward-blocking voltage. The bridge arms can on the other hand conduct current in both directions as soon as a GTO-valve is turned on.

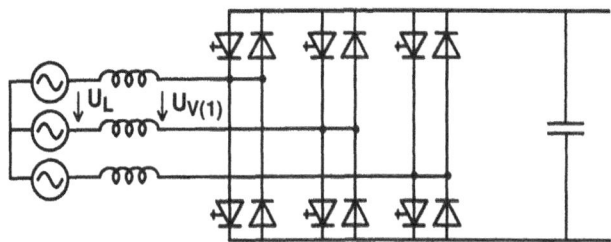

Fig. 7. Forced-commutated voltage-source converter.

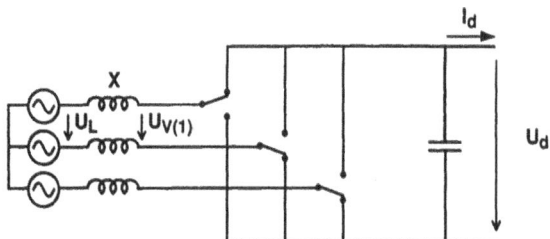

Fig. 8. Equivalent circuit for a forced-commutated voltage-source converter.

For the voltage-source converter, the commutation between the valves will take place between two valves connected to the same AC-phase. The function of the bridge may be more easily understood from the equivalent circuit diagram according to Fig. 8. As the voltages on the AC-terminals of the bridge are completely determined by the switching of the bridge arms and the voltage across the bridge, we will get the follow-ing relationship between the line-to-line fundamental frequency compo-nent of this voltage $U_{V(1)}$ and the DC-voltage U_d

$$U_{V(1)} = k_u \cdot U_d \qquad (3)$$

The quantity k_u is a constant and equal to $\pi / (2 \bullet \sqrt{6})$ for a square-wave six-pulse converter with one commutation between the upper and lower valves and vice versa per cycle, i.e. network frequency commutation. The quantity k_u can also be controlled and decreased to lower values by applying additional number of commutations per cycle, i.e. the so called pulse-width modulation, PWM.

The phase-position of the fundamental of the AC-voltage $U_{V(1)}$ is controlled by the firing of the valves. With the phase-shift as related to the line voltage U_L, denoted by δ, we get the following expression for the AC-tive power

$$P = \frac{U_L \bullet U_{V(1)} \bullet \sin\delta}{X} \tag{4}$$

where X is the total reactance per phase.

Setting the active power on the AC- and DC-side equal and using equation (3), we obtain

$$I_d = \frac{k_u \bullet U_L \bullet \sin\delta}{X} \tag{5}$$

The equations (4) and (5) indicate that the direction of the power flow and the magnitude and direction of the current are controlled by the angle δ. The direction of the reactive power flow is determined by the difference in the magnitude of the voltages U_L and $U_{V(1)}$. The behaviour of the forced-commutated voltage-source converter can also be illustrated by the equivalent circuit according to Fig. 9. Here, the phase and amplitude of the internal fundamental frequency voltage $U_{V(1)}$ is controlled by the firing of the valves and by the direct voltage across the bridge.

In so called square-wave operation, i.e. with the commutation frequency equal to the network frequency, the amplitudes of the internal voltage harmonics are inversely proportional to the harmonic number n. The magnitude of the harmonic voltages can be decreased either by the use of an additional number of commutations per cycle or by using a number of phase-shifted bridges connected in series or in parallel. To limit the harmonic currents to acceptable levels, usually the converter transformer is designed with a fairly large leakage reactance. In addition, two or more phase shifted bridges are usually connected in series on the AC-side.

However, in order to be able to reduce the cost of the converter transformers and also to avoid installation of AC-filters, it would be very attractive to increase the commutation frequency since this would permit both control of the factor k_u and also reduction at the magnitudes of the harmonics. From this point of view, a commutation frequency of the order of at least one kiloherz would be desirable. However, this is not possible with the existing type of high voltage GTO's and will probably also be difficult to obtain in the near future. The reason is that the switching losses in the semiconductors as well as the losses in the snub-ber circuits are large and will be proportional to the switching frequency. For a forced-commutated voltage-source converter, the commutation will either take place from the diode-valve in one branch to the GTO-valve in the other branch connected to the same phase or from the GTO-valve to the diode valve depending upon the polarity of the current at the mo ment of commutation.

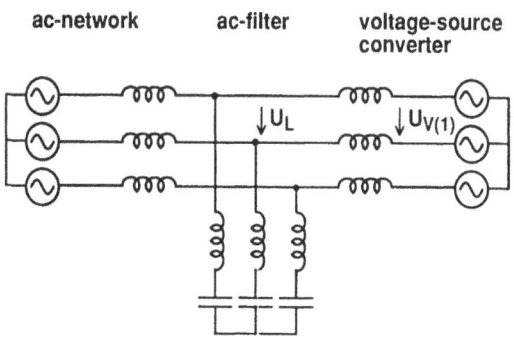

ac-network   **ac-filter**   **voltage-source converter**

Fig. 9. Equivalent cicuit for a forced-commutated voltage-source converter connected to an AC-bus with an AC-filter and an AC-network.

However, in both cases the commutation will occur at a voltage close to the full DC-voltage. The GTO-thyristors or other alternative thyristors with current turn-off capabilities, like the MOS Controlled Thyristor, MCT, as well as the diodes are built-up of silicon wafers with layers hav-ing different dopings. The layer which will take up the major part of the forward or reverse blocking voltage is a weakly doped n-layer. The thick-ness of this layer determines the maximum blocking voltage but also the on-state voltage drop in the semiconductor. To avoid a too high forward voltage drop or current dependent part of the on-state voltage in high voltage components, a fairly long life time for the charge carriers is cho-sen. As already mentioned, this will increase the recovery charge for thyristors. The same is valid for the diodes. Increased life time will also increase the tail current of GTO's and MCT's which together with high voltage across the thyristor after turn-off will generate turn-off losses. Additional losses are generated immediately after turn-off during the time-interval when the forward voltage is built-up in the thyristor. Losses will also be generated at turn-on, but those losses are usually smaller.

Beside the losses in the semiconductors, also the losses in the snub-ber circuits have to be considered. Capacitors have to be connected across the thyristors to limit the steepness of the voltage rise at turn-off and also to limit the spread in the voltage distribution between the thyristors in high voltage valves, where many thyristors are connected in series.

Inductors have also usually to be connected in series with the thyristors to limit the steepness of the current rise at turn-on. The energies stored in the capacitors and inductors will, at least to some extent, be dissipated in the resistors of the snubber circuits, either during energization or deenergization.

In order to limit the losses in the thyristors and the snubber circuits to acceptable values, the commutation frequency has probably to be limited to some hundred Hertz for high voltage components (> 6 kV), at least during the near future.

An interesting alternative converter circuit with regard to the requirements on control of the factor k_u, limiting the amount of harmonics, limiting the commutation frequency and using a simple transformer, is the three level or Neutral Point Clamping (NPC) converter as shown in Fig. 10. The disadvantage with the circuit is, that it requires six additional diode valves and that the control of the midpoint of the DC-capacitor can be critical at disturbances.

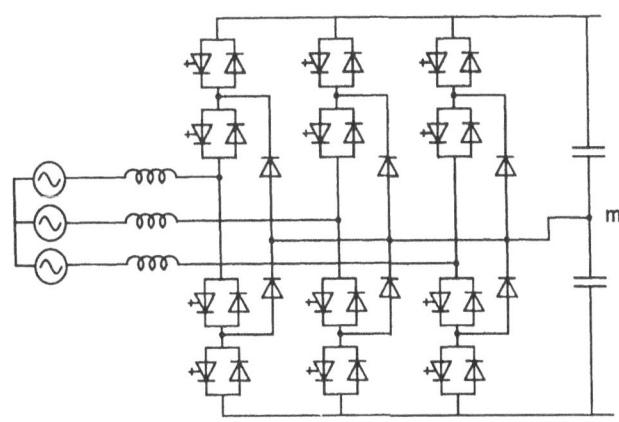

Fig. 10. Three-level (Neutral Point Clamping, NPC) converter.

The major advantage with the forced-commutated voltage-source converters as compared to line-commutated current-source converters are that they behave like synchronous machines, although without any inertia. This means that they can be controlled to both generate and consume reactive power and that they can operate in an AC-network with no other supply of AC-voltage.

The main disadvantages are that they are not as robust as line-commutated current-source converters and that the risk for high overcurrents is much higher, especially at faults on the DC-side. The costs of the forced-commutated valve bridges are also considerably higher than those of corresponding line-commutated bridges.

Much further development work is required until cost-effective valves with many series-connected thyristors can be built. The requirement for high energy turn-off pulses as well as gate signals during the whole conduction interval with today's GTO's results in rather expensive gate control units which are not suitable for valves on high potential. It is, however, to be expected that the future development of new thyristors, e.g. MCT's, will make it possible to build more cost effective valves.

5. EXPECTED FUTURE DEVELOPMENT OF THE HVDC TECHNIQUES

The further development of the HVDC techniques will most probably be mainly an evolutionary development of the existing techniques with line-commutated current-source converters. The major applications will in the future, as in the past, be long cable transmissions, long overhead line transmissions and back-to-back converter stations. It is also to be expected that some larger multiterminal HVDC systems will be built.

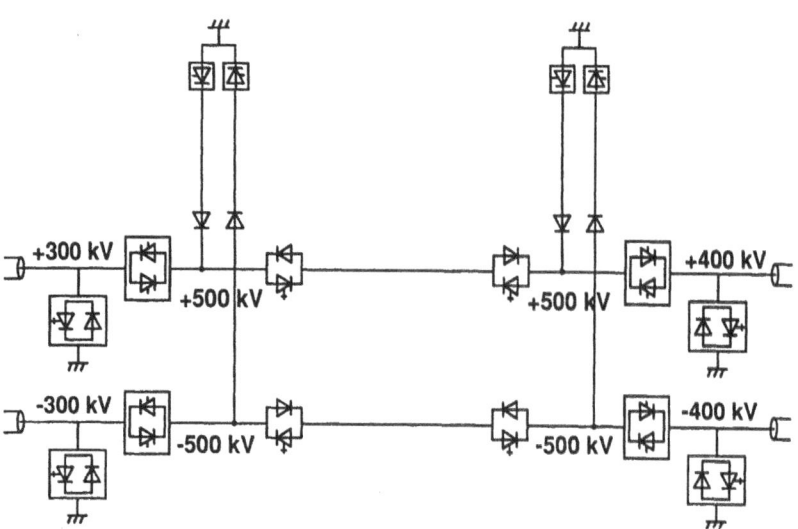

Fig. 11. Multiterminal HVDC system with three different voltages 500/400/300 kV.

The fast development of the semiconductor technology is expected to result in that the electrical rating of both conventional thyristors as well as of thyristors which can turn-off the current can be further increased with reduced requirements for the snubber-circuits. The thyristors will probably be light-triggered, at least for firing. The improved semiconductors will make it possible to build more compact valves which can be encapsulated and placed outdoors. The reduced costs of the valves will also make it attractive to use forced-commutated converters, at least for back-to-back converter stations, cable transmissions and special applications for which the properties to be able to operate in very weak AC-systems and to control the reactive power are essential. Converter stations for multiterminal systems or tapping stations, requiring that a disturbance in the local AC-network must not disturb the operation of the whole system, might also partly be built-up of forced-commutated converters.

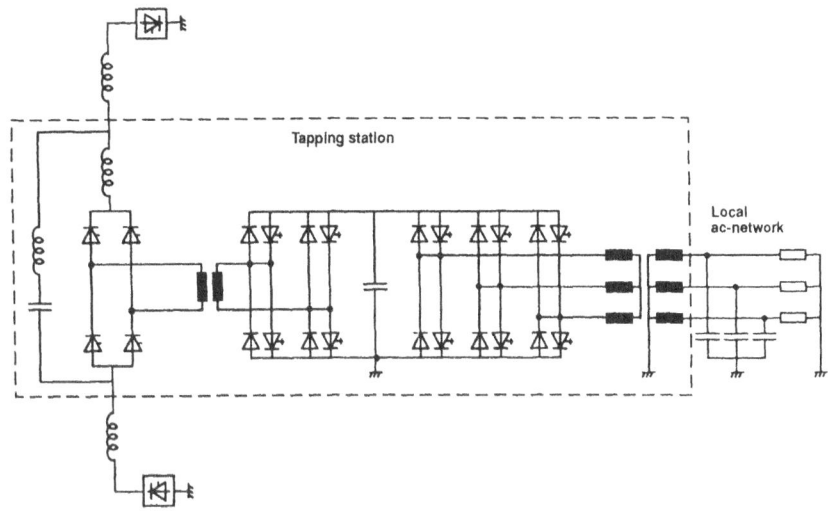

Fig. 12. DC transmission line with a series tapping station.

The application of forced-commutated converters for HVDC is further treated in [1,2,3]. As described in [2], it could also be expected that, in the future, it will be required that a fault in a part of an HVDC multiterminal system must not result in a short time interruption of the whole system. To prevent this, fast HVDC breakers with semiconductors might be used. It is also to be expected that multiterminal systems have to be built-up of different voltage levels as illustrated in Fig. 11.

In [3], a new concept for a tapping station according to Fig. 12 is proposed. The tapped-off power which is presumed to be only a few percent of the power rating of the main transmission line is converted via three converter bridges and an intermediate DC-stage from the DC-line to the local AC-network.

6. FLEXIBLE AC TRANSMISSION SYSTEMS

As mentioned before, EPRI has introduced the Flexible AC Transmission Systems concept, which could be understood as collective concept for a number of system or equipment concepts, each having the ability to improve the utilization and control of the existing AC-system. The major concepts which also will be treated in this paper will all use high power electronics. They are thyristor controlled shunt compensation, thyristor controlled series compensation and thyristor controlled phase-shifters. The use of back-to-back stations could also be considered as a Flexible AC Transmission Systems concept. It should be noted that all the individual Flexible AC Transmission Systems concepts were first presented a number of years before the general Flexible AC Transmission Systems concept was introduced. The major advantage with the introduction of the Flexible AC Transmission Systems concept is that it has stimulated a great interest among utilities to introduce the different concepts. The major hindrance for a more general introduction is probably the costs and losses in the different systems and also uncertainty how the devices shall be controlled. However, besides the existing conventional SVC stations, a number of pilot installations of so called Advanced Static Var Compensators, ASVC, controlled series capacitors and energy storage systems have already been built.

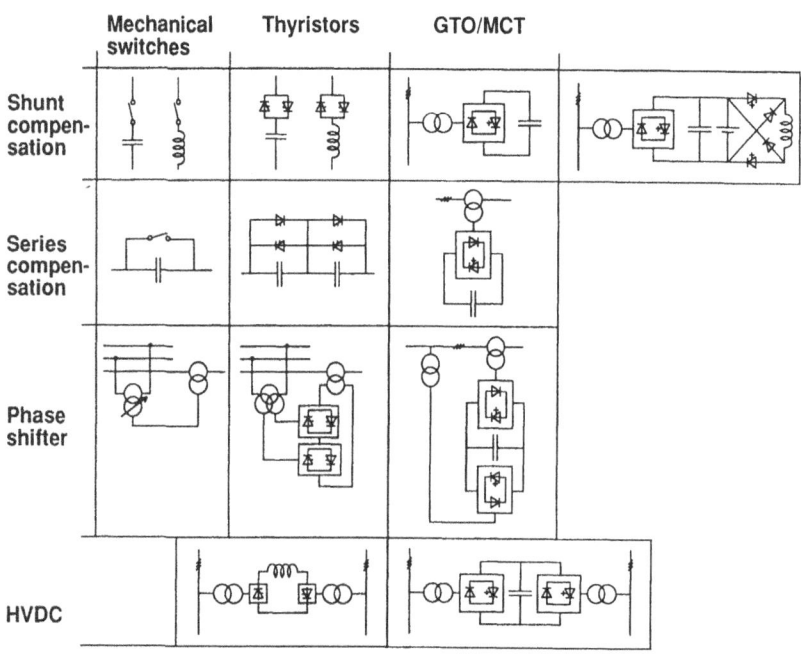

Fig. 13. Summary of different FACTS and HVDC concepts. The rows give the different applications, i.e. shunt compensation including energy storages, series compensation, phase-shifters and HVDC including back-to-back stations.

A summary of the different concepts including HVDC is presented in a matrix form in Fig. 13. The columns give the different switching elements used, i.e. mechanical switches, which also might be new superfast switches, thyristors for which only the turn-on can be controlled and finally GTO or MCT thyristors for which both the turn-on and turn-off can be controlled.

7. SHUNT COMPENSATION

As previously described, the technique of using Static Var Compensation, SVC, with thyristors instead of rotating synchronous compensators is today a commonly used and accepted technique. The SVC stations are used to control reactive power, either by generation or absorption of reaction power or both. SVC's are primarily used to control the AC-voltage level and minimize the transient voltage changes, e.g. to limit temporary overvoltages. As the transmitted active power on an AC-transmission line is determined both by the amplitudes and the phase-angle difference δ between the voltages at the terminals, it is of course also possible to use an SVC installed in one of the terminals to damp power oscillations in an AC-system. However, to be most effective, the SVC unit should if possible be placed at locations for which the largest voltage variations will occur at power changes, i.e. for a single transmission line in the middle of the line. The transmitted power on a transmission line is often illustrated in a simple way by a two-machine model with the terminal voltages U_a and U_b connected via a loss-free reactance X_{ab} as shown in Fig. 14.

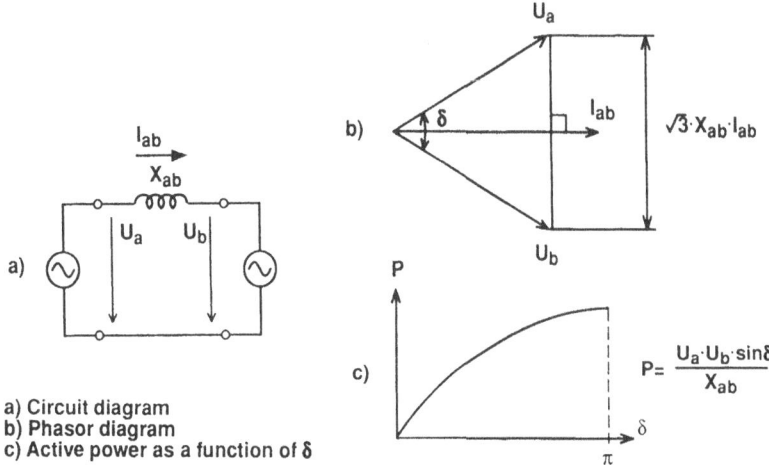

a) Circuit diagram
b) Phasor diagram
c) Active power as a function of δ

Fig. 14. Simplified two-machine model: a) circuit diagram; b) phasor diagram; c) active power as a function of δ.

The SVC stations are usually built-up of Thyristor Controlled Reactors (TCR), Thyristor Switched Capacitors (TCR) and Fixed Capacitors (FC) which also may be built-up of filters as illustrated in Fig. 3.

The TCR unit is used to obtain continuous fast control of the reactive power. The disadvantage with the TCR unit is that it generates harmonics which usually will require installation of filters. Because of that it could be advantageous to use switched reactors instead of thyristor controlled reactors when no continuous control is needed. To obtain fast switching, anti-parallel thyristors might be used as switching elements. This is usually called Thyristor Switched Reactors (TSR).

During the last years, a number of pilot installations have been built with conventional SVC's replaced by forced-commutated voltage-source converters for reactive power control. These systems are often called Advanced Static Var Compensator (ASVC). It has been claimed that an ASVC could be built more compact and possibly also cheaper than a conventional SVC, at least if it is required that the maximum generation is almost equal to the maximum absorption of reactive power. The reason for the reduced cost should be that the same equipment is used for both generation and absorption of reactive power and that the rated reactive power does not correspond to energies stored in capacitors and reactors. Hence, savings can be made in reactors and capacitors. However, most manufacturers seem now to agree that an ASVC will be more expensive to build than a conventional SVC, mainly because of the increased costs of the valves and the transformers. The costs of the forced-commutated valve-bridges are today 2-3 times the costs of the conventional thyristor valve-bridges.

However, a decision about which technique should be chosen must not be based only on costs but also based on performance. An ASVC might of course be justified if this technique could offer special valuable advantages not offered by a conventional SVC. One such advantage, usually not sufficient to justify the usage of an ASVC, is that the maximum reactive current is independent of the voltage while the maximum current will decrease with decreased voltage for a conventional SVC. For very weak AC-systems it might also be of importance that an ASVC, in the contrary to the case of a conventional SVC, will not decrease the resonance frequency of the system. On the other hand the conventional SVC is more robust. The operation and performance of the ASVC at faults in the AC-network have also to be further investigated.

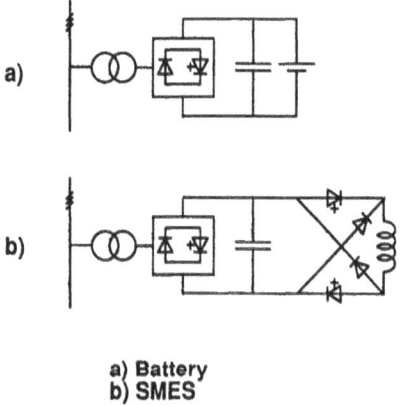

a) Battery
b) SMES

Fig. 15. ASVC combined with energy storage: a) battery; b) SMES.

One interesting application for which an ASVC will offer special advantages is an ASVC, combined with an energy storage element, e.g. in the form of a battery, as in Fig. 15a, or in the form of a superconducting magnetic energy storage, SMES, as shown in Fig. 15b.

Energy storages for short time power supply might be of interest to use to avoid short interruption of power supply, for stabilization purposes and to reduce the need for dimensioning the transmission system capacity for short duration high power peaks, e.g. for the supply of electric railway systems.

8. SERIES COMPENSATION

As shown in Fig. 14 for the two-generator case, the transmitted power is inversely proportional to the impedance X_{ab}. An efficient way to increase the power transmission capability of a long-overhead line involves installing a series capacitor as illustrated in Fig. 16. Here, the total line reactance is split into two inductive parts, X_a and X_b, and a capacitive part, X_c, connected in series. The influence of the series capacitor can either be considered as a reduction of the total line reactance or, for a given line current, as an addition of a voltage perpendicular to the line current and with opposite sign as compared to the voltage drop across the line reactances X_a and X_b.

It is also obvious from Fig. 16 that the transmitted active power on the line can be controlled by control of the reactance X_c. This control can be obtained as illustrated in Fig. 17. As shown in Fig. 17a, the whole or part of the series capacitor can be switched in or out with mechanical switches or breakers. Faster and improved control in smaller steps can be obtained with anti-parallel thyristors as shown in Fig. 17b. A pilot project with outdoor thyristor valves according to this concept is going on in cooperation between ABB Power Systems and the power utility AEP in USA.

If a continuous control is needed, it might be performed with a combination of thyristor controlled reactors and capacitors or possibly also by a scheme according to Fig. 17c. The thyristors are here assumed to be of the turn-off type and switched with a fairly high switching frequency. The drawback with this concept and all other alternatives giving continuous control is however increased costs and generation of harmonics.

Figure 17d shows a concept which has certain similarities with a controlled series capacitor as it generates a longitudinal voltage perpendicular to the line current. The advantage of this concept is that the magnitude of the voltage can be directly controlled, as the forced-commutated voltage-source converter is connected in series in the overhead line via a booster transformer. However, the costs of the transformer as well as the costs of filtering have also to be considered.

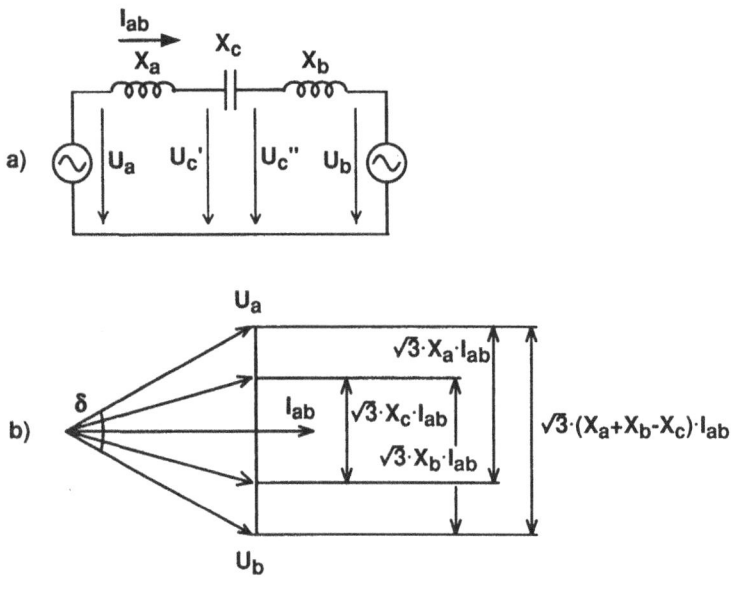

a) Circuit diagram
b) Phasor diagram

Fig. 16. Simplified two-machine model with capacitive series compensation: a) circuit diagram, b) phasor diagram.

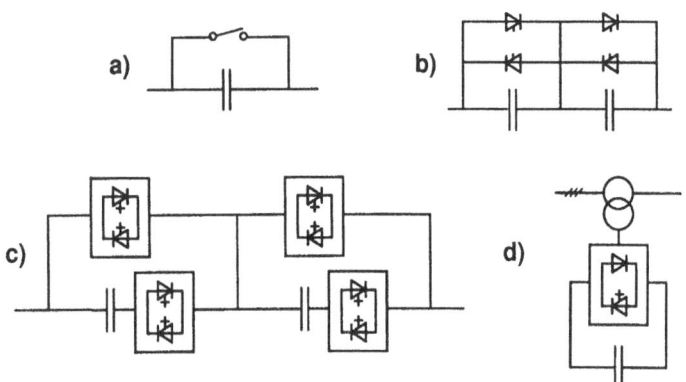

a) Breaker switched capacitor
b) Step-wise thyristor switched capacitors
c) Continuously controlled GTO-switched capacitors
d) Forced-commutated voltage-source converters

Fig. 17. Different alternatives for capacitor series compensation: a) breaker switched capacitor; b) step-wise thyristor switched capacitor; c) continuously controlled GTO-switched capacitors; d) forced-commutated voltage-source converters.

9. PHASE-SHIFTERS

As the transmitted power on a transmission line is directly dependent upon the phase-shift angle δ, as illustrated in Fig. 14, the control of
power can of course also be obtained by control of the phase-shift. The
principle is illustrated in Fig. 18, showing a simplified two machine
model with a phase-shifter and the corresponding phasor-diagram. The
phase-shifter gives a positive or negative contribution, Δδ, to the total
phase-shift angle between the voltages U_a and U_b. An important
difference as compared to the controlled series capacitors is that for the
phase-shifter the phase-shift or the additional longitudinal voltage is
directly controlled. For the series capacitor, the magnitude of the
longitudinal voltage is on the other hand directly proportional to the
magnitude of the line current and the phase position is perpendicular to
the line current. Figure 19 illustrates different possible concepts for
phase-shifters.

Figure 19a shows the conventional phase-shifter with mechanical
tap-changer controlled transformers. The basic principle is to connect in
series in each phase a controlled part of the voltages from the two other
phases. If this additional phase voltage is denoted ΔU_a and the current is
I_a, the minimal installed rating is $3 \; \Delta U_a \bullet I_a$. This means, to obtain a
maximum phase-shift of $\pm 60°$, the minimal installed rating of the
transformer must be equal to the maximum power on the line. However,
it should be noted that adding a voltage perpendicular to the phase-
voltage does not only give a phase-shift but also gives a change of the
voltage level. This is minimized with the arrangement according to
Fig. 19a, see [4].

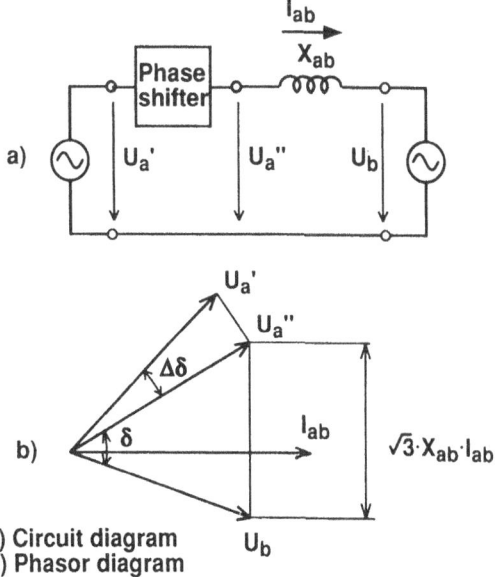

a) Circuit diagram
b) Phasor diagram

Fig. 18. Simplified two-machine model with phase-shifter: a) circuit diagram; b) phasor
diagram.

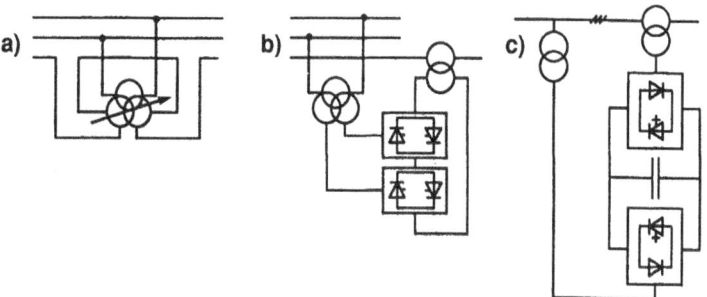

a) Conventional phase-shifter with transformers and mechanical
tap-changers.
b) Phase-shifter with anti-parallel thyristors.
c) Phase-shifter with forced-commutated voltage-source
converter.

Fig. 19. Different alternative concepts for phase-shifters: a) conventional phase-shifter
with transformers and mechanical tap-changers; b) phase-shifter with anti-parallel
thyristors; c) phase-shifter with forced-commutated voltage-source converter.

The phase-shifter is often combined with a separate transformer for
change and control of the input/output voltage. However, it should be
noted that the combined transformer and phase-shifter only can control
the reactive power flow but not generate reactive power. This type of
phase-shifter will always consume reactive power. It is mainly used for
slow power control. Most of the installations are to be found in USA.

Figure 19b shows a variation of the phase-shifter with the mechanical
tap-changers being replaced by bridges using anti-parallel connected
thyristors. The principle is further described in [5].

The secondary side of the transformer consist of a number of
windings, e.g. with the turn ratios 1:3:9. These windings can via thyris-
tor-bridges be connected to a booster transformer connected in series in
the AC-line. The connection and the polarity of each winding is controlled with the thyristors. In principle it might be possible to connect
the thyristor bridges directly in series into the line and thus avoid the
booster transformer, but in practice it will probably be cheaper to use the
arrangement according to Fig. 19b. This will, however, require that the
total ratings of the two transformers will be twice as large as for a
conventional transformer phase-shifter. The separate transformer for
voltage level control will probably be needed in both cases.

The cost of the thyristor valves will probably also be substantial for
the concept according to Fig. 19 b. If it is assumed that the cost is proportional to the r.m.s. value of the current times the voltage, it is found
that the costs of the valves according to the concept according to
Fig. 19b will be twice as high as the costs of the valves for a conventional
symmetrical SVC unit with the same rating. The additional costs and
losses for the configuration according to Fig. 19b as compared to Fig. 19a
will probably not justify the improved performance with regard to fast

control. It should, however, be noted that both concepts have the advantages of not generating any harmonics.

Finally, Fig. 19c shows a more advanced but probably also a more expensive concept, based on the usage of forced-commutated voltage-source converters. It is very similar to the concept according to Fig. 17d for series compensation. The major difference is that the addition of the shunt-connected forced-commutated voltage-source converter now makes it possible to obtain a longitudinal voltage which no longer must be perpendicular to the line current. The shunt-connected converter will also act as an ASVC which means that, besides control of the flow of active and reactive power, also the generation and absorption of reactive power can be controlled. From this point of view, this advanced phase-shifter could be compared with the behaviour of an HVDC back-to-back station, built-up of forced-commutated voltage-source converters. It should, however, be noted that the total power ratings of the two forced-commutated converters according to Fig. 19c will be lower than for the back-to-back stations if the maximum phase shift is smaller than about $\pm 70° - 80°$. It should also be noted that the rating of the shunt-connected converter usually can be chosen substantially smaller than that of the series-connected converter.

The total ratings and costs of the transformers will probably be lower for the concept according to Fig. 19c than for the concept according to Fig. 19b. The control will be continuous but, as for all other converter applications, the additional costs of the filters have also to be considered.

Finally, it should be noted that, most probably, it often will be found advantageous to use a combination of two or more different concepts in the future e.g. Thyristor Switched Capacitors and a small ASVC for the continuous control of reactive power.

10. CONCLUSION

As illustrated in Fig. 13, quite a number of new interesting high power electronic concepts have been proposed besides the conventional SVC and the line-commutated current-source converters for the control and conversion of reactive and active power in electrical power transmission. The conventional SVC and HVDC techniques are now widely accepted and a number of new installations are commissioned every year. It is to be expected that, also in the future, the conventional SVC and HVDC will dominate, although new concepts, e.g. based on the usage of forced-commutated voltage-source converters, will find interesting applications. One major factor preventing even faster and more widely used applications of the presented new concepts are the costs and sometimes also the losses. The fast development of power semiconductor devices such as diodes, thyristors, GTO's and MCT's, will be decisive, offering possibilities of reducing the costs and increasing the commutation frequency in the future.

REFERENCES

1   Å. Ekström, "Line-commutated and forced-commutated converters for future HVDC transmission systems", *Int. Symposium on Electrical Energy Conversion in Power Systems*, Capri, 1989, T-C.5.

2   L. Knudsen, B. Hansson, Å. Ekström, "Description and perspecitive applications of new multi-terminal HVDC systems concepts", *CIGRE, 1990*, 14-201.

3   Å. Ekström, P. Lamell, "HVDC tapping station - power tapping from a DC transmission line to a local AC network", *IEE London*, 1991, pp. 126-131.

4   D. Perce, "Special transformer with control power flow between Ontario, Manitoba", *Electrical News and Engineering*, July 1972, pp. 22-25.

5   G. Güth, R. Baker, P. Eglin, "Static thyristor controlled regulating transformer for AC-transmission", *IEE London*, publ. 205, 1981, pp. 69-72.

DISCUSSION

T. P. Lipo (U. Wisconsin, Madison, WI)
Have you given any thought to using resonant converters in your applications ?

Å. Ekström
I think this is a very interesting concept. First of all, however, we need some sort of device which can switch very high power at high frequency, probably at about 1 kHz or more. These components should exhibit a high blocking voltage, say 10 kV, and have a low forward voltage drop to reduce the losses. The carrier lifetime required will make it difficult to obtain a fast switch.

I would expect that losses in your high power resonant converter applications might be fairly high. Is that correct ?

T.P. Lipo
With a resonant converter, you should be able to remove the switching losses and therefore improve the efficiency by quite a bit.

Å. Ekström

Apart from switching losses, the conduction losses are very essential, leading to components with very high carrier lifetime. Basically, we are interested in all concepts which can both control active and reactive power flow.

M. Campagna (ABB Corp. Research, Baden)

Would it be posible for you to make a proper analysis of the basic topology and then, by looking at the various cost contributions including the losses, predict the kind of characteristics your device should have ?

Å. Ekström

Together with universities, we are speculating along that line to see what happens 10 years from now. We could build HVDC valves with forced commutated converters using existing GTOs. But using very expensive, large gate units which need high auxiliary power and which must be placed at a potential of 500 kV is not very attractive.

H.Ch. Skudelny (RWTH, Aachen)

Are there already real industrial installations of AC power flow control by means of static converters or is this just a concept ?

Å. Ekström

Apart from the existing line-commutated type of a reactive power control, some forced commutated converters have been installed in Japan, in the US and in Finland. The problem is that they are too expensive.

H. Ch. Skudelny

Are these experimental type converters ?

Å. Ekström

They are mostly pilot installations in order to gain experience. We have not yet studied what kind of performance could be obtained by using that kind of converters instead of the conventional type.

H. Stemmler (ETH, Zürich)

For your information, the first 80 MVA GTO reactive power compensator is in operation in Japan since spring of this year.

H. Ch. Skudelny

My point was power flow control ; can you comment about sharing of power flow in parallel lines ?

Å. Ekström

The installation of a thyristor controlled capacitor according to Fig. 17 b, which I shortly described, will be installed next month in the US and will, in a first step, be just for one phase. This will be the first application of real power flow control.

NEW MATERIALS BEYOND SILICON FOR POWER DEVICES

B. Jayant Baliga

North Carolina State University
Raleigh, NC, USA

ABSTRACT

At present, semiconductor based power electronic systems rely extensively upon silicon power devices. The introduction of Metal Oxide Silicon (MOS) technology to power devices in the 1970s and 1980s has resulted in a displacement of the bipolar power transistors with power MOS Field Effect Transistors (MOSFET's) and Insulated Gate Bipolar Transistors (IGBT's). This change is motivated by the simplification in the gate control circuit due to the high input impedance of these MOS-gated power devices. In the 1990s, the Gate-Turn-Off Thyristor (GTO) is also expected to be challenged by the development of MOS-gated thyristors.

Looking into the future on a longer time scale, it is anticipated that the silicon devices could be replaced by devices based upon other semiconductor materials. Recent analysis has shown that the specific on-resistance of power MOSFETs can be reduced by one order of magnitude using Gallium Arsenide, by two orders of magnitude using Silicon Carbide, and three orders of magnitude using Semiconducting Diamond. It has also been shown that a 5000 V Silicon Carbide based MOSFET could have a forward drop lower than that of a 5000 V Silicon GTO, which indicates the possibility to replace the entire spectrum of Silicon devices with Silicon Carbide devices. This paper will describe the reasons for these conclusions and outline the technological challenges that must be overcome to fabricate these devices.

1. INTRODUCTION

At present, silicon based power devices and integrated circuits are extensively used in power electronic applications. Looking back to 50 years ago, germanium based devices were favored over silicon. When the semiconductor industry began to capitalize upon the high insulating quality of the natural oxide film that could be thermally grown on silicon, it rapidly overtook germanium as the material of choice. Although, interest in other semiconductor materials continues to exist, their commercial success has been restricted to cases where the physical and electronic properties of silicon cannot provide the same function. One such case is the field of optoelectronics which has supported the development of process technology for III-V compound semiconductor materials.

Recent progress in silicon power device technology has pushed its performance close to the theoretical limits for silicon. This has created an interest in the investigation of other semiconductor materials that would provide superior intrinsic performance, based upon their unique electronic properties. This paper reviews the criteria to be used in the selection of these new semiconductor materials that would provide superior power device characteristics. The state of the technological developments for each material is also discussed, providing a rationale for determination of the time scale within which power devices based upon these new materials would become available.

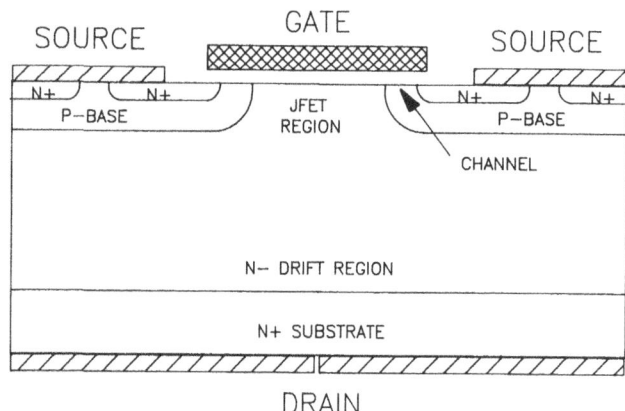

Fig. 1. Cross-section of power DMOSFET structure.

2. SILICON POWER DEVICES

In this section of the paper, a review of the status of silicon based power semiconductor technology is provided. Although the bipolar power transistor has been extensively used for power electronic applications in the past, it is rapidly being replaced by Metal-Oxide-

Semiconductor (MOS)-gated power devices. This revolutionary trend has taken place due to the simplification of the gate control circuit for the MOS-gated power devices. In fact, it is possible to integrate the gate control circuit of these devices, resulting in a tremendous reduction in cost, volume, and weight in applications.

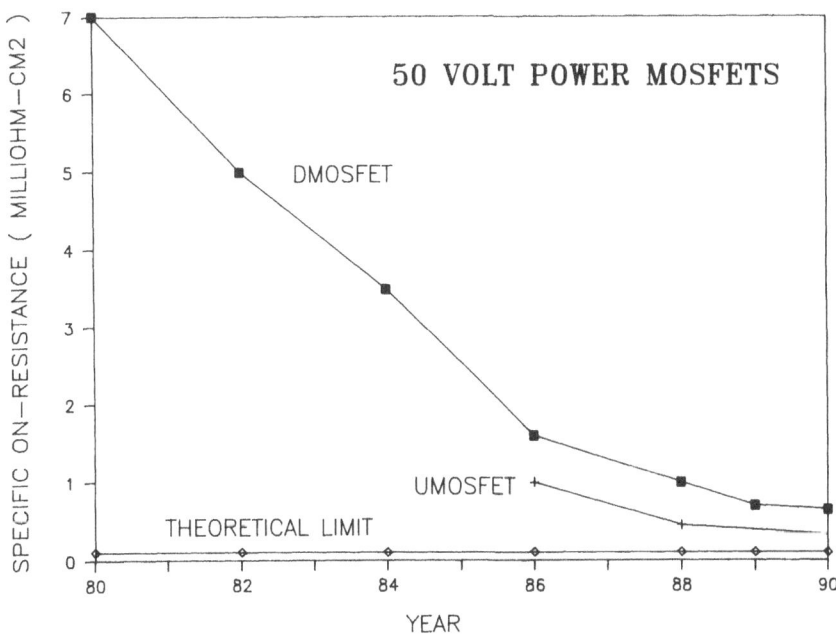

Fig. 2. Progress in reducing the specific on-resistance of silicon power MOSFETs.

Although many possible MOS-gated power device structures have been proposed, only the power MOS-Field Effect Transistor (MOSFET) and the Insulated Gate Bipolar Transistor (IGBT) have become commercially available at high volume on a world-wide basis. A cross-section of the power MOSFET is shown in Fig. 1. The power MOSFET is a unipolar device in which the current flow is controlled by the application of a gate voltage[1]. The gate voltage creates an inversion layer (electrons in n-channel devices) at the surface of the p-base region. This results in the formation of a current flow path between the source and drain terminals. Since the current flows only by majority carrier (electron) transport, it is limited by the resistances within the device structure. The largest resistance contributions arise from the channel (inversion layer), the Junction FET (JFET) region between the p-base diffusions and the drift region. The drift region is required to support the device blocking voltage and its resistance increases very rapidly as the breakdown voltage of the MOSFET is increased. Advances in device design and process technology have been aimed at reducing all the internal resistances in the power MOSFET structure. The impact of this effort on the specific on-resistance (resistance for a 1 cm^2 active area) is shown in Fig. 2 for the case of devices with breakdown voltage of 50 volts[2]. It can be seen that an improvement by over one order of magnitude has been achieved. By 1990, devices have been reported with specific on-resistances below 0.3 mΩcm^2. This value is very close to the theoretical limit for silicon (0.15 mΩcm^2) based upon just the drift region resistance. From this, it can be concluded that further advances in the performance of the power MOSFET structure will require the use of alternate semiconductor materials.

Another problem that has limited the application of the silicon power MOSFETs is a very rapid increase in the resistance of the drift region with increase of the breakdown voltage, as shown in Fig. 4. This degrades the on-state characteristics of these devices making the power losses excessive for many high current applications. A solution to this problem was developed by the creation of the IGBT structure shown in Fig. 3[1]. In this structure, minority carrier injection into the drift region during the on-state reduces its resistance. Although IGBTs have lower on-state voltage drop than power MOSFETs, their switching speed is much slower. The creation of high voltage power MOSFETs with low specific on-resistances from new materials is consequently of interest for high voltage applications.

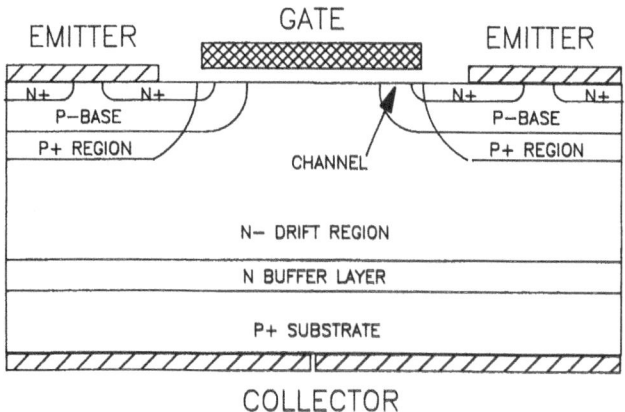

Fig. 3. Cross-section of insulated gate bipolar transistor (IGBT) structure.

A high performance power electronic system requires not only improved switching devices (transistors) but also improved rectifiers. For low blocking voltages (< 100 volts), silicon Schottky rectifiers are available with excellent characteristics[1]. High voltage silicon Schottky rectifiers are not viable because of poor reverse blocking characteristics due to the Schottky barrier lowering effect and a high on-state voltage drop arising from the large series resistance of the drift region. Consequently, for high blocking voltages, the silicon P-i-N rectifier is being used. The P-i-N rectifier operates with a low forward voltage drop in the on-state due to high level injection. However, its reverse recovery behaviour produces large power losses in applications[1]. Consequently, it would be advantageous to develop Schottky rectifiers from new materials that have smaller drift region resistances.

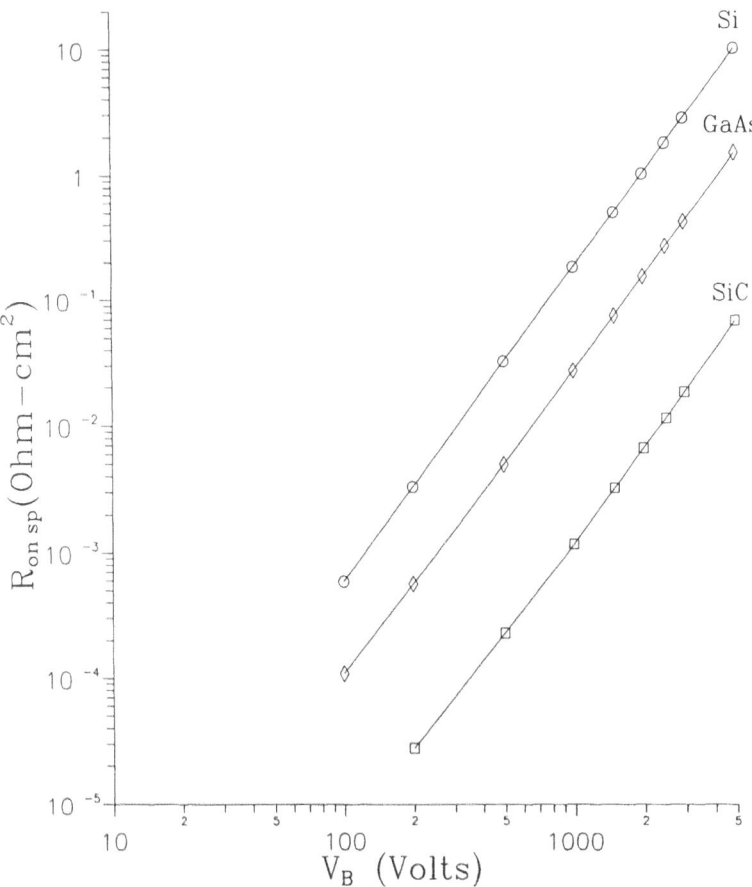

Fig. 4. Comparison of the specific on-resistance of gallium arsenide and silicon carbide based power FETs with that for silicon devices.

3. FIGURE OF MERIT FOR NEW MATERIALS

In order to make a judicious selection of new materials that would be candidates for the development of improved power devices, it is necessary to define a figure of merit based upon the fundamental electronic properties of semiconductors. It has been shown by Baliga[3] in 1982 that the specific on-resistance of the drift region is given by the following relationship :

$$R_{on,sp} = K / (\epsilon \bullet \mu \bullet E_c^3)$$

where K is a constant, ε is the dielectric coefficient, μ is the mobility, and E_c is the breakdown field strength. Since a small specific on-resistance is desirable, Baliga's Figure Of Merit for power devices has been defines as:

$$BFOM = \varepsilon \bullet \mu \bullet E_c^3$$

This figure of merit has been analysed for all the III-V compound semiconductors and other wide band gap materials, such as silicon carbide and semiconducting diamond. It has been found that gallium arsenide has the largest figure of merit among the III-V compounds. A comparison of Baliga's figure of merit for the best materials is provided in Table I. From this table, it can be concluded that gallium arsenide devices will have one-tenth of the drift resistance of silicon devices, silicon carbide devices will have one-hundredths of the drift resistance of silicon devices, and diamond based devices will have even lower drift region resistances.

Table I. A generalized comparison of semiconductor materials (All values normalized to silicon)

Semiconductor	Maximum Electric Field	Dielectric Constant	Mobility	BFOM
Si	1.00	1.00	1.00	1.0
GaAs	1.29	1.09	5.70	13.3
SiC	8.10	0.85	0.20	106
Diamond (n-type)	18.90	0.47	1.27	8574
Diamond (p-type)	18.90	0.47	1.00	6751

The above table is intended to provide a macroscopic view of the improvements that can be expected from the use of new materials. A more detailed analysis of the specific on-resistance of the drift region has been done for the case of silicon carbide and gallium arsenide de vices by taking into account the dependence of the electric field and mobility upon the doping concentration in the drift region[4]. Insufficient information is available for semiconducting diamond to be included. From this analysis, the variation of the specific on-resistance with breakdown voltage has been calculated and plotted in Fig. 4. Based upon these values, it can be concluded that silicon carbide power MOSFETs with on-state drops comparable to those for a 5000 volt silicon Gate-Turn-Off (GTO) thyristor may be achievable.

The use of these new materials with much smaller drift region resistances will also allow the development of high voltage Schottky rectifiers with a low forward voltage drop. As an example, the on-state characteristics of 1000 volt Schottky rectifiers made from gallium arsenide and silicon carbide are compared with those for a silicon device in Fig. 5. It can be seen that, at a given current density, the forward voltage drop for the Schottky rectifiers made from these new materials is much lower than for the silicon Schottky rectifier and comparable to that for a good silicon P-i-N rectifier. Of course, the switching characteristics of the Schottky rectifiers made from these new materials will be much superior to those for the silicon P-i-N rectifier.

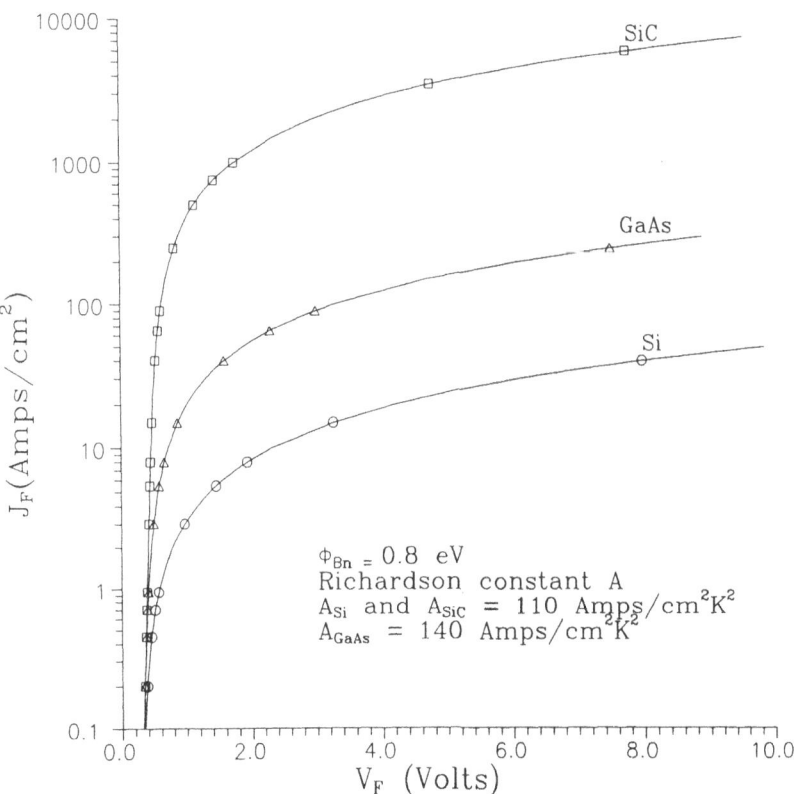

Fig. 5. Comparison of the forward conduction characteristics of gallium arsenide and silicon carbide Schottky barrier rectifiers with that for a silicon device (1000 volt reverse blocking case).

4. TECHNOLOGY ASSESSMENT

The development of gallium arsenide based power devices on the basis of the improved figure of merit was first reported in the early

1980's. The first technological challenge was the growth of lightly doped drift regions on highly doped substrates. This was accomplished using both liquid phase epitaxy and vapor phase epitaxy with high breakdown voltages[5,6]. Using this material, high voltage Schottky barrier rectifiers and vertical channel JFETs and Metal Semiconductor FETs (MESFETs) were fabricated[7,8,9]. Devices with breakdown voltages of 200 volts have been reported. Recently, gallium arsenide based high voltage Schottky barrier diodes have become commercially available, based upon an improved resistive field plate edge termination[10]. These devices are attractive for high frequency power electronic applications where the blocking voltage of the diode exceeds 100 volts.

In the case of silicon carbide, the process technology is not as advanced as that for gallium arsenide. One of the serious limitations for silicon carbide device development is the high cost and small wafer size for the substrates. For this reason, it would be advantageous to use silicon substrates with heteroepitaxial silicon carbide layers. Although the heteroepitaxial growth of silicon carbide layers on silicon substrates has been demonstrated, these layers have been found to contain a large number of defects due to the significant lattice mismatch (20 %) between the two materials[11]. The effect of these defects upon the characteristics of power devices is yet to be evaluated. However, it has been already demonstrated that enhancement mode MOSFETs can be fabricated in this material by the growth of a gate dielectric layer by thermal oxidation[11].

5. CONCLUSIONS

Advancements in silicon power device design and process technology have pushed their performance close to the theoretical limits based upon the fundamental electrical parameters of this material. In a pioneering study performed a decade ago, it was shown that superior intrinsic performance can be obtained in power devices fabricated from materials with higher breakdown field strength and high free carrier mobility. The most promising semiconductor materials were identified to be gallium arsenide, silicon carbide and diamond. Progress in gallium arsenide technology have now made the fabrication of power devices from this material commercially viable. Recent progress with the technology for the growth of homoepitaxial and heteroepitaxial silicon carbide layers have made the investigation of power device structures possible at this time. Due to the large number of process unknowns, the demonstration of power devices from silicon carbide will require research over a period of several years. The current research activity on the growth of semiconducting diamond films should lead to device quality material on the longer time frame.

REFERENCES

1   B.J. Baliga, "*Modern Power Devices*", John Wiley & Sons (1987).

2   B.J. Baliga, "Impact of VLSI technology on power devices", *Int. Conf. on Solid State Devices and Materials*, Abstr. A-1-2, pp. 5-9 (1990).

3   B.J. Baliga, "Semiconductors for high voltage vertical channel field effect transistors", *J. Applied Physics*, 53, pp. 1759-1764 (1982).

4   M. Bhatnagar and B.J. Baliga, "Analysis of silicon carbide power device performance", *Int. Symp. on Power Semiconductor Devices and ICs*, pp. 176-180 (1991).

5   B.J. Baliga, R. Ehle, J.R. Shealy, W. Garwacki, "Breakdown characteristics of gallium arsenide", *IEEE Electron Device Letters*, EDL-2 (11) pp. 302-304, (1981).

6   B.J. Baliga, A.R. Sears, P. Menditto, P.M. Campbell, "Extended measurements of gallium arsenide breakdown characteristics using punchthrough structures", *IEEE Electron Device Letters*, EDL-5 (9) pp. 385-387, (1984).

7   B.J. Baliga, et al, "Gallium arsenide Schottky power rectifiers", *IEEE Trans. Electron Devices*, ED-32, pp. 1130-1134 (1985)

8   P.M. Campbell, et al, "150 Volt vertical channel GaAs FET", *Int. Electron Devices Meeting Digest*, Abstr. 10.4, pp. 258-260 (1982).

9   P.M. Campbell, W. Garwacki, A.R. Sears, P. Menditto, B.J. Baliga, "Trapezoidal groove Schottky gate vertical channel GaAs FET", *IEEE Electron Device Letters*, EDL-6 (6) pp. 304-306 (1985).

10   K. Ohtsuka and Y. Usui, "Improvement of high speed blocking voltage by means of metal field plate for GaAs Schottky power rectifiers", *Int. Symp. on Power Semiconductor Devices and ICs*, pp.159-163 (1991).

11   H. Matsunami, "Crystalline SiC on Si and high temperature operational devices", in *'Amorphous and Crystalline Silicon Carbide and Related Materials II'*, Ed. M.M. Rahman, C.Y. Yang, and G.L. Harris, pp.2-7 (1989).

DISCUSSION

H.Ch. Skudelny (RWTH Aachen, Aachen D)

One of the important parameters is the maximum junction temperature a device can withstand. Are there remarkable differences in the materials you mentioned ?

B.J. Baliga

When people first looked at silicon carbide, this material was considered mainly for high temperature applications and some of the early and even of current work on FETs has been done to demonstrate high temperature operation. There is an example of a device running at 650 °C. In my analysis for a power device, I do not use such high junction temperatures because I am concerned about the package development effort required to tolerate these kinds of temperatures.

D. Metzner (TU Munich, Munich D)

Do you see any problems in transferring your argumentation to the case of conventional bipolar switches ? In my understanding the width of the drift zone would also influence the trade-off curve which you see in switching behaviour and on-state losses in those devices.

B.J. Baliga

Rather than bipolar switches, it is my preference to work in the direction of FETs because we expect much faster switching times, less storage time as well as a better safe operating area.

T. Ohmi (Tohoku U., Sendai J)

Do you have an idea how to treat the edge contour of this kind of high voltage devices ?

B.J. Baliga

Answers to these questions will have to be developed over the next many years. However, the material intrinsically has the ability to withstand very high electric fields and this is why we feel the realization is possible. The surface passivation of the material is going to be a critical issue and we must not only look at the center part of the device but also on how to treat the edges.

H.J. Krokoszinsky (ABB Corp. Research, Heidelberg D)

Where do you see the limits regarding electromigration effects on the Schottky barrier for the new materials ?

B.J. Baliga

In the case of the curves for the new materials, approaching 1000 A/cm^2, you are still well below the electromigration point and its not a limiting issue.

M. Campagna (ABB Corp. Research, Baden)

Twenty years ago, gallium-arsenide was predicted to be the material of the future. You have made similar remarks about gallium-arsenide as well as silicon carbide. Which of the critical issues such as epi on large wafers, ohmic contacts etc. do you consider most important, particularly for the various polytypes for silicon carbide ?

J.B. Baliga

First of all, it is no longer true to say that gallium-arsenide is a material of the future because commercial ICs as well as a gallium-arsenide power device are available. In terms of silicon carbide, we are targeting two different polytypes because these are the ones on which most of the work has been done. The first is 6H for the bulk material and the other is the 3C which we grow on silicon. I first picked the 3C material is because of the cost of the initial wafers. Recent inputs from manufacturers and increasing activity in growing silicon carbide indicate that the 6H material may become cost effective on the long run. So we intend to look at both of these materials in our research. As far as other critical technologies, I think silicon carbide has advantages over gallium-arsenide like the ability to grow a high quality thermal oxide.

R. Sittig (TU Braunschweig, Braunschweig D)

At present, there is a strong development in vacuum microelectronics and vacuum exhibits a very good breakdown strength and ultimate mobility. Hence the Baliga figure of merit should be very high for vacuum. Did you compare vacuum to silicon carbide ?

B.J. Baliga

I promised to talk about new materials and there is no material in a vacuum. Personally, I am attracted to vacuum microelectronics, but I was disturbed to learn at IEDM'91 that the forward voltage drop is as high as 150 Volts. This is not the limit and better emitters are being worked on. Prof. Ohmi told me last night about some of the work being done in Japan, and we also have work going on at the Microelectronic Center in North Carolina on vacuum microelectronics. If progress is made in developing better emitters, we will start looking at power devices with this approach.

R. Dutton (Stanford U., Stanford CA)

Can you comment about the wafer cost of silicon carbide in the US and in Japan ? Maybe Prof. Ohmi would like comment too. What is driving the economics for that field ?

B.J. Baliga

There is a company in the U.S. selling 1.0 inch wafers and, being an exclusive source, they can demand whatever price they want from the marketplace. That situation is changing because there are a lot of companies interested in entering this market.

T. Ohmi

In Japan, we are very eager to realize blue LEDs by using silicon carbide substrates. Their quality is rapidly improving and we should easily overcome the economical difficulties, I believe. There is no commercial source as yet, but I think 3 or 4 inch wafers of silicon carbide substrates will become practical. So we see two very attractive targets for semiconductor engineers : one is the blue LED and the other is a very excellent high power semiconductor device thanks to the effort at NCSU.

D. Silber (TU Bremen, Bremen D)

Do you think, apart from the power market, the new materials will also be used in other fields of semiconductor products ?

B.J. Baliga

I think, until now, the development has been driven by LEDs and high temperature electronics and not by power devices. In the US, some of the support has come from defense agencies which were interested in all the high temperature aspects and I think that will nurture the technology for a while, but eventually, we will have to find a high volume application. The highest volume that I see now are the LEDs and power devices. I do not think that there are enough reasons to use it in main stream IC technology.

André A. Jaecklin

Recieved his diploma and doctorate degree in electrical engineering from the Swiss Federal Institute of Technology (ETH) in Zurich, Switzerland. After several years in magnetics research at Ampex Corp., Redwood City, California, he joined the Brown Boveri Research Center in Baden in 1968. He started and headed the power semiconductor device activity there. Later, he joined the Power Semiconductor Department as Head of Basic Development. In 1990 he transferred to ABB Corporate Research. He is also a Visiting Professor at the ETH.

INDEX

The manufacturer's authorised representative in the EU is Springer
Nature Customer Service Centre GmbH, Europaplatz 3, 69115 Heidelberg,
Germany. If you have any concerns regarding our products, please
contact ProductSafety@springernature.com

Printed and bound by CPI Group (UK) Ltd, Croydon, CR0 4YY
23/04/2026
02095625-0013